SOTHIS PRESS

Sothis Press aims to completely reimagine the public domain book with an emphasis on universal access to captivatingly designed print editions. We believe that technology will enable every single book ever published to be perpetually accessible in beautiful print form. Sothis is beauty in ubiquity.

www.sothis.press
1 Waterhouse Square, London, ECIN 2ST

The Metallographist, 1901

A Quarterly Publication Devoted to the Study of Metals, With Special Reference to Their Physics and Microstructure, Their Industrial Treatment and Applications

Albert Sauveur

SOTHIS PRESS

ISBN 978-1-0365-0680-3

The Metallographist, 1901
is in the public domain.

THE METALLOGRAPHIST

A Quarterly Publication devoted to the Study of Metals, with Special Reference to their Physics and Microstructure, their Industrial Treatment and Applications

Edited by ALBERT SAUVEUR

VOL. IV

1901

PUBLISHED BY THE

BOSTON TESTING LABORATORIES,

BOSTON, MASS., U.S.A.

CONTENTS

JANUARY, 1901

APRIL, 1901

JULY, 1901

OCTOBER, 1901

Contributors to "The Metallographist"

H. BEHRENS, Professor at the Polytechnic School, Delft, Holland.

T. A. BRINELL, Chief Engineer, Fagersta Bruks Aktiebolag, Fagersta, Westanfors, Sweden.

G. CHARPY, *Docteur ès sciences*, Engineer, Forges de Chatillon et Commentry, Montluçon, France.

J. A. EWING, F.R.S., Professor of Mechanism and Applied Mechanics, University of Cambridge, Cambridge, England.

G. GUILLEMIN, Metallurgical Engineer, Paris, France.

E. HEYN, Professor, Royal Technical Experimental Station, Charlottenburg, Germany.

H. M. HOWE, A.M., S.B., Bessemer Medallist, Professor of Metallurgy, Columbia University, New York, U.S.A.

BARON HANNS JÜPTNER VON JONSTORFF, Instructor at the Royal School of Mines, Leoben, Austria, and Chief Chemist of the Oesterreichisch-Alpinen Montan-Gesellschaft, Donawitz, near Leoben, Austria.

H. LE CHATELIER, Chief Engineer of Mines, Professor of Chemistry, Collège de France, Paris, France.

A. MARTENS, Director of the Royal Mechanico-Technical Experimental Station, Charlottenburg, Germany.

F. OSMOND, Engineer, Paris, France.

Sir W. ROBERTS-AUSTEN, C.B., F.R.S., Assoc. R.S.M., F.I.C., F.C.S., President Iron and Steel Institute, Professor of Metallurgy, Royal College of Science; Chemist of the Mint, London, England.

J. E. STEAD, F.C.S., F.I.C., Bessemer Medallist, Analytical Chemist and Metallurgist, Middlesborough, England.

D. TSCHERNOFF, Professor of Metallurgy, Imperial Academy of Artillery, St. Petersburg, Russia.

H. WEDDING, Ph.D., *Geheimer Bergrat*, Bessemer Medallist, Berlin, Germany.

The Metallographist

A QUARTERLY PUBLICATION DEVOTED TO THE STUDY OF METALS, WITH SPECIAL REFERENCE TO THEIR PHYSICS AND MICROSTRUCTURE, THEIR INDUSTRIAL TREATMENT AND APPLICATIONS.

Vol. IV JANUARY 1901 No. 1

NOTES ON THE TECHNOLOGY OF MICROSCOPIC METALLOGRAPHY*

By H. LE CHATELIER

MICROSCOPIC metallography constitutes, for the study of alloys, a method of immediate analysis extremely valuable. It has opened to chemistry a new field of investigations. All the improvements introduced into a method of such general application, have their interest, and this is my apology for publishing the following remarks.

My first metallographic researches were carried on under the direction of Mr. Osmond, who had the kindness of initiating me personally to his methods. Without hoping to do better, I have sought to increase the speed of the manipulations. Not only in industrial work, but in scientific work also, much importance should be attached to economy of manual labor. The scientific production of every worker is necessarily limited by the amount of time which he has at his disposal.

Metallographic manipulations include four distinct parts:—

I. Preparation of the polished surface.
II. Etching of the polished surface.
III. Microscopical examination.
IV. Preparation of alloys.

* Received September 24, 1900.

I. Preparation of the Polished Surface

This operation may be further divided into three parts: —

1. Cutting of the section.
2. Filing and grinding.
3. Polishing.

1. Cutting of the Section. — Bulky samples, such as pieces of rails, pigs of cast iron, etc., should be sent to the laboratory in small fragments. The attempt should not be made to machine in the laboratory samples weighing more than one kilogram, otherwise the loss of time will be considerable. The dimensions of the final sample to be subjected to examination should be in the neighborhood of ten millimeters.

The cutting of these small samples may be accomplished very satisfactorily by means of a hack saw, in the case of all metals whose hardness does not exceed that of unhardened steel,

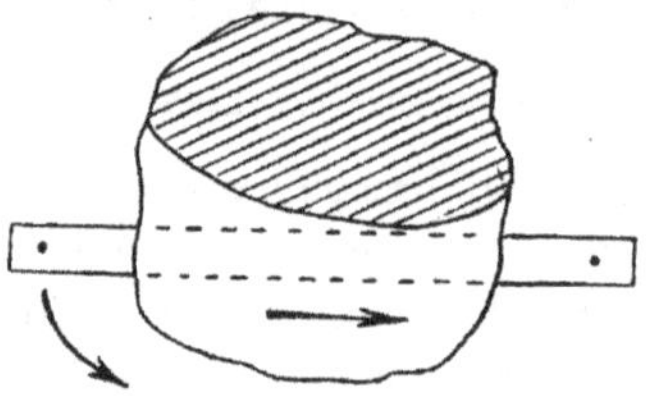

Fig. 1.

that is in the case of nearly all metallic alloys. So-called American blades, one millimeter thick, give very good results. A certain experience in holding the saw is, however, necessary, to avoid a rapid destruction of the blade. The saw should be held in such a way that the metal will first be attacked at the center of the mass, never near the outside surfaces. The cut made by the saw (Fig. 1) is then curved; the blade, being held horizontally, is in contact with the central part which is first attacked; as the saw is pushed forward, it is inclined more and more, until it comes in contact with the edge of the sample; it is then again placed in a horizontal position, thus coming in contact once more with the central part, and, while being drawn, is inclined more and more in the opposite direction until it comes in contact with the other edge of the sample.

In the case of hard and brittle metals, such as white cast iron, the easiest way is to use a hammer to obtain the desired fragment. The objection to this method is, that it yields irregular pieces. If power is available, it is preferable to cut the sample by means of a thin disc of emery. One-fourth of a horse-power is required to carry on the operation with sufficient speed. It is the only method that can be used in the case of hardened or tempered steels which are too hard to be sawed, and not sufficiently brittle to be broken.

2. FILING AND GRINDING. — This operation is generally started with a file or an emery wheel, and continued with emery-paper of increasing fineness. It should yield a smooth surface, somewhat specular, perfectly flat, and free from coarse scratches.

This operation is long and laborious, requiring much care, on account of its influence upon the final polishing. The experiments which I have made to shorten it have not so far yielded any results worthy of being reported here.

3. POLISHING. — I succeeded in shortening and simplifying greatly the polishing operation by means of a proper preparation of the polishing powders. This operation, which has been, until now, considered the most tedious of the three, has become the quickest. The most important point in the preparation of polishing powders, and also of emery powders, is to obtain an absolutely accurate classification with regard to size. The levigation method used commercially is altogether defective; even when carried on in the laboratory with all the precautions indicated by Mr. Osmond, the results are not very satisfactory. A perfect classification, on the contrary, is obtained by the washing method used by Mr. Schlœsing for the analysis of kaolins.

The pulverulent matters in suspension in water contain some salts, especially calcium salts, which have a tendency to cause the formation of lumps containing both coarse and fine grains. A very small amount of salt is sufficient to produce this result, the carbonate of lime present in ordinary water being more than sufficient. Ordinary water, therefore, can only yield a very imperfect levigation. It is the formation of such lumps that causes the rapid precipitation of argillaceous matters in suspension in calcareous waters.

Mr. Schlœsing treats the powders, which he desires to classify, by means of water containing one part of nitric acid in one thousand parts of water, in order to dissolve the carbonate and sulphate of lime and other salts which might be present. After a few hours, and occasional stirring, the mixture is allowed to settle. The powder falls rapidly to the bottom, and the clear liquid above it can be easily decanted. This liquid is now replaced by distilled water, and the mixture stirred; it is then again allowed to settle, and decanted. After a few similar operations, when all the acid has been removed, the settling takes place more slowly, and the liquid remains milky. The conditions are now favorable to proceed with the levigation, and the operation is further facilitated by adding two cubic centimeters of ammonia to each liter of water, as it helps the suspension of the finest particles. Decantation is now resorted to, at stated intervals, by means of a syphon forming a hook at its extremity, so as to avoid disturbing the portions which have settled.

By heating ten grammes of powder in a one-liter flask, nine-tenths of the liquid may be syphoned without fear of disturbing the deposit.

The decantations are made at the following intervals: fifteen minutes, one hour, four hours, twenty-four hours, eight days.

The deposit left after the first decantation contains all the coarse grains unsuitable for polishing.* The second deposit, resulting from the first decantation, after one hour's settling yields a substance which is not yet very homogeneous, but which may serve to start the polishing. The third deposit constitutes a good polishing powder for hard metals such as iron. It is, however, the deposit collected between the first and the eighth day which constitutes true polishing powder.

Instead of waiting eight days for the formation of the last deposit, it may be precipitated immediately after the removal of the twenty-four-hours powder, by adding some acetic acid, in order to saturate the ammonia. The totality of the powder, still remaining suspended, is then deposited after a few hours. The only objection is that some very fine particles, which would have remained suspended after eight days, are also precipitated.

It is not necessary to insist upon the necessity of conducting these operations with great cleanliness. The flasks, syphons, and the operator's hands must be washed with great care, and

* This first deposit may be subjected to a second washing to recover some of the fine particles which it may contain. This operation is necessary if a greater amount of powder is used, say 50 grammes per liter, in order to reduce a little the amount of distilled water required.

the flasks always kept covered. Indeed, these precautions constitute half of the art of polishing.

Once prepared, the powders must be carefully kept, and the contamination by any foreign matter avoided. A method, which has given me very good results, consists in converting them into a paste by mixing them with soap. To do this, a piece of very dry castile soap should be shaved by means of a very clean knife. The powder of soap* thus obtained is added to the polishing powder, when the latter is still wet, in the proportion of one part of dry soap to ten parts of the wet powder. The mixture is melted in a water bath, and allowed to cool, stirring all the while, until the mass begins to thicken; it is then poured into tin tubes similar to those used for keeping oil paints. After complete cooling, the tubes are closed, and they are now filled with a paste which can be used and preserved very conveniently.

This process of levigation may be applied to powders of any composition, provided the shape of the grain is quite regular. Lamellar substances, such as infusory silica, carborundum, etc., remain much longer in suspension, and the deposits, obtained under the conditions described, are much too coarse to be used for the final polishing.

The substances which have been found the most satisfactory are given below in their order of excellence.

Alumina derived from the calcination of ammoniacal alum.

Commercial flour-emery.

Oxide of chromium derived from the combustion of bichromate of ammonium.

Oxide of iron derived from the calcination in air of iron oxalate.

The last two substances must first be pulverized, in a mortar or crusher, in order to break up the soft lumps formed under the influence of heat, and which would not remain suspended in the water.

Alumina gives by far the best results; it is with this powder that the polishing can be accomplished in the quickest time. The proportion of fine powder resulting from the levigation is quite satisfactory, fully 25 per cent of the amount treated.

Flour-emery yields only a small proportion of powder suit-

* The powdered soaps on the market frequently contain some hard particles which scratch the metals.

able for the final polishing. The first deposits may be utilized for the beginning of the polishing operation.

The sesquioxide of chromium yields a very large proportion of fine powder, at least 80 per cent. For the polishing of iron and steel it is not quite as satisfactory as alumina. It is better suited to the polishing of softer metals, such as copper alloys. In that case the softer parts, which still remained suspended after three or four days, may also be utilized.

Oxide of iron is far inferior to the chromium oxide for polishing iron, its action being very slow. The percentage of fine powder is quite high, but smaller than that yielded by oxide of chromium.

The supports upon which these substances are used must be sufficiently flexible to come in contact with the entire surface of the sample, and yet not too yielding, otherwise the softest parts of the alloy are dug out more deeply, and, instead of a polished surface, a relief structure is produced, which cannot be easily examined with high powers. The most important requirement for these supports is, that they will not contain any hard particles capable of scratching the preparation. Substances such as albumine, leather, etc., which have been subjected to very few manipulations, are preferable. Other supports, however, such as paper, cloth, velvet, felt, etc., may also be used. These manufactured products should be examined with great care; it should be ascertained whether they do not scratch a polished surface of metal, such as bronze, for instance.

For the polishing of sections of iron and steel, previously rubbed upon commercial emery-papers, including the finest grade, I use the following:—

1. An emery-paper prepared with albumine, according to Osmond, with the deposit obtained in between a quarter of an hour and one hour in the ammoniacal washing of flour-emery.

2. A felt disc, covered with some soap paste prepared with the deposit of alumina or of emery, obtained in between one and three hours. The outside circumference of the disc is used, there being no need to fear that the center will be worn too deeply, as the sample is generally somewhat convex after the treatment with the emery-papers.

3. A flat disc made of wood, metal or ebonite, covered with cloth, velvet or leather which is strongly glued upon it. Upon

this covering the soap preparation, obtained with the deposit of alumina after twenty-four hours, is spread.

The last two discs are rotated by some mechanical device producing a great speed. With the polishing machines made by Grauer of Paris, which are run by foot power and are provided with ball bearings, a good speed may be obtained without much fatigue. The discs must be frequently moistened with a brush or sponge, for they become quickly dry under the action of the centrifugal force, which first eliminates the excess of water drop by drop, and then causes the remaining water to evaporate quickly. The vertical position of the discs has the advantage of preventing more readily the settling of dust upon their surfaces. It is, however, important to cover them carefully when not in use.

In the last stage of the polishing it is important to constantly turn the specimen so as to avoid the production of furrows starting from the minute cavities in which the polishing powder accumulates. The easiest way consists in rotating the sample in a direction opposite to the rotation of the disc. Under these conditions the polishing proper may be finished in about five minutes.

II. Etching of the Polished Surface

The method which appears to me most rational, and which I have already described, consists in etching the specimens when subjected to the influence of an electrical current, in a solution which has no action upon the metal. It is then possible to control the etching by regulating the intensity of the current. Such method yields excellent results in the case of copper alloys immersed in diluted solutions of caustic potash or of ammonia. The intensity of the necessary current varies between one thousandth and one hundredth of an ampere per square centimeter, the action lasting from a few seconds to a few minutes, according to the intensity of the current.

With iron and steel samples I have not obtained good results by this method. A neutral solution of potassium sulphate, or a warm neutral solution of hyposulphite of sodium was found to work better. The liquids are neutralized by means of a little calcium carbonate. In the present state of our knowledge, however, the best etching method for iron and steel specimens is undoubtedly the combined polishing and etching recommended by Mr. Osmond. The following simple treatment answers in almost all cases: some commercial tincture of iodine, diluted with four times its volume of alcohol, is rubbed over the polished surface with the tip of the finger, and the operation repeated two or three times, until a gray film appears over the surface, which, however, must not be so deeply etched that it ceases to be specular.

Fig. 2. Gray Cast Iron of good quality, etched with tincture of iodine. Magnified 660 diameters.

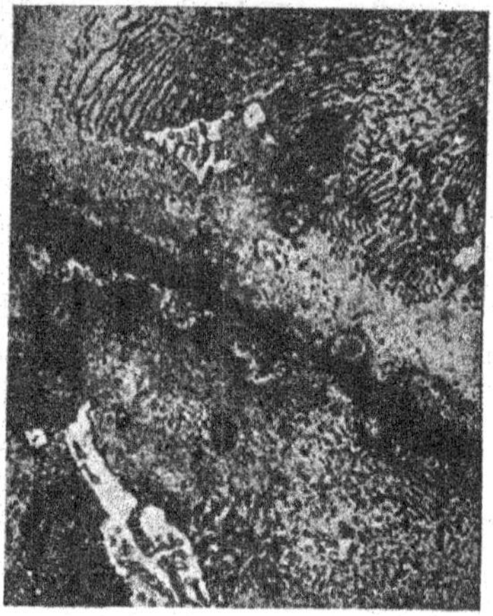

Fig. 3. Gray Cast Iron of good quality, after reheating, etched with tincture of iodine. Magnified 660 diameters.

I give further on, as an illustration, some photomicrographs of some samples of gray cast iron prepared by this method. The magnification is 660 diameters. Fig. 2 represents some gray foundry iron of good quality, Fig. 3 the same iron after annealing, and Fig. 4 shows the structure of an abnormal sample of gray cast iron.

By rubbing gently the polished surface while being treated with iodine, the etching is uniform all over the preparation, the constituents which are not acted upon remain brilliant, and the filling up of the minute parts, such as pearlite,* by some deposit derived from the etching of other parts, is prevented.

The etched sample is washed and dried as usual. To preserve the preparation unaltered for a long time, the simplest way consists in varnishing them with some "zapon," a solution of gun cotton in amyl acetate. The very thin coat of varnish is

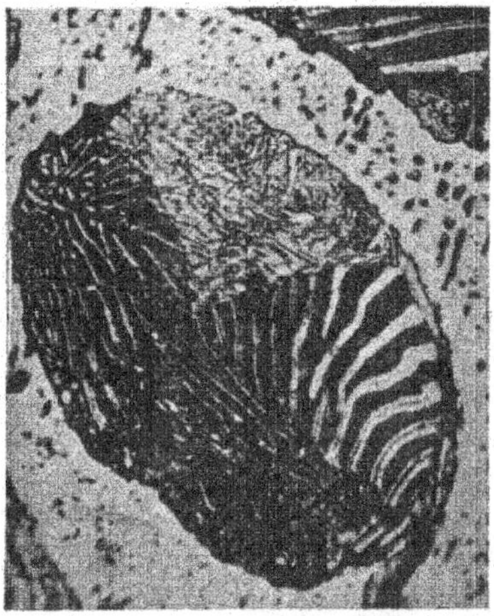

Fig. 4. Gray Cast Iron, abnormal structure, etched with tincture of iodine.
Magnified 660 diameters.

very transparent and allows examination with the highest powers; the samples may thus be kept untarnished for several months. It is not, of course, as effectual as an immersion in oil, but it is infinitely more convenient.

---

* In previous numbers of *The Metallographist*, the name of this constituent has been spelled "pearlyte," because such was the orthography proposed by Prof. H. M. Howe, with whom, as is well known, the name originated. Since, however, Prof. Howe himself has now adopted the more generally used spelling "*pearlite*," we shall also abandon the original orthography. — Ed.

III. Microscopical Examination

The microscopical examination and the photomicrography of the specimens require, at present, costly outfits, which I have endeavored to simplify as much as possible. Mr. Pellin, of Paris, constructed, according to my instructions, a microscope shown in Figs. 5, 6, and 7. *A* is the metal being examined, *B*, the support, *C*, fine adjustment for focussing the objective, *D*, objective, *E*, totally reflecting prism, *F*, eye-piece, *G*, a movable tube carrying the prism, *H*, tube carrying the eye-piece,

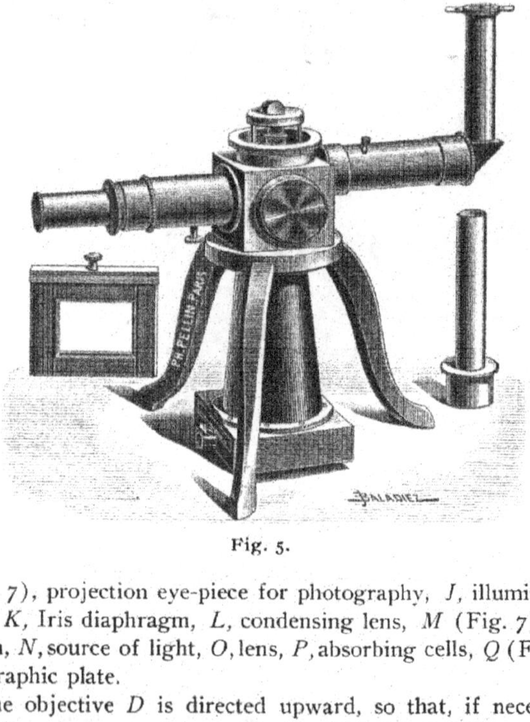

Fig. 5.

I (Fig. 7), projection eye-piece for photography, *J*, illuminating prism, *K*, Iris diaphragm, *L*, condensing lens, *M* (Fig. 7), diaphragm, *N*, source of light, *O*, lens, *P*, absorbing cells, *Q* (Fig. 7), photographic plate.

The objective *D* is directed upward, so that, if necessary, very bulky pieces may be examined. I have been able to examine, with this arrangement, a steel plate 250 millimeters wide. The greatest advantage, however, is to be found in the fact that it is only necessary to have one flat surface in the samples examined, which is placed upon the support taking the place of the stage of the microscope. The necessary adjustment to bring the polished surface in a plane perpendicular to the axis of the microscope is done away with, or the still more tedious operation of cutting two parallel surfaces on each specimen may be omitted. In either case a notable saving of time is gained.*

The eye-piece is placed horizontally, and receives the image by the reflection of a totally reflecting prism *E* placed under the

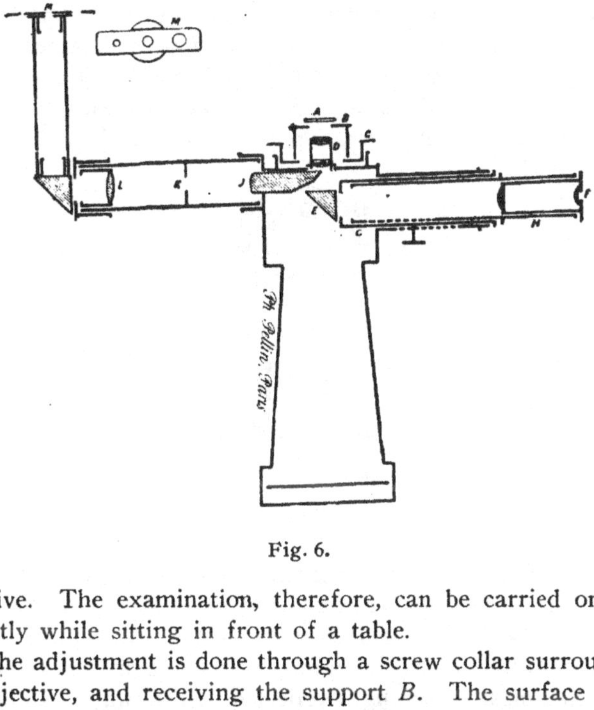

Fig. 6.

objective. The examination, therefore, can be carried on conveniently while sitting in front of a table.

The adjustment is done through a screw collar surrounding the objective, and receiving the support *B*. The surface under examination must be absolutely perpendicular to the axis of the objective. The supports, which rest upon three points, have a very uniform thickness, so that the preparation is in the proper position. In a microscope for the examination of metals, the illuminating arrangement is the most delicate to regulate. It is

* The specimen holder universally used by American workers accomplishes the same purpose as the special arrangement of Prof. Le Chatelier. — Ed.

generally accomplished tentatively by adjusting a number of movable pieces. I have endeavored to reduce these adjustments to a minimum, and to give to each one a well-defined effect. In my arrangement, two adjustments only are required, the proper *opening* and proper *position* of a single diaphragm *M* placed in such a way that its *opening* regulates the angle of the beam of light which falls upon the preparation. This angle must vary with the nature and the quality of the objective. The greater the angle, the greater the effect of the spherical aberration, and, on the contrary, the smaller the angle the more marked the chromatic aberration. The best definition is obtained with a certain mean angle which must be found tentatively in each case. The *position* of the diaphragm *M* regulates the mean inclination of the beam of light falling upon the preparation. To obtain

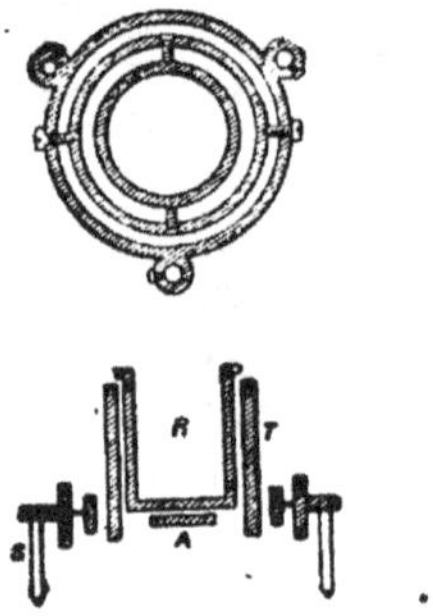

Fig. 6*a*.

the greatest clearness, its direction should be nearly vertical, but in order to reduce the amount of light reflected by the lenses, and sent back to the eye, it is necessary to give a more inclined direction to the beam of light. If the inclination is what it should be, most of the light reflected in this way is stopped by the illuminating prism itself.

The diaphragm *M*, which regulates the illumination, is placed at the principal focus of the complex optical system made of the objective *D*, the illuminating prism *E*, and the lens *L*. To ascertain its proper position, the image of a luminous point, placed at a distance of two or three meters above the objective, is located by means of a magnifying glass. It may be more convenient to place the luminous point at the proper distance horizontally, but only a little above the objective, and to reflect the rays in the objective by means of a totally reflecting prism, placed above it. This determination is done once for all, for each

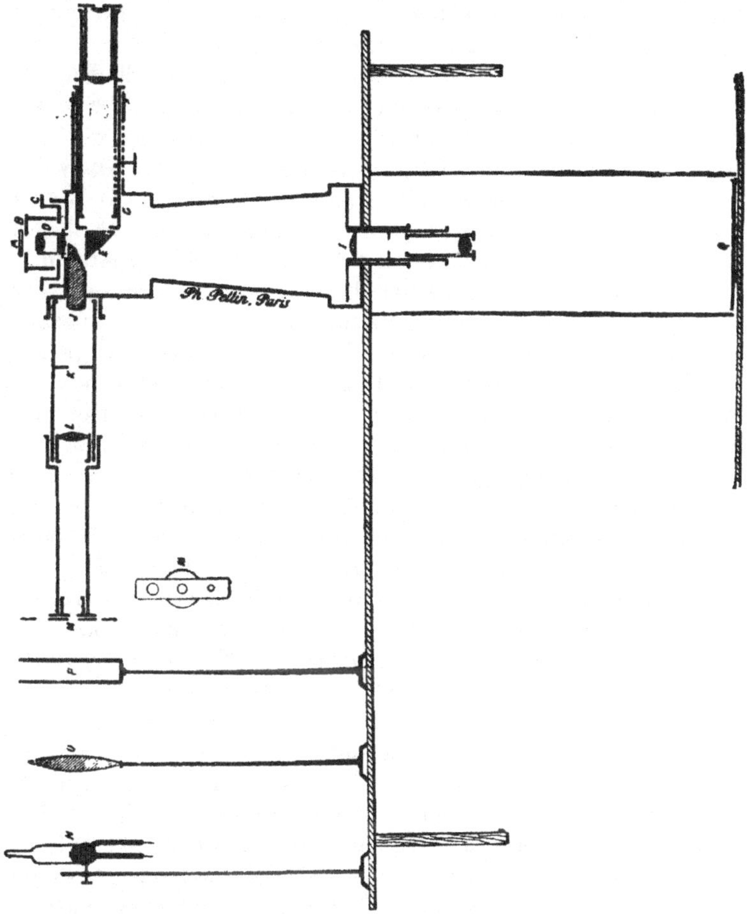

objective, and the corresponding positions, in each case, of the draw tube carrying the diaphragm, is noted.

The purpose of the diaphragm *K* is to stop the light which would fall upon parts of the preparation outside of the portion examined. Its opening is regulated so that its circumference falls within the field of the microscope, and it need not be altered as long as the eye-piece is not changed. This elimination of the useless rays is very important in order to reduce the amount of light reflected back by the lenses of the objective.

The tube carrying the diaphragm *M* may, according to the source of light used, be straight (Fig. 7), in which case the source of light must be placed at the level of the diaphragm, or it may be bent by means of a totally reflecting prism (Fig. 6), which allows the source of light to be placed at any height. It is generally preferable to locate the source of light, so as to be able to keep in a horizontal position the tube carrying the diaphragm.

As a source of light, if the object is only to be viewed with the eye, a Welsbach lamp is the most convenient. Its intensity may be regulated by passing the light through pieces of ground glass, or through thin sheets of paper. The mantel should not be placed at a greater distance than ten centimeters from the diaphragm *M*, so as to utilize the whole of its surface.

Microscopic photography has a very great importance, for it enables us to preserve some durable impressions of the various metallic preparations. It is especially for the purpose of photography that I have adopted the arrangement just described.

The whole of the photographic outfit is placed vertically below the objective (Fig. 7). During direct observation it is covered by the totally reflecting prism *E*. This prism, however, is carried by a movable tube which may be drawn so as to clear the field.

The camera receives at its lower part plate holders measuring 4.5 by 6 centimeters; the distance is such that when the image formed by the objective is sharp on the plate holder, it is equally sharp for a mean position of the eye-piece, when viewed by the eye. This distance is determined tentatively by taking a series of photographs after focussing with the eye-piece in different positions. The latter is carried by a graduated tube, and its positions may therefore be recorded.

The distance of the plate to the objective is, of course, that for which the aberrations of the objective have been córrected, i.e., 170 millimeters in the case of French or German objectives, and 250 millimeters in the case of English lenses.

The resulting image measures 15 millimeters in diameter. It

is extremely sharp, and could be subsequently enlarged ten diameters, were it not for the grain of the plates. As a matter of fact, a greater enlargement than 3 diameters is not advisable, and this yields photographs 45 millimeters in diameter, a very convenient size for projections.

It is preferable, however, to magnify the image before receiving it on the plate. To do this, a projection eye-piece, magnifying two diameters, takes the place of the small plate holder, and the photographic plate, measuring for instance 9 by 12 centimeters, is placed one meter further down. It is convenient, for this purpose, to set the microscope on a table, or, better still, on a shelf fastened to the wall, under which is placed the camera and plate holders. The projection eye-piece passes through a hole in the table or shelf as shown in Fig. 7. The focussing is done, as in the previous case, by means of the eye-piece after having ascertained tentatively its proper position. It is, however, necessary, owing to the high magnification, to use a very high-power eye-piece, so that the focussing will be sufficiently sharp.

With the above arrangement and a three-millimeter objective, very sharp images are obtained, magnified 500 diameters.

In order that the images be equally sharp all over the field, the polished surface must be accurately perpendicular to the axis of the objective. It is not always sufficient to place it on the support. It is preferable to use the support shown in Fig. 6*a*. The object is fastened, by means of sealing-wax, at the end of a cork which fits a central tube. This central tube may rotate around two axes perpendicular to its length, and carried by two concentric rings. The external ring carries, moreover, three projections by means of which the whole arrangement rests upon the stage of the microscope, or upon one of the supports.

The surface under examination must be absolutely parallel to the three supporting projections. To do this, the contrivance is reversed, and a sheet of glass, of uniform thickness, is placed upon the three legs. A luminous point placed at a great distance is then observed by reflection upon the sheet of glass, and the central cylinder is inclined gradually, until the image reflected by the metallic surface occupies upon the glass exactly the same position as that reflected by the glass; the parallelism is then perfect.

From the point of view of photography, the character of the light used is of vital importance. A Welsbach lamp, an acetylene flame or an oxyhydrogen light are very satisfactory, if white light is deemed sufficient. The luminous beam must completely cover the diaphragm *M*. If the source be very small, as is the case with a pearl of magnesia in an oxyhydrogen flame, its image must be thrown upon the diaphragm by means of an additional lens *O*, which is placed, when necessary, so as to obtain a magnified image of the source of light. It is indispensable when an electric spark is used, between two points of magnesium, for instance.

It is, however, preferable to use a monochromatic source of light, because so-called achromatic and apochromatic objectives are never completely free from chromatic aberration. It is even probable that it would be advantageous to use objectives in

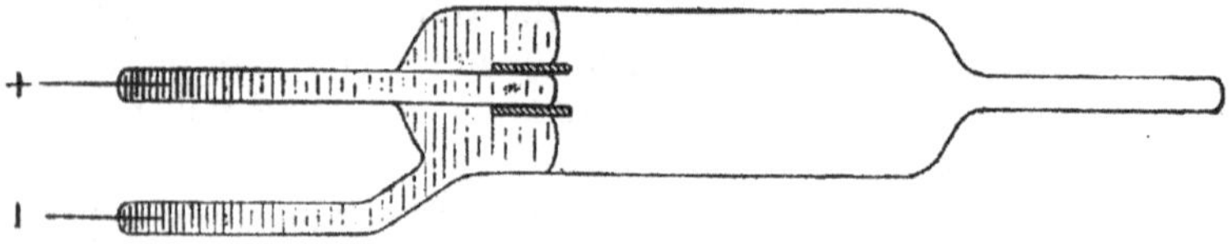

Fig. 8.

which the chromatic aberration was altogether overlooked, but which would be more perfectly corrected with regard to spherical aberration. At any rate, in using only one kind of glass, it would be possible to select a stronger glass, and the construction of the objectives would be much cheaper.

An excellent monochromatic light for photographing is obtained by means of a mercury arc lamp in vacuum. The one I use (Fig. 8) presents the following features: The upper vacuum space is very large, so as to increase the radiating surface, and therefore decrease the heating of the lamp. The central tube is covered with a refractory coating which greatly lengthens the life of the lamp. With a projection eye-piece, like the one mentioned above, it requires an exposure of from one to two minutes.

According to the researches of Messrs. Parrot and Fabry, the radiations emitted by this arc are as follows: —

Two yellow rays 0.μ577 and 0.μ579
One green ray 0.μ546
One indigo ray 0.μ436
Several violet and ultra violet rays.

The green radiations, very desirable for ocular examination, may easily be isolated as well as the indigo ray so effective for photography.

In order to use the green ray only, a glass cell five millimeters thick, and containing some bichromate of potassium, is placed between the source of light and the diaphragm. Such a cell stops the indigo, violet, and ultra violet radiations. Another cell, containing a saturated solution of chloride of didymium, is used to absorb the yellow rays.

To isolate the indigo ray, an acid solution of sulphate of quinine is used to absorb the violet and ultra violet rays, and an ammoniacal solution of copper to stop the yellow and green radiations. For the purpose of photographing, however, the latter solution may be dispensed with, as yellow and green rays have no action upon ordinary photographic plates. This lamp should preferably be worked by a continuous current. The difference of electro-motive force between the two poles varies from 15 to 25 volts as the lamp becomes heated. The current intensity required to start it is four amperes; it may be lowered to two amperes after the lamp is hot. Three amperes is the best intensity for normal work.

IV. Preparation of Alloys

The object of microscopic metallography may be to study some known alloys, industrially prepared, for instance, but it may also be the chemical study of the various combinations which two metals may form. It is necessary, then, to prepare a great number of alloys of various compositions, so as to be sure not to overlook some of these combinations. Numerous trials are required before the alloys of some interest are produced. I have obtained much better results by superposing the two metals so that they will mix only in the vicinity of their contact. By preparing then a vertical section of the resulting mass, the metals may be observed alloyed in all proportions from one pure metal to the other.

This superposition, it is true, is not easily obtained, and

ing alloys in the case of alloys of copper and zinc whose constitution has been ascertained by the work of Mr. Charpy.

Figs. 15 and 16 show the crystals of a definite compound whose formula is supposed to be Zn_4Cu, Fig. 15 after a slight etching, and Fig. 16 after a deep etching. This difference in appearance indicates that the crystals are not homogeneous.

Fig. 17 shows the eutectic alloy of Zn_4Cu and Zn_3Cu. The latter compound contains numerous cracks running in all directions.

Fig. 18 exhibits some crystals of Zn_2Cu, surrounded by a eutectic alloy poorly defined.

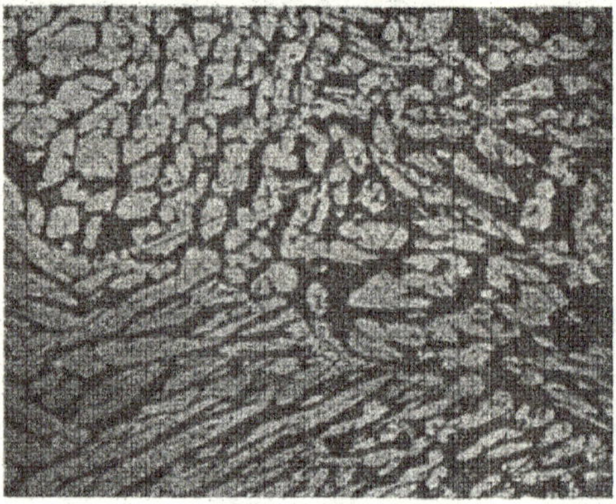

Fig. 19. Crystals of ZnCu.
Magnified 150 diameters.

Finally, Fig. 19 shows some crystals of ZnCu which constitute the essential part of all forgeable brasses. It will be noted that many parallel bands belong to the same crystalline element.

My experiments have not revealed the crystals of $ZnCu_2$ discovered by Mr. Charpy, because they are formed only on reheating. The bottom of the metallic button was composed of crystallites, similar to those found in alloys rich in copper prepared by simple fusion. I have not thought it necessary to reproduce them here.

NOTE ON SOME INSTANCES OF SUPERFICIAL HARDENING*

By F. OSMOND

DAUBRÉE, in his experiments on the synthesis of meteorites, was led to place some thin sheets of steel in a closed vessel, in which some powder was lighted by means of a voltaic current. The steel was found partly melted: the portion which was not melted and which was originally ductile, was now found to have a high elastic limit, or even to be brittle, as if it had been subjected to a hardening treatment. In another experiment, in which the fusion had been complete, a section of the small ingot showed, after polishing and slight etching, that the structure of the metal was not homogeneous: the parts nearest the surface corresponded to a narrow band which had assumed a peculiar tint, very different from the appearance of the rest of the section. "This difference," the author says, "conveys the idea of some kind of hardening produced by a very rapid cooling of the surface of the metal. Some chemical action of the surrounding gases, however, may also contribute to this result." The fragments of unmelted steel left, after the explosion of some dynamite, exhibited also a similar appearance.*

I had the opportunity of confirming Daubrée's observations. Mr. Vieille had given me five steel cylinders containing, parallel to their axes, a narrow channel for the escape of the gases of various powders. In examining the microstructure of a longitudinal section of these cylinders, I noticed at once, running the whole length of the channel which had been corroded by the gases, a band decidedly hardened.

As the heating and cooling of these bands had taken place under peculiar conditions of pressure and speed, it was interesting to study this phenomenon carefully.

The five samples will be designated by the letters A, B, C, D, and E.

A was made of soft steel; B, C, D, and E were of a harder grade, such as is commonly used for the manufacture of guns.

---

* Received September 10, 1900.

* *Études synthétiques de Géologie expérimental.* Paris, Dunod, 1879, pages 624 to 649.

A and B were simply rolled or forged; C, D, E were subjected to some treatment after forging, probably to a quenching in oil followed by tempering, or to some similar treatment.

Sample A. — Upon being rubbed on the last emery-papers, the hardened band assumed a more brilliant polish. By polishing in relief, hard grains are revealed corresponding to the dark areas of Fig. 1, on a light background or ferrite. In the har-

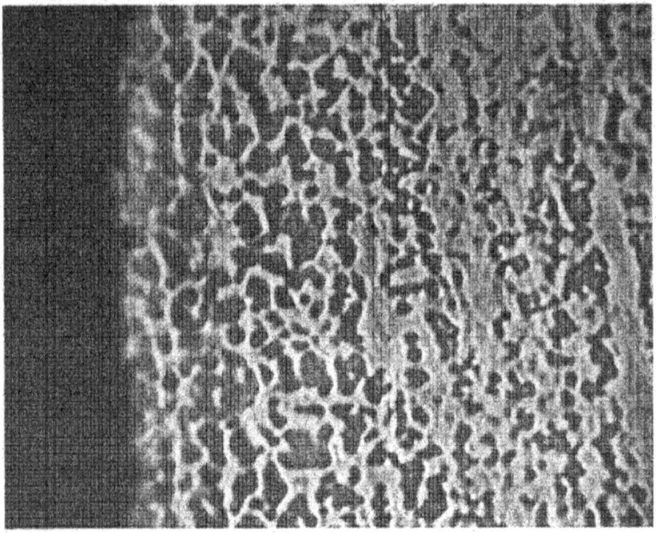

Fig. 1—Sample A. Polished in relief.
Magnified 200 diameters.

dened band, however, which actually measures about 0.25 mm., the hard grains take a brighter polish and stand more in relief. Their size increases as the corroded end is approached and, very near the edge, they show a tendency to weld. A slight etching with tincture of iodine (Fig. 2) imparts a dark-brown color to the hard grains of the preparation, which consists of pearlite changing somewhat to sorbite, but, in the hardened band, the grains of pearlite are suddenly replaced by corresponding grains of martensite (or of hardenite, if that term is retained to designate saturated martensite). By the etching method used, hardenite is only slightly colored. As we proceed from the inside towards the corroded surface, the grains of hardenite are found to be surrounded by a brown border and to increase in size, as may be seen in Fig. 3, magnified 1000 diameters. These brown borders or membranes, situated as they are between the martensite and the ferrite, correspond to the constituent troostite,

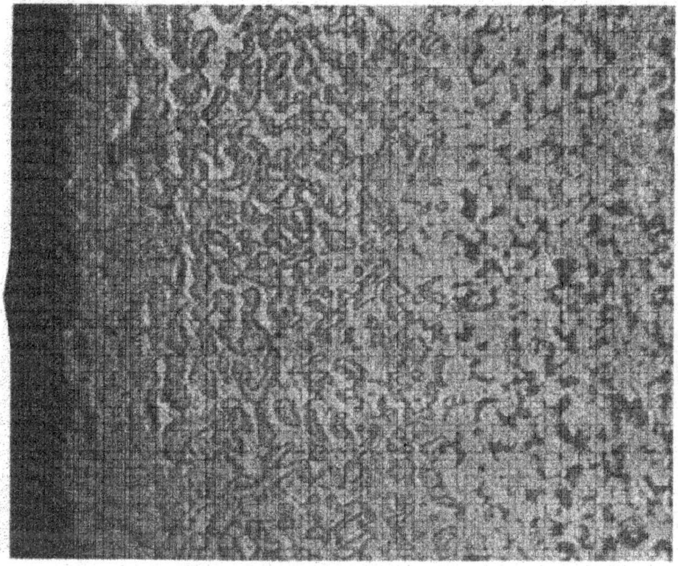

Fig. 2—Sample A. Slightly etched with tincture of iodine. Magnified 200 diameters.

which, however, exhibits here a new peculiarity, that of assuming the shape of needles, not easily seen in the photographs, and which recall the structure of martensite. We are, therefore, here in presence of a secondary form of transition, which had not yet been described, between martensite and troostite, and which corresponds to the beginning of the diffusion of the carbon of the hardenite into the surrounding ferrite. At the extreme edge, the hardenite, which formed the center of the grains, becomes softer owing to the increasing diffusion of its carbon and exhibits, although minute and somwhat confused, the characteristic forms of martensite. The distribution of the carbon is not yet, however, uniform, as may be seen in comparing Figs. 1 and 2. Fig. 2, which was etched, does not show any ferrite near the edge, while Fig. 1, which was polished in relief, exhibits in the same region some relatively soft furrows (light on a dark background). It follows from these appearances that the martensite

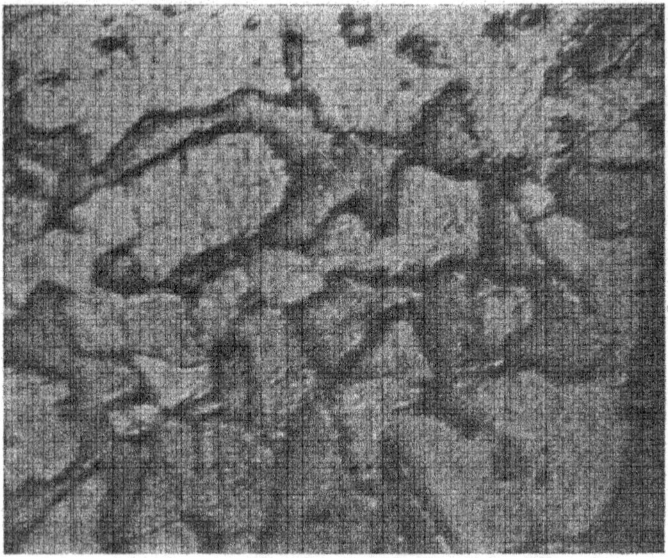

Fig. 3—Sample A. Etched with tincture of iodine. Magnified 1000 diameters.

is not homogeneous and that its carbon content still exhibits some maxima in the vicinity of the center of the original grains.

The above observations are easily interpreted. The gases, as they escape at a high temperature and under high pressure, heat suddenly the walls of the channel, which are then cooled also very suddenly, either owing to the large mass of steel surrounding them and which had not had time to become heated, or owing to the expansion of the lost gases. The line of demarcation between the hardened band and the unhardened metal corresponds to the regions of the cylinders which had reached, when the cooling began, the point $Ac_{3\,2\,1}$ of the pearlite: the transformations of the iron and carbon took place *in situ*, but along this line the cooling followed so closely the heating that the diffusion of the carbon had not even time to begin. As we approach the edge we find layers of metals which were subjected to gradually increasing temperatures during gradually

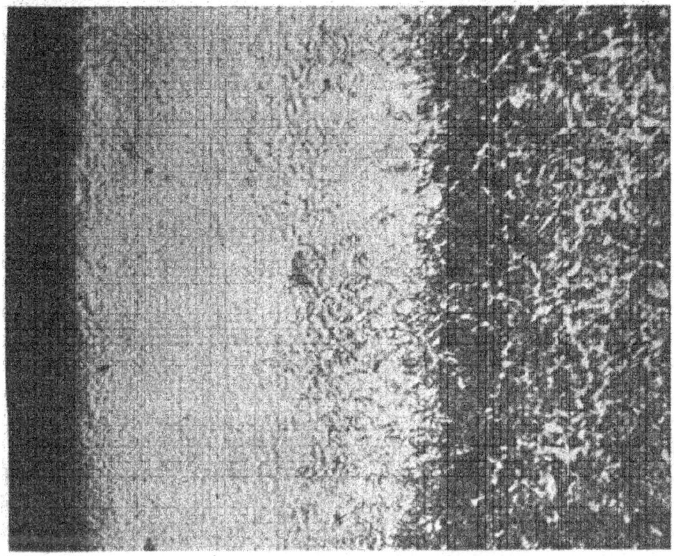

Fig. 4. Appearance of Samples C, D and E, slightly etched with tincture of iodine. Magnified 200 diameters.

increasing lengths of time, which, however, always remain very short. The carbon is found to have diffused more and more, without, however, reaching a homogeneous distribution even in the extreme edge, since the original network of ferrite is still indicated by the presence of soft martensite. The hottest layers were carried away by the gases.

Sample B. — The results were similar to the preceding ones.

Samples C, D, E. — These three samples exhibited, beyond the hardened band, the normal structure of guns quenched in oil and tempered, i.e. a distorted network of ferrite upon an ill-defined background representing the transition forms between pearlite and martensite, troostite and sorbite, without sharp demarcation. The corroded surface of the channel was lined, as in the previoüs instances, with a hardened layer, but as the distribution of the carbon was, in these samples, quite uniform before the escape of the gases, owing to the previous treatment, the greatest portion of the hardened band consisted of martensite.

These observations are illustrated in Fig. 4, magnified 200 diameters. The sample was slightly etched with tincture of iodine. The martensite, occupying about 0.25 mm., remained clear and unresolved, while the metal which was not affected by the action of the gases was intensely colored, with the exception of the particles of ferrite. · The latter extend into the martensite band for about one-third of its width. A darker band will be noted between the hardened band and the unaltered region. It represents the layer, whose temperature resulting from the passage of the gases was exactly below the critical points. It could not, therefore, be hardened. On the contrary, the structure resulting from the previous treatment was effaced to a certain extent, the metal assuming the characteristics of pure sorbite.

The very exceptional conditions to which the above samples were subjected are not the only ones capable of producing a superficial hardening of steel.

I have had occasion to examine a steel rail of English origin which broke in several fragments and produced a serious accident. The portion of the head upon which the wheels bear was found to be hardened to a small depth in a way identical in the smallest details to the hardening produced by the passage of gases in the channel of samples A and B. The reason for such hardening is not known. In pressing a piece of steel upon an emery wheel, and without using water, I did not succeed in heating it above a blue color. It is hardly to be doubted, however, that under favorable conditions a piece of steel may be heated to a red heat by friction, resulting in the hardening of a very thin outside layer. At any rate, the fact which has now been ascertained has probably happened more than once, and, on account of the dangers to which it may lead, it was desirable to bring it to light.

CASTING ON MICA OF METALLOGRAPHIC PREPARATIONS*

By H. J. HANNOVER

IT is well known that it is difficult to obtain good preparations of soft alloys for microscopical examination. In polishing, for instance, an alloy of tin and lead, the lead spreads itself upon the tin in such a way that it is impossible to distinguish clearly the structure under the microscope. For these reasons an other method of obtaining smooth surfaces was sought. Messrs. Ewing and Rosenhain proposed to cast such alloys upon plates of glass or of polished steel,† and I know that Prof. H. Le Chatelier has tried casting on glass by inserting vertically in a glass tube a strip of glass reaching almost to the bottom, so that two samples, easy to detach, were produced at each operation.

Glass, however, frequently breaks, as noted by Messrs. Ewing and Rosenhain. Having mentioned this fact to Mr. C. Christensen of Copenhagen, Denmark, he suggested the use of mica in place of glass, and by following his advice I obtained some very good results. The accompanying photograph shows an alloy containing 20 per cent of antimony and 80 per cent of lead, cast on mica. A background, made up of a eutectic alloy, may be easily distinguished as well as large crystals corresponding to an excess of antimony. In preparing alloys of tin and lead, the two metals are first weighed and then melted together in a small crucible under a cover of cyanide of potassium. A small funnel cut in a piece of charcoal is then placed upon a freshly detached sheet of mica, and the alloy poured in the funnel. Such methods do not yield perfectly polished surfaces in the case of alloys of lead and antimony. In order to obtain the preparation seen in the accompanying photograph, the small mass of alloy, after being cast upon mica, was placed upon another sheet of mica, and covered with some potassium cyanide. It was then heated by means of a blow-pipe, and when it began to melt, another sheet of mica, previously heated, was placed upon it and pressed until solidification was complete. The upper surface remained perfectly smooth, as may be seen from the photograph.

---

* *Bulletin de la Société d'Encouragement*, August 1900, page 211.

† *The Metallographist*, Vol. III, page 94, April 1900.

No good results are obtained by heating from below the sheet of mica upon which the alloy rests, because the mica is distorted by the action of the heat. It will be necessary to modify

Lead-antimony alloy—lead, 80 per cent; antimony, 20 per cent—cast on mica.

the method to suit the various alloys, but, in general, casting on mica will possess material advantages over casting on glass. It is on that account, that, upon Prof. Le Chatelier's request, and with Mr. Christensen's permission, I published these notes which may be of service to metallographists.

THE EFFECT OF ANNEALING UPON THE PHYSICAL PROPERTIES, AND THE MICROSTRUCTURE OF A LOW CARBON STEEL*

By HENRY FAY and STEPHEN BADLAM

THE enormous demand for low carbon steel for structural purposes, plate, tubing, and its many varied uses, offers great temptation for the manufacturer to allow almost any kind of material to pass through the mills without paying particular attention to the condition in which it appears upon the market. That the manufacturer does occasionally submit to the temptation to work the mills beyond their capacity is shown in the failure of the metal to meet certain physical requirements. That these failures are not more frequent is to be greatly wondered at, and it goes to show under what excellent control the steel plants are held. This occasional failure, however, makes it necessary to study carefully the conditions which, if not strenuously adhered to, will produce steel which will not only give trouble to the manufacturer himself, causing loss of time and perhaps material, but may also be a source of trouble to the buyer, and perhaps a source of danger to the public. The buyer, in order to protect himself, usually subjects the material to certain physical tests which tell him whether the steel has the requisite strength. This determination of tensile strength was considered for a long time as entirely satisfactory and all that was necessary, but the presence in the steel of certain elements, such as phosphorus and sulphur, which were supposed to have an injurious effect on the physical properties, led the cautious buyer to test his steel both physically and chemically. It is a well-known fact that two pieces of steel having identically the same composition may have entirely different physical properties. This marked difference in physical properties was found to be due to the difference in heat treatment. We are able to say of any piece of steel that its physical properties depend primarily on its chemical composition, and secondly, on the heat treatment to which it has been subjected. If we know, then, the chemical composition and the heat treatment, we should be able to predict the physical properties.

---

* *Technology Quarterly*, December 1900. The illustrations accompanying this article are reproduced here through the courtesy of the *Technology Quarterly*.

The query now arises, if we have two steels of the same composition but of different physical properties, is there any means of knowing the heat treatment? Fortunately this can be answered in the affirmative. It has been shown by Sorby, Osmond, Stead, Ridsdale, Sauveur, and other investigators, that there is a direct relation between the heat treatment and the microstructure of steel. A study of the microstructure, together with the physical properties of metals, has given rise to the science which has been designated by the name Metallography.* For some varieties of steel and many alloys these relations have been very carefully worked out, and this information has added enormously to our knowledge of the properties of steel and other alloys. It was the object of this investigation to start with a steel of known composition and to study carefully the heat treatment, and how it would effect the physical properties and the microstructure. Having this information established for this class of material under normal conditions, we have certain standards from which to draw our conclusions. This, however, does not give all of the information which we desire, as it is well known that work, such as rolling, hammering, etc., has a decided influence on the physical properties. This portion of the work, however, was impossible for us to carry out, and we present the paper in its present form with the hope that the subject will be of sufficient interest to be completed along similar lines by some practical metallurgist. The investigation is incomplete in another sense, viz., that we have heated our specimens to a certain definite temperature and allowed the metal to cool from that temperature, but have not held the steel for any length of time at the temperatures to which it was heated. Our cooling was carried out slowly enough, as will be shown later on, for complete crystallization to take place, but according to Stead's† experiments, coarse granulation takes place by holding the sample for some time at about 700° C. This portion of the work is now in progress, and it is hoped to bring forth our completed results later. The material selected for this purpose was a low carbon steel which was kindly furnished by the National Tube Company, of McKeesport.

---

* For the elements of the subject see *The Metallographist*.

† "The Crystalline Structure of Iron and Steel." J. E. Stead. *Journal Iron and Steel Institute*, 1898, No. 1, 145. *The Metallographist*, Vol. I (1898), page 289.

The steel which we used is generally classified as "extra mild steel." Many investigations have been conducted on mild steel, but none of the investigators have followed exactly the same manner of work. Before proceeding to the experimental work it will be necessary to review briefly the work which has already been done on mild steels.

Until the publication, in 1898, of Mr. Stead's paper on "The Crystalline Structure of Iron and Steel," it was tacitly assumed that all grades of steel, whether soft or hard, followed the general laws of structure, as enunciated by Tschernoff and Brinnel. These two investigators were the first to formulate any laws connecting the heat treatment and the structure. Both of them experimented principally upon steels of medium hardness, about 0.3 per cent of carbon. Their conclusions in regard to the structure were based upon the appearance of the fractures.

Tschernoff* summarized his conclusions as follows: Steel on solidifying has a crystalline structure. This may be prevented:

1. By simply reheating to a cherry red when an ingot with a coarse structure will suddenly be transformed into a fine-grained and strong product.

2. By hammering at a high temperature, and continuing the work down to a low temperature, to a certain point, varying with different steels, below which the metal preserves its whole power of resistance.

3. Rapid cooling of the cast metal destroys the crystalline structure.

Tschernoff points out the important fact that the higher the percentage of carbon, the lower is the point below which no change in structure occurs.

Brinnell's† conclusions follow along the same line as those of Tschernoff, but are expressed more in detail. A few abstracts from his work, having a special bearing upon the subject, are given below:

1. Whenever steel loses its crystalline structure, other than by mechanical work, the change of texture will be found to occur

* "The Structure of Cast Steel Ingots." *Proc. Inst. of Mech. Eng.*, January 1880; translated from the Russian by Mr. W. Anderson.

"The Manufacture of Steel and the Mode of Working it." *Proc. Inst. of Mech. Eng.*, April 1880; translated by Mr. W. Anderson.

† "The Changes in the Texture of Steel on Heating and on Cooling." Abstracted in *Journal Iron and Steel Institute*, 1886, I, 365.

simultaneously with the transformation of cement into hardening carbon, or *vice versa,* and the change will be complete only in the event of the carbon being converted by heat from cement into hardening carbon.

2. In order to effect the transformation of hardening into cement carbon, in a steel heated to a white heat, it must be allowed to cool slowly to a lower temperature than that necessary to change cement into hardening carbon.

3. If hardening carbon, in heating or cooling, is converted completely or largely into cement carbon, a crystallization occurs, yielding a fracture the coarser, as the former texture of the steel was crystalline.

4. The crystallization of steel requires a certain amount of time as well as a certain amount of heat. If the duration of the cooling process is cut short by quenching in water, the development of crystallization is partly or wholly prevented.

With regard to the development of a coarse crystallization, Brinnell says that on heating the steel to a point W, say 750°, the size of the granules increases with the temperature, the coarsest being obtained by heating to the highest temperature, and the structure remains coarse whether the steel is quenched or cooled slowly from the temperature.

Within the last few years several important articles have appeared dealing with the effect of heat treatment upon soft and extra soft steels, the results of which seem to be quite different from those of the earlier experimenters upon medium and hard steels. These later researches are also characterized by the use of the microscope for the examination of the structure instead of the naked eye and fractured surfaces.

From his microscopic work upon soft steels and pure iron, Mr. Osmond* came to the following conclusions in regard to heat treatment:

1. Steel with 0.02 per cent C. "It would seem that the structure of pure iron is almost independent of the temperature and the rate of cooling."

2. Steel with 0.14 per cent C. "In proportion as the temperature of reheating is raised, other things being equal, the polyhedrons of ferrite enlarge, slowly at first, and keeping their

* *Bulletin de la Société d'Encouragement pour l'Industrie Nationale* (5) 10, 476.

forms, then more rapidly, until at 1000° they become irregular and tend to lengthen into groups of juxtaposed bands. The pearlite generally remains exterior to the arrangement of ferrite, and stratifies in seams."

The most important articles on this subject that have yet appeared have been by Mr. J. E. Stead.* His first paper on "The Crystalline Structure of Iron and Steel" revolutionized our ideas upon the effect of heat treatment on soft steels and showed that they followed laws of structure which so far have not been found to apply to carbon steels. Among the conclusions that Mr. Stead draws, we have selected the following, which have a bearing on the present work:

1. Granules and crystals should not be confused, for although a granule is built up of crystals, its external form is not crystalline, as it takes its shape from its surroundings. It is an allotrimorphic crystal, or more simply, a grain.

2. Grains formed in the solidification of liquid metals are large or small according to whether the freezing is rapid or slow, and they take their form from their surroundings, and this is most irregular, owing to natural interference.

3. In carbonless pure irons or steels of fine grain produced by either forging or certain heat treatment the grains increase in size, slowly at 500° C., and more rapidly at between 650° to 750°, and it is possible, by heating at about 700° for a few hours, to develop granular masses of exceeding coarseness. At 900° the granules again become small, and heating to 1200° apparently does not change them.

4. Granules in carbon steels 0.20 per cent to 1.20 per cent carbon do not grow in size by long continued heating at 700°, but they increase in size slowly at first, and then more rapidly with each increment of temperature above 750° C., the change point varying with the carbon. When, however, coarsely granulated steel is reheated again a little above 700° to 750°, the coarse structure vanishes.

5. In steels of from 0.10 per cent to 0.15 per cent carbon containing the pearlite in widely separated areas, on heating and quenching from about 750° the large ferrite grains are not broken up, and the carbon apparently does not expand or diffuse beyond

* *Bulletin de la Société d'Encouragement pour l'Industrie Nationale* (5) 10, 476.

the original area, but when heated to 850° and slowly cooled the carbon areas are found as a number of smaller segregations, and where the carbon has diffused the ferrite is found in small grains.

Mr. Stead's second paper on "Brittleness Produced in Soft Steel by Annealing,"* follows the line of his previous paper, but brings out a number of new facts. Among other points he distinguishes two kinds of weakness, viz.:

1. Intergranular weakness which occurs when the grains are imperfectly cohesive or are surrounded by a brittle envelope.

2. Cleavage weakness which occurs where the grains are so oriented that the component crystals are all in the same phase. He explains the brittleness sometimes observed in annealed plates by saying that the rolling causes a tendency to such orientation, and that annealing develops this tendency.

Mr. Stead suggests that at about 900° for soft steel *Gamma-iron* is formed, and coincident with it the structure is refined. The results of the present investigation bear out this conclusion. Mr. Stead also makes the pertinent suggestion that the banded or laminated structure of the ferrite, observed in overheated steel, provided there is no intergranular separation, is a remarkably strong one.

Shortly after the publication of Stead's papers, Ridsdale published an article on "Brittleness in Soft Steel,"* containing much valuable material. He draws the following conclusions, "In order to avoid brittleness:

1. "There should not be too high an initial temperature nor 'soaking' at a high temperature for a long time."

2. "Work should be continued down to red heat, but not to a blue heat."

3. "If work has been continued unavoidably to near blue heat, there should be no chilling, but slow cooling, and where possible subsequent heating, if only for a short time, to cherry redness. Also there should be absence of jar or vibration whilst cooling through the blue heat."

A year later, in 1899, Mr. Ridsdale published a very comprehensive paper on "Practical Microscopic Analysis for Use in

---

* *Journal Iron and Steel Institute*, 1898, II, page 137. *The Metallographist*, 1899, Vol. II, page 85.

† *Journal Iron and Steel Institute*, 1898, I, page 220.

the Steel Industries."* He emphasizes strongly the importance of soft steels, and the neglect which they have suffered at the hands of the metallographists. His conclusions, which are derived from the study of numerous specimens of steel made by various processes and makers, are illustrated by photographs at 50 diameters.

His conclusions may be stated as follows:

1. Normal steel, work continued to a low red. The structure is regular and homogeneous, the grains small and ill-defined.

2. Normal treatment and abnormal composition. The structure is in general the same as the preceding. Phosphorus tends to enlarge the grain slightly.

3. Normal steel heated too long or to too high a temperature. This shows an outer carbonless band with large grains, or if actually burnt, a separation of the grains.

4. Normal steel worked at too low a temperature or through a blue heat. This shows a strained structure with distinct flow lines. There is a breaking up of the surface grains and a loosening or actual rupture of the grain junctions.

"In general," Ridsdale says, "outside of normally treated specimens, the effect of heat treatment far outweighs that of composition."

Morse,† who worked on a steel containing 0.34 per cent carbon, finds a diminution in tensile strength when held between 650° and 750°, an increase from the latter point up to 900°, and a falling off from 900° to 1200°. Beyond 1200° the tensile strength rises again. The area of the grain increases up to 600°, falls to a minimum at 900°, and then increases rapidly. The author notes the occurrence above 1000° of the banded or ramified structure noted by Stead, and he reaches the same conclusion as to its being a strong structure.

In regard to the existence of a point of change in the physical properties at a temperature above 1000°, there are several articles which deserve mention. Ball published an article "On the Changes in Iron Produced by Thermal Treatment,"‡ in which

---

* *Journal Iron and Steel Institute*, 1899, II, page 102. *The Metallographist*, 1900, Vol. III, page 64.

† *Trans. Am. Inst. Min. Eng.*, 1900, 29. *The Metallographist*, 1900, Vol. III, page 130.

‡ *Journal Iron and Steel Institute*, 1890, I, page 85; 1891, I, page 103

he investigated the effect upon the tensile strength and magnetic properties of a steel containing about 0.10 per cent carbon when slowly cooled, and when quenched from various temperatures. The steel shows three critical points, one at 660°, another at 900°, and a third at 1200°, at which latter point the magnetic and physical properties reach a maximum.

Recently Roberts-Austen* has published a cooling curve of electrolytic iron which shows an evolution of heat at a temperature of 1132°, and this he identifies with the point of maximum tenacity observed by Ball. This upper critical point coincides fairly closely with the second point of maximum tenacity observed in the present series of experiments.

Experimental

The material which was used for these experiments was a low carbon, acid Bessemer steel of a commercial grade, and was kindly furnished by the National Tube Company, of McKeesport, Pennsylvania. It represents the average material of this grade turned out by the company for the manufacture of pipes and tubes.

The chemical analysis was as follows:

C = 0.07. S = 0.058. P† = 0.10. Mn = 0.32. Oxides = 0.25.

The carbon and manganese were determined by colorimetric methods.

The heat treatment in the mill was approximately as follows: The ingots were subjected, in the soaking pits, to a temperature of 1000° to 1200°. They entered the roughing train at probably over 1000°, and the mechanical work was continued to a full red heat. The specimens, thirty in number, were all cut from plates rolled from the lower two-thirds of the same ingot, and came in the shape of bars 18 inches long, 1¾ to 2¼ inches wide, and ⅜ inch in thickness. The edges were planed. In order to heat the specimens uniformly and to a temperature which could be accurately measured, the usual forms of furnaces were found in-

---

* *The Metallographist*, No. II, 186.

† Although high, the phosphorus seems to have had no effect on the size of the grain.

convenient. At the suggestion of Mr. C. L. Norton,* a furnace heated by electricity was constructed and was found to answer all of the requirements most admirably.

The general principle on which the furnace is based is that when a conductor is traversed by a current, heat is developed according to Joules's law, $W = C^2R$. The resistance of the wire, however, increases as the temperature rises, and as the loss of heat by radiation also increases with the temperature, it may be

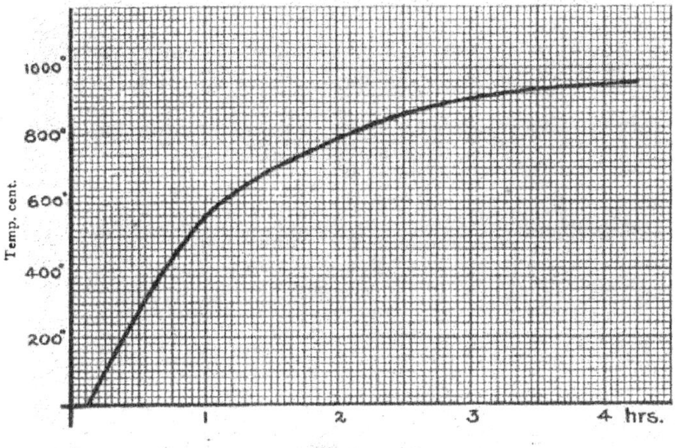

Fig. 1. — Heating Curve of Furnace.

seen that for each difference of potential between the terminals there is a certain maximum temperature which may not be exceeded. The rise in temperature for 110 volts is shown by the curve in Fig. 1. By the use of 220 volts, however, the temperature may be readily raised to 1300°.

The furnace consisted of a cylindrical muffle 20 inches long and 3¼ inches in diameter wound with about 72 feet of No. 29 platinum wire and inclosed in an insulating jacket of about 15 inches external diameter and 30 inches long. The muffle was made by taking three clay porous cups, sawing the bottoms off

* Complete details in regard to this and other forms of electric furnaces used in the heat laboratory, will appear in a separate article by Mr. Norton.

two of them and placing the three in line so as to form a hollow cylinder closed at one end. The bottom of the third cup was perforated by a ¾-inch hole, so that a tube could be inserted, and a neutral gas, such as nitrogen or carbon dioxide, drawn through to prevent oxidation of the specimens. It was not found necessary, however, in these experiments.

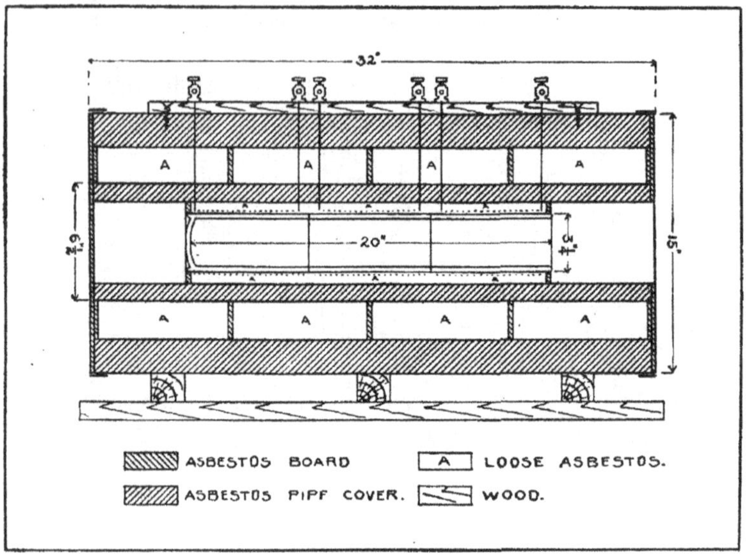

Fig. 2
Section of Furnace.

The three sections of the muffle were wound independently, each with about 24 feet of No. 29 platinum wire, and the ends were brought out to six binding posts on the outside of the furnace. The insulating jacket consisted of two pieces of asbestos pipe cover, one 12-inch piece outside, and a 4-inch piece inside. The spaces between the two pipe covers, as well as between the inner pipe cover and the muffle, were packed with fibrous asbestos. The whole furnace was supported in a wooden cradle, and is shown in section by the accompanying sketch, Fig. 2.

The electric current was taken from the three wire mains of the incandescent circuit by the three coils in parallel. For temperatures below 800°, 110 volts were used, and for temperatures above 800° the furnace was heated to 800° on the 110 volt circuit, and then shifted to the 220 volt circuit.

The efficiency of the furnace was very good indeed. With a difference of potential of 110 volts the current passing varied from 10 amperes when cold to 6.5 amperes when hot, and the furnace could be heated to 800° in about one and one-half hours. (See Fig. 1.)

The pyrometer used was of the Le Chatelier thermo-electric type. The junctions consisted of a platinum, and a platinum 10 per cent·iridium wire, and were protected by perforated clay tubes. The junction was simply laid on top of the bar in the furnace, but the heating was so gradual that this represented very closely the actual temperature.

The galvanometer was a four-coil, astatic instrument suspended from the ceiling to prevent vibration. The source of light was a 32 C. P. lamp, a narrow adjustable slit being interposed between it and the mirror. The image of the slit was thrown onto a ground glass strip graduated in millimeters.

The pyrometer was calibrated against the following substances:

Boiling water, 100° C. Boiling naphthalene, 219°. Boiling sulphur, 445°. Freezing aluminum, 660°. Freezing gold, 1,072°.

The temperatures above 1072° were obtained by extrapolation on the calibration curves. The general arrangement of furnace, galvanometer, and wiring is shown in the photograph of the apparatus, Fig. 3.

The treatment of the specimens was carried on as follows: A bar, numbered at each end with its serial number, was placed in the furnace. Upon it, and extending to a point one-third the length of the bar was placed the pyrometer junction. The mouth of the muffle was then closed with a wad of loose asbestos, and the heating was begun. The current was turned on at 110 volts with the coils in parallel until a temperature of 800° was reached, when 220 volts were turned on and the furnace heated to the desired temperature. The circuit was now broken and the maximum temperature recorded. The bar was allowed to cool in the furnace, and after 24 hours, at the end of which time

Fig 3. Furnace, Pyrometer, etc.

it was still at a temperature of 80° to 100°, was removed. All the bars except Nos. 23 to 27 inclusive were allowed to cool in the furnace. The latter bars were removed from the furnace just after reaching the maximum temperature and thrown into a box of sand and allowed to cool naturally, reaching the atmospheric temperature in a couple of hours. The furnace was so well protected from radiation that the specimens which were allowed to remain in the furnace were cooled extremely slowly.

The bars, after the heat treatment, were machined for a space of about 12 inches in the middle of the bar to remove the scale and allow an accurate measurement of the cross section of the specimen. After measuring the cross section, the bars were marked for the elongation in 5 inches, and broken in a testing machine.

The testing machine used was a 50-ton Olsen arranged to be run by an electric motor. With the motor running at a constant rate the load rose rapidly at first, and then slowed up suddenly as shown by a drop of the beam. The stress at which this occurred was taken as the elastic limit, though its accuracy is probably questionable to the extent of 1000 pounds. After the elastic limit was passed, the piece began to elongate more rapidly and the load to rise more slowly up to a maximum stress, beyond which the piece began to draw down under a diminishing load. The maximum load was recorded as the ultimate strength of the material. As the piece drew down, two lines of greatest stress, forming a St. Andrew's cross, showed themselves by faint depressions across the bar. These depressions enlarged until the fracture occurred, beginning at the center and extending to the sides. This mode of fracture, which was characteristic of all the specimens, was beautifully shown by specimen No. 10, with which the machine was stopped just after the middle had parted and before the break had reached the edges. The fractured bars were placed end to end, and the elongation was measured and recorded. Owing, however, to the way in which the bars parted, the ends could not be placed in perfect contact, and there is consequently some uncertainty in the accuracy of the percentage of elongation.

From the facts obtained in the above described manner the following data were calculated:

1. Temperature from which annealed.
2. Maximum load, pounds per square inch.
3. Elastic limit, pounds per square inch.
4. Elongation, per cent of in 5 inches.

The tabulated results are given in the table, and are shown graphically in Fig. 4.

The specimens for microscopic examination were cut from the ends of the fractured bars and numbered with the serial number.

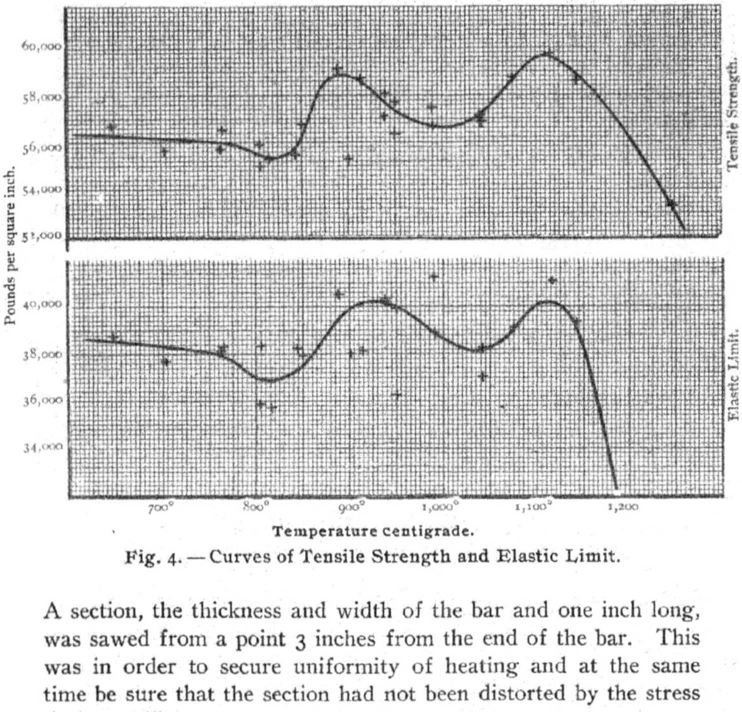

Fig. 4.—Curves of Tensile Strength and Elastic Limit.

A section, the thickness and width of the bar and one inch long, was sawed from a point 3 inches from the end of the bar. This was in order to secure uniformity of heating and at the same time be sure that the section had not been distorted by the stress during pulling.

The small block so cut was then planed and filed down one-sixteenth of an inch on the flat surface to remove all surface injuries. The specimen was then bevelled on the upper edges and ground on three grades of emery cloth, Nos. 80, 100, and 120, and four grades of "Rupert's" French emery paper, Nos. 10,

1, 0, and 00. This left the specimen with a dead smooth surface ready for the final polishing, which was done with the finest jeweller's rouge and water upon wash leather stretched on a square tile. The polishing was continued until the specimen was entirely free from scratches and had attained a mirror-like surface.

Tabulated Results Shown Graphically in Figure 4.

Specimen number	Temperature (Degrees C.)	Area	Tensile strength (Lbs. per sq. in.)	Elastic limit (Lbs. per sq. in.)	Elongation in 5 inches (Per cent.)
1	647°	·5670	56,660	38,790	31·0
18	703	·5827	55,620	37,760	32·2
2	764	·5767	55,690	38,150	31·2
6	765	·5740	56,570	38,320	29·0
4	805	·4953	55,880	38,370	29·8
16	806	·6829	55,000	35,870	28·4
15	817	·6149	55,310	35,780	31·6
8	844	·5768	55,470	38,140	30·6
19	850	·6063	56,790	37,930	30·6
3	889	·5923	59,120	40,520	30·0
11	901	·5920	55,280	38,010	33·2
17	914	·4388	58,750	38,130	—
14	939	·7077	57,110	40,260	33·6
5	939	·5706	58,120	40,310	—
9	950	·6002	57,760	39,980	32·0
22	951	·5244	56,300	36,230	30·6
7	988	·4363	57,530	41,260	28·4
21	990	·5907	56,690	38,920	28·2
28	1,041	·6078	57,190	37,020	31·2
10	1,042	·6538	56,840	38,230	31·0
12	1,075	·6134	58,760	39,130	29·8
29	1,117	·5599	59,770	41,080	30·6
13	1,143	·4831	58,600	39,340	29·8
20	1,247	·5667	53,280	21,170	31·2
*23	695	·6764	58,880	43,610	32·5
*24	805	·7256	58,530	38,600	32·5
*27	891	·6089	60,150	39,000	29·0
*26	990	·6355	60,530	36,980	—
*25	1,247	·5855	58,100	35,870	30·6

*** Specimens 23 to 27 inclusive, were rapidly cooled.**

During the progress of the polishing the specimens were examined at intervals under the microscope. When the scratches were all removed they showed a ground mass of ferrite with small yellowish markings of pearlite unevenly distributed over the field. The pearlite stood out slightly in relief due to its superior hardness. This is shown in Figs. 5 and 6. The uneven distribution of the pearlite was quite noticeable even to the naked eye, the block presenting a banded or moiré appearance. Under the microscope the carbon was seen to be highly segregated, some areas being thickly strewn with the pearlite, while others showed only scattered particles.

When the polished specimens had been examined they were etched. The best results were obtained by immersing one end of the specimen in 10 per cent nitric acid for 30 seconds, and washing first with running water and then with alcohol. The surface was dried by warming gently. The etched specimens showed very clearly the granular structure of the steel, the joints between the grains being well developed. It is well known that erroneous results are likely to follow the use of nitric acid as an etching medium, but after trying many other agents with indifferent success it was decided to use nitric acid uniformly throughout the series.

The microscope used was one made by Reichert, of Vienna, especially for metallographic work. It was fitted with an adjustable mechanical stage, vertical illuminator, and iris diaphragm. The field was illuminated by means of a 16 C. P. incandescent lamp placed 18 inches from the microscope. A 3-inch bull's-eye condenser was interposed between the two.

After the preliminary examination the specimens were photographed. The camera for this purpose was the vertical type and was used with the eye-piece in the microscope just as for eye examination. The ground glass screen was provided with a small circle of clear glass in the center, and the microscope was focussed by placing a low-power eye-piece, with the ocular lens removed, upon it, and focussing its smaller field. The exposure varied from 20 to 25 minutes. The photographs were taken at 84 diameters and are reproduced at 73 diameters.

Discussion of Results

After the specimens had been heated they were, as a rule, covered with a greater or less amount of scale. Those heated to below 750° had a thin powdery coating of red oxide upon them, readily removable with the finger. The specimens heated to above 750° had a coating of black magnetic oxide varying in thickness from the thinnest possible film at 800° to a thirty-second of an inch at 1200°.

The machining of the specimens gave no indication of the temperature to which they had been heated; even the 1247° specimens were apparently uninjured. The first difference in behavior was noted when the tensile tests were made. The specimens heated to below 1200° pulled in a perfectly normal manner, having a high elastic limit and ultimate strength, and, when finally breaking, showing a smooth face and fibrous structure. The two specimens heated to 1247° behaved differently. The elastic limit was decidedly lowered, to 21,170 pounds in the case of the slowly cooled bar. When the elastic limit was passed, however, the bars held out remarkably well, the slowly cooled bar breaking at 53,280 pounds and the rapidly cooled bar at 58,100 pounds per square inch. After fracture had occurred it was seen that the surface of these two bars had roughened, having the appearance of a piece of putty when pulled apart. This was much more noticeable, in. the case of both bars, upon one side than upon the other, and as the two bars were heated in the furnace together it may be that the rougher sides were more exposed to oxidizing gases than the other sides which were in contact with each other. To settle this point will require further experimental work. The fractured areas of these two bars showed some bright crystalline spots resembling the grain of overheated carbon steel.

The results of the tensile tests as shown by the table and the curves plotted from it seem to indicate:

1. That there is a fair degree of constancy in the tenacity of the bars heated to below 850° with possibly a slight falling off between 800° and 850°.

2. That at 890° there is a sudden rise in the tensile strength from 56,790 pounds to 59,120 pounds per square inch.

3. That beginning at 900° there is a slight falling off reaching a minimum of 56,690 pounds at 1000°, from which point it rises again.

4. That at 1125° the strength reaches a second maximum of 59,000 pounds.

5. That from 1143° there is a very decided falling off in the strength, reaching 53,280 pounds at 1247°, beyond which point the experiments were not carried.

The concordance of the results obtained is seen to be quite good, most of the observations lying within 500 pounds of the mean curve. One result only, No. 11, lies widely off, showing a discrepancy of 3600 pounds. It lies, however, in a portion of the curve where the strength changes rapidly, so a small error in the temperature would make a large apparent error in the strength.

The results of the elastic limit experiments are very similar to those of the ultimate strength, but show much larger variations, and consequently the curve is more doubtful; but it seems to be of the same nature as the ultimate strength curve.

The results of the elongation determinations were very unsatisfactory, owing, as already stated, to the impossibility of matching the ends together. As well as could be determined, however, it seems to show:

1. A decrease in elongation to 800°, at which point it is about 29 per cent.

2. A rise to 950°, where it reaches 32.5 per cent.

3. A sudden decrease to 1000° at which point it again reaches 29 per cent.

4. A gradual increase up to the limit of the experiments 1247°, at which temperature the elongation is 31.2 per cent.

The rapidly cooled specimens show a tensile strength quite uniformly from 2000 pounds to 2500 pounds higher than the corresponding slowly cooled specimens, with the exception of the 990° rapidly cooled bar which broke at 60,530 pounds, or even higher than the 900° bar. The stretch limit shows a gradual decrease from 43,000 pounds at 695° to 36,000 pounds per square inch at 1247°. The variations in the elongations seem to be less than the errors of determination.

The microscopic examination of the different specimens has been disappointing. On account of the very small amount of carbon it is extremely difficult, if not impossible, to distinguish the different specimens which have been etched only by polishing in relief. There is in all specimens a ground mass of structurally free ferrite through which are disseminated particles of another constituent which we would expect to be pearlite, but which looks very much more like structurally free cementite (Fig. 5). In no case was it possible to observe the characteristic pearlite markings. The appearance is usually that of an amorphous granular mass which stands in relief. With increase of temperature there is a greater tendency for this constituent to segregate in masses.

Fig. 5. Etched by polishing. Specimen No. 1. Cooled slowly from 647°. Tensile strength, 56,660 lbs. Magnified 73 diameters.

Fig. 6. Etched by polishing. Specimen No. 20. Cooled slowly from 1247°. Tensile strength, 53,230 lbs. Magnified 73 diameters.

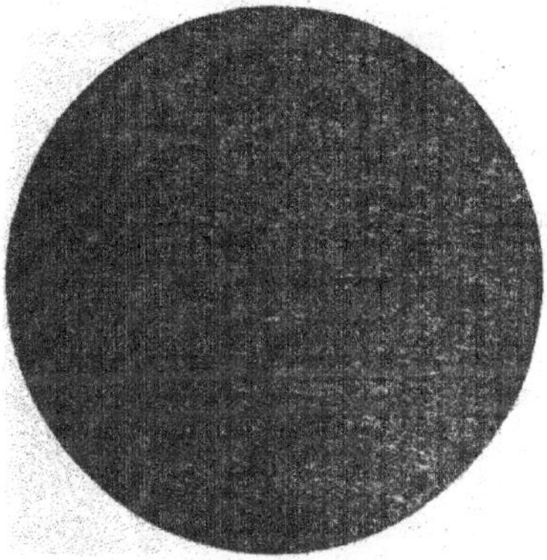

Fig. 7. Etched 30 seconds $\frac{1}{10}$ HNO_3. Specimen No. 3. Cooled slowly from 889°. Tensile strength, 59,120 lbs. Magnified 73 diameters.

Fig. 8. Etched 30 seconds $\frac{1}{10}$ HNO_3. Specimen No. 7. Cooled slowly from 988° Tensile strength, 57,530 lbs. Magnified 73 diameters.

The scratches which appear on the specimens from polishing would also seem to indicate the presence of cementite, as the course of the scratch is usually interrupted and continues beyond the particle. Figs. 5 and 6 show the general appearance of the microstructure of the specimens which have been etched by polishing in relief.

The examination of specimens etched with dilute nitric acid shows that the grain is practically uniform in size in those specimens annealed from 850° and below. When we come to specimen No. 3, Fig. 7, slowly cooled from 889°, however, we see a marked change in the structure. The granular structure has disappeared and has been replaced by a fine close structure resembling porcelain. Specimens Nos. 17, 5, and 9, slowly cooled from 914°, 939°, and 950° respectively, show a fine structure but with grains, visible under a magnification of 84 diameters.

Specimen 7, Fig. 8, shows again a larger development. Specimens 12 and 13 show a grain of a size not greatly different, but with a structure very different. Specimen 13, Fig. 10, shows parallel striations or bands appearing as alternating dark and light streaks. These bands occur in groups in which the component lines are parallel, but in the different groups they are nearly at right angles. Specimen 20, which was cooled from 1247°, shows an enormous increase in the grain and also a marked development of these bands or striæ through the grains. This is well shown in Fig. 11.

A comparative study of the results of the tensile tests and the microscopic examination show in general a fairly close relation between the size of the grain and the tensile strength. The larger the grain the lower is the tensile strength and elastic limit. The latter seems to be affected by the grain more than the former. The elongation does not show any definite relation to the size of the grain.

The bars which were cooled slowly from below 850° show a tensile strength of from 55,500 to 56,500 pounds and a grain of fairly large diameter. Passing from 850° to 900° there is a very marked increase in the tensile strength, 59,000 pounds, and an equally sudden diminution in the size of the grains, the latter becoming the finest noted in the series. From 900° to 1000° the gradual increase in the grain is accompanied by a decrease in the strength, the two being quite proportional.

Above 1050° there is an increasing size of grain to a slight extent, but the structure here is changing to the banded type. The strength increases through this range to 1143°, showing that this structure is a strong one and masks the effects of the grain structure. Above 1143°, however, the grain increases rapidly and the strength decreases correspondingly.

In general the following tentative conclusions may be drawn:

1. For this particular steel annealing seems to give two points of maximum tenacity and elastic limit, one at 890°, and

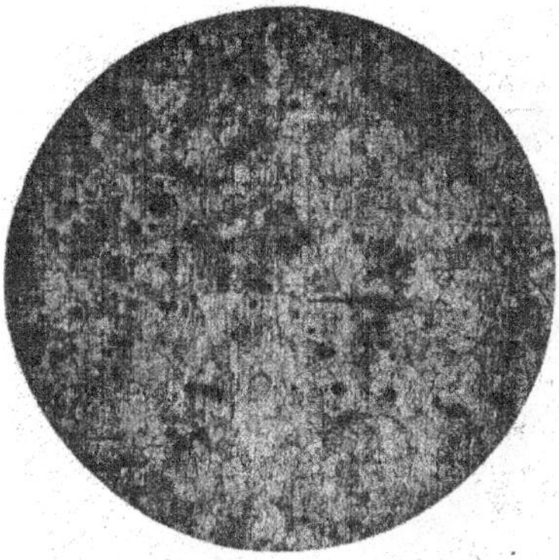

Fig. 9. Etched 30 seconds $\frac{1}{10}$ HNO_3.
Specimen No. 28. Cooled slowly from 1041°.
Tensile strength, 57,190 lbs.
Magnified 73 diameters.

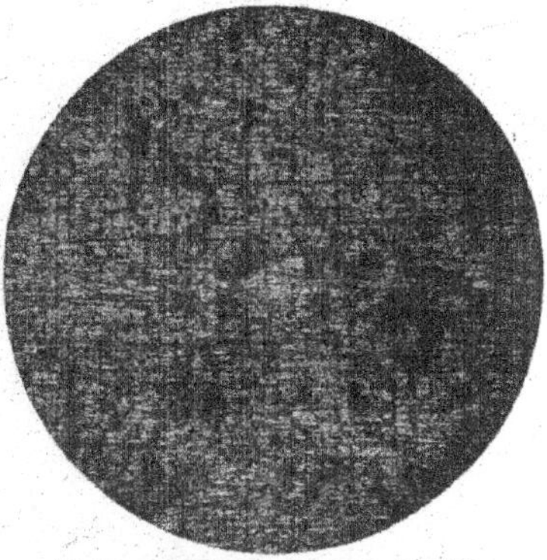

Fig. 10. Etched 30 seconds $\frac{1}{10}$ HNO_3.
Specimen No. 13. Cooled slowly from 1143°.
Tensile strength, 58,600 lbs.
Magnified 73 diameters.

another at 1125°. Below 850° the properties of the steel do not change to any appreciable extent, and between 890° and 1125° the steel shows a minimum value at 1000°. Annealing from temperatures above 1125° causes a decrease in strength with increasing temperature.

2. The size of the grain is apparently constant or increases but slowly at temperatures below 850°. At 890° the grain is suddenly refined, and the structure becomes close and almost porcelain-like. From temperatures above this, annealing seems to cause a gradual increase in size slowly up to 1125°, and rapidly from that point up to the limits of the experiments.

3. Above 1100° a new type of structure appears which seems to consist of parallel bands and striations sometimes crossing the grain junctions. This type of structure has not, however, been carefully studied, and we are not prepared at present to state its significance.

4. Below 1000° there is a fairly close relation between the size of the grain and the strength, the latter falling off as the former increases, and *vice versa.* A fine grain indicates a high

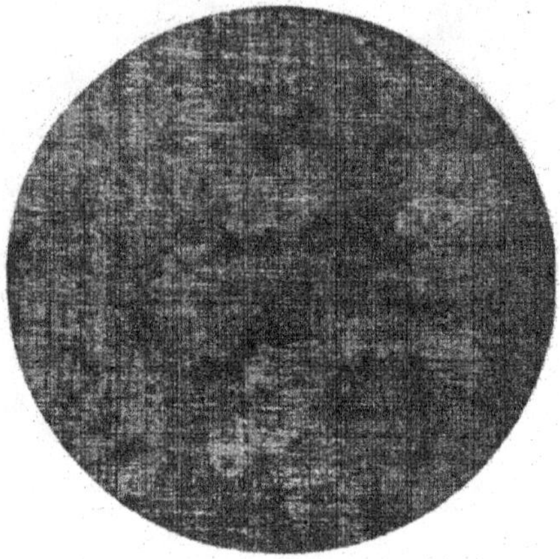

Fig. 11. Etched 30 seconds $\frac{1}{10}$ HNO_3. Specimen No. 20. Cooled slowly from 1247°. Tensile strength, 53,280 lbs. Magnified 73 diameters.

tensile strength and elastic limit. The latter is affected more than the former.

5. Above 1000° the banded structure appears to increase the strength and also somewhat retard the grain development, but at 1125° the latter takes the ascendency as indicated by the decrease in strength from that point to 1247°.

The banded structure when not accompanied by an excessive grain appears to be a very strong form.

6. Rapid cooling, as in the open air, gives a higher tensile strength than does slow cooling from the same temperature.

7. The size of the grain of the more rapidly cooled specimens (23-27) follows the same general laws as for the slowly cooled,

except that the grain for the same temperature is uniformly smaller.

In connection with the point of maximum tenacity and the related change in structure it may be said that its position 1140° corresponds closely with the highest critical point observed by Roberts-Austen in his cooling curve of electro-iron, and also with the point of maximum tenacity observed by Dr. Ball.

It is desired to express the thanks of the authors to Mr. C. L. Norton, without whose aid and kindly advice much of the work would have been impossible; likewise to Professors Miller and Schwamb for the use of much valuable apparatus.

SOME EXPERIMENTS ON THE METHODS OF ANNEALING STEEL*

By A. CAMPION

A YEAR ago the author had the honor of presenting to this Institute a paper in which he endeavored to set forth some of the changes of physical properties and of microstructure in various grades of steel, brought about by heat treatment.† He also gave the results of some preliminary experiments which he had made, in an investigation of the process of annealing steel, in order to determine in what manner and to what temperature, also the length of time of heating to, and of cooling from that temperature to below the critical range Ar_1, the material should be subjected, so as to give the lowest possible tensility and greatest possible ductility, accompanied by fine grain for different grades of steel.

The experiments have been continued during the past twelve months on similar lines to those described in the previous paper; and the present contribution deals with the results which have been obtained.

Although a considerable number of experiments have been performed, the progress has not been as rapid as the author would have liked; there have been no startling developments or

* *Journal West of Scotland Iron and Steel Institute*, 1901. The original paper was illustrated by a large number of photomicrographs.

† *The Metallographist*, 1900, Vol. III, page 168.

facts new to science to report; nevertheless, it was thought advisable to publish the details of the experiments performed and results obtained up to date, in order that they might be subjected to criticism in the hope of producing suggestions for future work; and also to give any one who desired to do so an opportunity of repeating the work, with a view to confirmation or otherwise, as it is of the utmost importance that the results should be verified, by a number of investigators working independently, before it is possible to establish with certainty the particular conditions under which one must work in order to obtain the best possible results from annealing steel. It is also very desirable that any interpretation of the results to which exception might be taken should be discussed and inquired into at the present early stage of the investigation.

The author has made many repetitions of the experiments described in the last paper under series 1, on soft steel containing between .16 and .20 per cent of carbon, and has made microscopical examinations and bending tests of the samples after treatment; but, so far, has been unable to obtain results differing to any appreciable extent from those previously published.

As it was not possible to give the results of bending tests in the paper read last year, it was considered advisable to include in the present paper the results of a number of experiments which have since been made on round steel bars of similar composition and of the same dimensions as those used in series No. 1. The results are given below of bending tests and microstructure of steel after various treatment of the following composition:

Carbon (by combustion) . .	0.174 per cent
Silicon	0.075 " "
Sulphur	0.033 " "
Phosphorus	0.057 " "
Manganese	0.560 " "
Arsenic	0.036 " "

The bars used were 1.25 inches diameter, 15 inches long, and weighed about 5½ pounds.

The times taken to heat up and to cool down to 550° C., as suggested by Mr. Stead, have been noted and are also given.

No. 51. — Heated to 840°–860° C. for half an hour, and allowed to cool in the closed furnace. Time taken to heat up thirteen minutes, and to cool to 550° C. sixty-nine minutes.

Sample bent almost close double and cracked slightly.

Microstructure fairly fine grained. Ferrite in small meshes.

No. 52. — Heated to 840°–860° C. for fifteen minutes, and covered with sand to cool. Time taken to heat to 840° C. seventeen minutes, and cooling down to 550° C. took fifty-seven minutes.

This bar just started to crack after bending close double.

Microstructure fairly fine grained. Ferrite meshes rather smaller than in No. 51.

No. 53. — Heated as in experiments 51 and 52, but held at that temperature for one hour forty-five minutes, and slowly cooled in sand. Time of heating to 840° C. sixteen minutes, and of cooling to 550° C. sixty-two minutes.

Sample bent nearly double and began to crack.

Microstructure coarser than in the case of No. 51, with rather large ferrite meshes.

No. 54. — Heated to 940° C. for fifteen minutes, and covered with sand to slowly cool. Heated up in seventeen minutes, and cooled to 550° C. in one hour and nineteen minutes.

Bent almost double and broke.

Structure coarse, with rather large grains of ferrite.

No. 55. — Heated to a temperature between 730° C. and 760° C. for one hour and slowly cooled in sand. Time taken to heat up twelve minutes, and in cooling to 550° C. forty-eight minutes.

Bend test gave excellent result, as bar bent quite close double and then stood several heavy blows with steam hammer without showing the slightest sign of fracture. The microscope shows a very fine structure, the pearlite and ferrite being very intimately mixed and evenly distributed throughout the section.

No. 56. — Treated as in No. 55, but held at that temperature for only half an hour. Took thirteen minutes to heat, and cooled in fifty-three minutes to 550° C.

Bent quite close double with the same treatment as the last experiment. No signs of cracking.

The microstructure almost identical with that of No. 55.

No. 57. — Heated slowly to 620°–670° C. for about three hours, allowed to cool in sand. Time of heating sixty-two minutes. Time of cooling to 550° C. forty-five minutes.

Bent close double showing small crack. Microscope showed

structure to be rather coarse and inclined to be crystalline. Coarser than No. 51.

No. 58. — Heated for six and a half hours at about 750° C. and allowed to cool in the air. Heating up took fourteen minutes. Time of cooling was not noted.

Structure very crystalline, and carbide has apparently separated into masses. Bar broke after bending through 150 degrees with crystalline fracture.

No. 59. — Heated at 910°–940° C. for two hours and five minutes, covered with ashes to cool. Time of heating to 910° C. twenty-one minutes. Cooled to 550° C. in fifty-four minutes.

Bent double and cracked.

Microstructure very coarse and crystalline, but not so much as No. 58. Coarser than No. 54.

No. 60. — Heated for two hours at 750° C., then allowed to cool in closed furnace till below Ar_1. Again heated for two hours at 750° C. and covered with sand to cool. Took twenty-two minutes to heat up the first time, and cooled in furnace to 650° C. in fifty-nine minutes, reheated to 750° C. in seven minutes, cooled to below Ar_1 in sixty minutes.

Bent test exactly the same as in the cases of No. 55 and No. 56. Hammering began to flatten the bars; not the slightest sign of crack appeared.

Microstructure very fine. Ferrite meshes not quite as fine as in 55 and 56. Heating was probably too long.

No. 61. — Heated for about two hours at 950° C. Allowed to cool in air. Heating up occupied seventeen minutes, and cooling to 600° C. took thirteen minutes.

Structure coarse and crystalline. Bent about double and then broke in two places.

No. 62. — Steel as rolled.

Bent close double and showed signs of cracking.

Microstructure fairly coarse and granular, but not as coarse as 54, 57, and 59.

No. 63. — Heated to 850°–900° C. for four hours, cooled in air. Heated up in sixteen minutes. Cooled to 550° C. in eighteen minutes.

Bent through only about forty-five degrees, and broke with a very crystalline fracture.

Microstructure very coarse, with large ferrite crystals.

The above-mentioned experiments have been repeated a number of times on the 1¼-inch bars, and the results obtained by the author working under conditions practically constant have been concordant; and it will be seen that the results given above agree with those given as series 1 of the previous paper.

Of course, in considering these results it must be remembered that bars of such small dimensions are naturally of a not very coarse grain owing to the large amount of work they receive in being reduced to so small a size, and also owing to their small mass and rapid cooling from the finishing temperature, which is usually very little, if at all, above the critical point Ar_1.

The amount of work which the metal has received, its mass and shape, are most important factors in determining the size of grain. In order to determine the effect of these under various conditions the author intends to carry out similar experiments to those already given on steel of different carbon content and shape, and of gradually increasing size. With this object in view a number of bars of rather larger diameter were treated in various ways as described below.

The analysis of the steel was as follows:

Carbon (by combustion)	0.180
Silicon	0.121
Sulphur	0.050
Phosphorus	0.073
Manganese	0.590
Arsenic	0.026

The bars were 1⅝ inches in diameter and 14 inches long, weighing about 8¼ pounds each.

No. 66. — Steel as rolled.

Bent close double but showed signs of cracking.

Microstructure coarse and crystalline, the pearlite present as granular variety.

No. 67. — Heated for two hours at 750° C. Took twenty-five minutes to heat up, and cooled to 550° C. in fifty-nine minutes in furnace, reheated to 750° C. in eight minutes and held at that temperature for another two hours, and finally cooled in sand to 550° C. in seventy-one minutes.

Bar bent quite close double, and showed no signs of crack even after a considerable amount of hammering.

The structure is very fine. Ferrite meshes very small and intimately mixed with the pearlite.

No. 68. — Heated rather slowly to 840°–870° C. for one hour and thirty-five minutes, and slowly cooled in sand. Heating to 850° C. took thirty-four minutes. Cooling in sand to 550° C. took sixty-two minutes.

Test bar bent double and then snapped with a crystalline fracture.

Microstructure coarse and crystalline, more so than in the case of untreated steel No. 66. The heating was too long and at too high a temperature. Probably also too slow.

No. 69. — Heated fairly rapidly to 840°–860° C. for forty minutes, slowly cooled in closed furnace. Heating to 850° C. took twenty minutes, and cooled to 550° C. in one hour and eighteen minutes.

Structure appears fairly fine, but under a magnification of 150 diameters is distinctly crystalline.

Test bar bent double, but started to crack after two or three blows from hammer.

No. 70. — Heated to 840° C. for about twenty minutes. Cooled in sand. Took tweenty-three minutes to heat up, and sixty-eight minutes to cool to 550° C.

Bent close double.

Structure coarser than No. 69, but not as crystalline.

No. 71. — Heated for one hour at 760°–790° C. and slowly cooled in sand. Heating took twenty-eight minutes, and cooling to 550° C. sixty-one minutes.

Bent close double, no signs of cracking after hammering.

Microstructure pretty fine, but not as fine as 67.

No. 72. — Heated very slowly to 740°–780° C. and kept at that temperature for an hour and five minutes, and slowly cooled. Took an hour and fifty-six minutes to heat to 750° C., and an hour and twenty-four minutes to cool to 550° C.

Bent double and cracked.

Structure coarse and crystalline, and somewhat resembling that of No. 66, but crystals scarcely as large.

No. 73. — Heated fairly quickly to 750°–780° C. for half an hour, slowly cooled in sand. Heated up in twenty minutes and cooled in fifty-eight minutes to 550° C.

Bent quite close double.

Structure fairly small, but not so fine as No. 71.

No. 74. — Heated to 690°–720° C. for about an hour, slowly cooled in sand. Heated up in twenty-three minutes, and cooled to 550° C. in fifty-seven minutes.

Bent close double.

Structure moderately fine, but not so fine as 67, 71.

No. 75. — Heated to 750° C. for a quarter of an hour, slowly cooled in sand. Heated up in twenty-two minutes, cooled to 550° C. in fifty-four minutes. This bar was kept at temperatures of 670°–640° C. for about twenty-minutes during the cooling.

Bent quite close double, began to flatten under hammer without showing any signs of crack.

Structure fine grained, but not as fine as No. 67. Ferrite in small meshes.

No. 76. — Heated to 760°–790° C. for about one hour, allowed to cool naturally. Took twenty-five minutes to heat, and eighteen minutes to cool to 550° C.

Bent close double.

Structure fine grained. Ferrite areas small but increased in size towards the centre of bar, a result which one would naturally

TABLE I. — *Showing results of tests for tensile strain, elongation and elastic limit of 1⅝ inch diameter soft steel bars annealed in various ways.*

No.	Size of Sample	Breaking Strain. Tons per sq. in.	Elongation per cent 3 in.	Elastic Limit. Tons per sq. in.	Reduction of area per cent.
66	·798 diam.	29·2	33·0	16·0	57·6
67	"	28·6	39·0	15·0	65·4
68	"	29·8	30·5	17·6	52·4
69	"	29·6	35·5	15·0	60·8
70	"	28·7	36·5	15·8	60·8
71	"	28·6	39·0	18·0	60·8
72	"	29·8	32·5	15·0	55·8
73	"	28·6	38·0	15·6	59·2
74	"	28·0	33·5	16·0	59·2
75*	"	29·2	32·5	15·8	52·4
76*	"	30·2	28·5	17·6	50·8

*** In these cases it was found after testing that a slag thread ran through the centre of the test piece. The breaking strains are possibly too high, whilst percentage of elongation and reduction of area may be less than the true amount.**

expect to obtain owing to the difference between the rates of cooling of the outside and center.

In the above series of experiments two pieces of equal size were in all cases treated together. One piece was used for bending tests, the other, after sawing off the section for microscopical examination, was turned down to ¾ inch diameter for a length of six inches and tested for tensile strain and elongation. The results of tensile tests are shown in Table I, and it will be seen that they are in accordance with what might be expected from the microscopical examination.

The sections for micro examinations were etched with nitric acid and photographed at a magnification of fifty diameters by vertical light. An objective having a focal length of 22 mm. and an angular aperture of 15° was used in conjunction with a Zeiss No. 5 projection eye-piece.

The etching was done by the method suggested by Mr. Sauveur, which consists of plunging the section in concentrated nitric acid of specific gravity 1.42 when the steel assumes the passive state; the piece is then washed under the tap in a good stream of water. The process can of course be repeated, and thus any desired depth of etching can be obtained. The author finds this method to give most excellent results.

Another set of six bars of still larger diameter, of a similar steel to the last, were treated at various temperatures as described below.

The steel had the following composition:

Carbon (by combustion)	0.167
Silicon	0.075
Sulphur	0.041
Phosphorus	0.055
Manganese	0.460
Arsenic	0.037

The bars were 2¾ inches diameter, 2 feet 3 inches long, and weighed about 45½ pounds.

No. 79. — Steel as rolled.

Bent double and just started to crack. Microstructure very coarse, with ferrite in large crystals.

No. 80. — Heated rapidly in billet-heating furnace to 950° C. for ten minutes, and slowly cooled by covering with ashes. Time taken to heat up about fourteen minutes, and cooling in ashes to 550° C. took two and a quarter hours.

Bent almost double and then broke with a crystalline fracture.

Microstructure is coarser and ferrite crystals larger than in the case of No. 79.

No. 81. — Heated rapidly to 890° C. in billet furnace for twenty minutes, and allowed to cool in the air. Time taken in heating to 890° C. was thirteen minutes, and cooling to 550° C. took about fifty minutes.

Bent double and broke.

Coarse and crystalline structure, very similar to that of No. 79.

No. 82. — Heated rapidly in billet furnace to 920° C. for ten minutes and allowed to cool in air. Heated up in a quarter of an hour, and cooling to 550° C. took about an hour.

Bent close double, but started to crack. Microstructure coarse and crystalline near the center of bar, but fairly fine near the outside.

No. 83. — Heated to 760° C. to 800° C. for about twenty minutes in small test annealing furnace and allowed to cool in closed furnace. Time taken in heating was twenty minutes, and in cooling to 550° C. about 3½ hours. The cooling was checked at about 700° C., and the bar kept at approximately that temperature for about 1½ hours, after which cooling was fairly rapid.

Bent quite close double, and showed no signs of cracking, even after repeated blows from steam hammer.

Microstructure shows the ferrite crystals to be fairly well broken up; but structure is not as fine as No. 84. Possibly the cooling was too prolonged.

No. 84. — Heated to 810° C. for five minutes, cooled in air to about 600° C. in twenty-five minutes, and reheated to 780° C. for fifteen minutes, and allowed to cool in air. Time taken to heat to 810° C., eighteen minutes, reheated from 600° C. to 780° C. in five minutes, and finally cooled in air to 550° C. in about three-quarters of an hour.

Bent quite close double, and showed no signs of cracking after a large number of blows from steam hammer.

Microstructure fairly fine, with well broken up ferrite. The pearlite and ferrite are pretty evenly distributed.

The sections were etched with nitric acid and photographed

at a magnification of fifty diameters by vertical light exactly as in the previous set. Tests for tensile strain and elongation were not made owing to the cost and labor of preparing test pieces from such large bars.

It appears from the above experiments that in annealing soft steel of from .16 to .20 per cent of carbon in bars of medium weight and round sections there are fairly wide limits of temperature to which the steel may be heated to give a fine grained microstructure, low tensile and ductility; provided that attention is paid to the rate of heating to, and to the length of time which the steel is kept at the particular temperature; and also to the rate of cooling of the piece to below the critical range Ar_1 or the V on the scale of Brinnel.

The experiments on steel of similar composition to that used in Series II of the previous paper have been repeated, and some of the results are given below.

Five pieces of 1½ inches diameter round bars of the same steel as No. 2 of last year's paper were treated as under:

The analysis of the steel was:

Carbon (by color)	0.360
Silicon	0.075
Sulphur	0.038
Phosphorus	0.039
Manganese	0.752
Arsenic	Not estimated.

The bars were 2 feet long and weighed about 12½ pounds each.

No. 27. — Heated to 700°–720° C. for half an hour and slowly cooled in ashes. Took twenty-one minutes to heat up, and cooled to 550° C. in fifty-four minutes.

Bent close double.

Fine microstructure, ferrite in small meshes. Examination under a high power showed pearlite to be finely striated.

No. 28. — Heated for two hours at 730°–760° C., cooled in furnace to 550° C., and reheated for an hour and forty minutes at 730°–760° C., and cooled in furnace. Took twenty-three minutes to heat to 750° C., and cooled to 550° C. in one hour ten minutes, reheated in nine minutes to 750° C., cooled the second time to 550° C. in an hour and a quarter.

Bent nearly double, and broke.

Structure not as fine as No. 27, but finer than Nos. 29, 30, 31. The percentage of elongation in 3 inches is less than in the case of untreated steel.

No. 29. — Heated for three hours at 610°–650° C., and cooled in lime. Heating up took thirteen minutes. The time taken in cooling to 550° C. was not noted, but was approximately one hour.

The structure is coarse and crystalline, ferrite in large meshes, but hardly as large as in the case of No. 31.

Test bar broke after bending through about 168°.

No. 30. — Heated to 900° C. for thirty-five minutes, and slowly cooled in sand. Time heating up, twenty-three minutes. Cooling to 550° C. in sand took one hour and seventeen minutes.

Structure coarse and crystalline.

Broke after bending through 130°.

No. 31. — Steel as rolled, untreated. Bent nearly double before breaking. Structure coarse, with ferrite in large meshes.

The bars, after treatment as above, were cut into two pieces, one portion being used for bending tests, whilst the other piece was turned to ¾-inch diameter and tested — breaking strain and elongation — the results of which are given in Table II.

TABLE II. — *Showing results of tensile tests of steel, containing .36 per cent carbon, annealed in various ways.*

No.	Size of Sample	Breaking Strain. Tons per sq. in.	Elongation per cent, 3 inches	Elastic Limit. Tons per sq. in.	Contraction of area, per cent.
27	·798 diam.	38·9	31·6	20·9	55·7
28	·798 "	39·3	28·6	21·0	52·6
29	·805 "	37·3	31·0	20·1	58·2
30	·800 "	38·4	30·3	20·9	52·8
31	·810 "	40·9	30·3	21·7	54·0

Another set of seven experiments was made on steel bars of similar composition to the last, as described below

The analysis of the steel was:

Carbon (by combustion)	0.313
Silicon	0.093
Sulphur	0.047
Phosphorus	0.063
Manganese	0.600
Arsenic	0.041

The bars were $1^{3}/_{16}$ inches diameter, 12 inches long, and weighed about 5 pounds each. The results of tests for tensile strains and elongation are given in Table III.

No. 37. — Normal steel as rolled. Cracked after bending nearly double. Microstructure rather coarse, with fair-sized ferrite grains.

No. 38. — Heated to 700° C. for twenty-five minutes and slowly cooled in sand. Took twenty minutes to heat up. Cooling to 550° C. took forty-three minutes.

Bent close double.

Microstructure very fine. Ferrite in small meshes. Pearlite finely striated.

No. 39. — Heated to 650°–670° C. for forty minutes, cooled in furnace outright, reheated to the same temperature for a similar length of time, and finally cooled in sand. Heating first time took thirteen minutes; cooling to 550° C., about one and a quarter hours. Second heating to 650° C. took seventeen minutes, and cooling in sand to 550° C. fifty-two minutes.

Bent close double.

Microstructure very fine, and similar to that of No. 38.

No. 40. — Heated to 800° C. for half an hour, and slowly cooled in sand. Heated up in twenty-one minutes; cooled to 560° C. in forty-seven minutes.

Bent close double.

Structure rather coarser, with ferrite in larger meshes than in the cases of Nos. 38 and 39; but not as coarse as that of No. 37.

TABLE III. — *Showing results of tests for breaking strain and elongation of steel containing* 0.313 *per cent of carbon annealed in various ways.*

No.	Size of Sample	Breaking Strain. Tons per sq. in.	Elongation per cent, 3 inches.	Elastic Limit. Tons per sq. in.	Contraction of area, per cent.
37	·750 diam.	35·2	30·0	19·9	54·8
38	·750 "	34·4	32·0	19·9	55·6
39	·750 "	34·1	31·0	19·8	56·6
40	·750 "	35·2	31·0	20·0	55·6
41	·750 "	35·2	32·0	21·7	57·2
42	·750 "	35·5	29·3	18·7	51·9
43	·750 "	35·0	30·5	18·3	52·1

No. 41. — Heated between 700° C. and 740° C. for one hour; slowly cooled in sand. Heating to 700° C. took thirteen minutes; cooling in sand to 550° C. took forty-eight minutes.

Bent close double.

Structure very fine, and almost identical with Nos. 38 and 39.

No. 42. — Heated for three and a quarter hours at about 900° C., and slowly cooled in sand. Took about a quarter of an hour to heat up, and sixty-seven minutes to cool to 550° C.

Broke when bent nearly double.

Structure very coarse, with large meshes of ferrite.

No. 43. — Heated to 800°–850° C. for two and a quarter hours. Slowly cooled in closed furnace. Heated up in nineteen minutes, and took one hour to cool to 550° C.

Test bar broke after bending through about 170°.

The microstructure is rather coarse. Ferrite grains fairly large, and increase in size towards the center of the bar. Coarser and more crystalline than in the case of No. 37.

Photograph of sections of the above pieces are shown as before at fifty diameters by vertical light. Nitric acid etched.

The pieces for bends and tensile test were heated and cooled together.

Another set of nine bars of steel of higher carbon content was treated in various ways, as described below.

The bars were 1⅝ inches diameter, 14 inches long, and weighed about 8½ pounds each.

The composition of the steel was as follows:

Carbon (by combustion)	0.440
Silicon	0.093
Sulphur	0.037
Phosphorus	0.074
Manganese	0.580
Arsenic	0.087

As in the experiments already described, two pieces were heated and cooled together in each case.

No. 89. — Normal steel as it left the rolls. Bent through 100° and broke with a crystalline fracture.

Structure very coarse, with ferrite in large meshes.

No. 90. — Heated to 800° C. and kept at that temperature

for a quarter of an hour. Took twenty-six minutes to heat up, and cooled in sand to 580° C. in sixty-four minutes.

Bar broke with a crystalline fracture after bending through 130°.

The structure is coarse, and ferrite meshes fairly large, although it appears to be rather more broken up than in the case of normal steel.

No. 91.—Heated to a temperature of 700°–740° C. and held there for one and a half hours, and then slowly cooled in sand. The samples took twenty minutes to heat up to 720° C., and the time taken in cooling to 600° C. was fifty-four minutes.

After bending through about 150° the bar broke with a fairly good looking fracture. The microstructure is moderately fine with ferrite pretty well broken up, although the grains are not very uniform in size.

No. 92.—Heated very slowly to a temperature of 720° C., held there for twenty minutes, and then cooled in ashes. Time taken in heating up one hour and thirty-five minutes, and time taken in cooling to 550° C. sixty minutes.

This bar gave a very good test, as it was not until it had bent double that it showed signs of cracking.

The structure is fine grained, and the ferrite meshes on the whole are small, although here and there larger patches of ferrite are shown in the photograph, but were much more plainly seen under the microscope.

No. 93.—Heated moderately quickly to 620°–650° C. for fifty-five minutes, and slowly cooled by covering with sand. Heating to 630° C. took twenty minutes, and cooling in sand to 550° C. occupied sixty-one minutes.

Bent test gave excellent result, bending quite close double without cracking.

The microstructure very fine, with minute ferrite grains. Under a magnification of 500 diameters the section prepared by polishing and straining on parchment with liquorice infusion showed a magnificently colored pearlite structure.

No. 94.—Heated rather quickly to 660° C. for about five minutes, and allowed to cool slowly in sand. Heating occupied fifteen minutes, and cooling to 550° C. sixty-three minutes.

The bar broke after bending through 135°. The structure is coarse, with ferrite in large meshes, and is very similar to the structure shown by No. 90, although the ferrite meshes are slightly smaller.

The heating was evidently of insufficient duration to break down the grain, thus demonstrating the importance of the part played by the time element.

No. 95. — Heated to 610°–650° C. for about two and a quarter hours, and allowed to slowly cool in sand. The temperature of 620° C. was attained in sixteen minutes. Cooling to 550° C. took about thirty-five minutes. The sand was heated to about 500° C. before the sample was put into it, in order that the cooling should not be too rapid.

The test bar bent close double, showing a very slight crack.

The structure is very fine, in fact, this and No. 93 are the two finest grain structures of the series. The ferrite and pearlite are very intimately mixed and evenly distributed throughout the mass.

No. 96. — Heated to 700°–730° C. and held at that temperature for half an hour. Time taken to heat up to 700° C. twenty-seven minutes, and to cool to 600° C. thirty-one minutes.

The bar bent through 165° before breaking. The microstructure is pretty fine, and very similar to that of No. 91.

No. 97. — Heated for about six hours at a temperature of 670°–710° C. and cooled in sand. The bars heated up in fourteen minutes. Time of cooling to 550° C. was not noted.

The bar broke after bending through 150°.

The structure is fairly fine, but not quite as fine as Nos. 91, 92, 93, and 95, but intermediate between Nos. 92 and 96.

The results of tests for tensile strain and elongation for this set of experiments are given in Table IV.

From a consideration of the above experiments it would appear that steel containing between .40 and .50 per cent of carbon requires to be handled with a considerable amount of caution in order to secure the greatest possible ductility by annealing, as it is only between comparatively narrow limits of temperature that it may be heated to obtain a fine grained microstructure.

The best results are shown in experiments 92, 93, 95 — cases in which the highest temperature recorded was 720° C. The same bars also show the finest grain structures. Experiments 91, 96, and 97 also gave very good results.

No. 96 section gave some considerable trouble with the nitric

TABLE IV.—*Showing results of tests for tensile strain and elongation of steel containing .44 per cent carbon annealed in various ways.*

No.	Size of Sample	Breaking strain. Tons per sq. in.	Elongation per cent, 3 inches	Elastic limit. Tons per sq. in.	Contraction of area, per cent.
89	·800 diam.	44·0	24·0	23·5	31·8
90	·800 "	41·2	25·0	22·5	38·0
91	·790 "	40·8	25·6	22·0	38·5
92	·800 "	41·4	26·0	22·1	40·0
93	·800 "	40·8	26·0	22·1	38·2
94	·800 "	38·6	21·6	21·1	41·8
95	·798 "	40·8	27·3	22·2	41·6
96	·798 "	38·0	22·6	20·6	47·2
97	·800 "	39·4	22·6	21·3	40·0

acid etching, as a thick, coherent, blackish-blue deposit of carbide formed on the polished surface, which required a good deal of rubbing on chamois leather to remove it before it was possible to photograph the structure. This fact suggested to the author that differential carbon determinations of the forms in which it existed in the different pieces after treatment would add considerable interest to the investigation. But, as the time was too far advanced when this suggested itself to the author to allow of the results being appended to the present contribution, he decided to leave it for a paper by itself, which he hopes to present to this Society at a future date.

A number of bars of steel of similar carbon content to the last set, and of larger diameter, have been treated in various ways, but, owing to the amount of time and labor required for cutting up and testing these large pieces, it was found necessary to leave the results for consideration in a future contribution.

There yet remains an enormous amount of work to be done (especially on large masses) before it is possible to say with certainty which is the most suitable method for obtaining the highest possible ductility and mineralogical softness of steels of different quality, shape, and size without seriously impairing the elastic limit, accompanied in all cases by fine grained microstructure.

Regarding this question of annealed steels possessing a finer grain structure than that of the steel in its original condition, the author is quite unable to agree with the views expressed by Prof.

J. O. Arnold in a paper* which he read before the Institution of Civil Engineers in 1895, in which he said: "The ideas prevalent among metallurgists on this subject are often very erroneous. It has been stated, both in text-books and in practical papers, that the action of annealing is to produce smaller crystals. *As a matter of fact, the crystals of an annealed steel are always larger than those existing in the metal before annealing.*"† ... "It has also been stated that, on annealing, the iron and the pearly constituent become more intimately mixed, for which Dr. Sorby has been quoted as the authority. Dr. Sorby's general conclusion on this matter was accurate, and was exactly opposite to that attributed to him. The mistake seems to have arisen from an imperfect knowledge of the meaning of the word 'segregate.' The true action of annealing on a moderately mild steel containing, say, 0.35 per cent. of carbon, is as follows: 1. The comparatively small and distorted crystals of the original metal become larger and more geometrical in form (they are therefore freer from internal stresses), and the inter-crystalline cohesion, if originally weak, is much strengthened. 2. The carbonized areas existing in the unannealed steel, chiefly in irregular elongated masses, gather together, during the slow cooling, into rounded or harp-shaped areas, in which form they favor the continuity of the iron crystals to a much greater extent than the original arrangement. The rounded or harp-shaped areas into which are concentrated the normal carbide of iron, split up during the slow cooling into plates of crystallized Fe_3C, separated by large interspaces of iron. Hence the latter become dove-tailed into the main body of the iron crystals. This continuity, however, is not perfect, being frequently broken by the sectional meshes partly environing the laminated areas. Thus long lines formed by a juxtaposition of two distinct constituents are broken up, and the iron becomes almost continuous throughout. In fact, the carbon is concentrated into small plates suspended in the iron, constituting only about 5 per cent of the total metal, instead of being distributed in large, more or less continuous areas forming about 40 per cent of the mass."

It may be advisable to point out that this opinion of Professor

---

* "The Influence of Carbon on Iron." *Proc. Institution of Civil Engineers*, Vol. CXXIII, Part. 1.

† The italics are the author's.

Arnold was based upon the results of subjecting a number of steels whose carbon content varied between 0.08 and 1.47 per cent to exactly the same treatment, which consisted of heating in quick-lime in a covered cast-iron box to a temperature of 1000° C. for a period of 72 hours, afterwards allowing them to cool in a closed furnace for 100 hours — a method involving both a prolonged heating and a heating to an excessive temperature, either of which (as has been shown by Mr. Stead, Mr. Sauveur, Mr. Ridsdale, the author, and others) will produce a coarsely crystalline or granulated structure. It is therefore not to be wondered at that Professor Arnold obtained larger ferrite crystals or grains in steel after such treatment than it possessed before. The ductility, moreover, of Professor Arnold's samples was not increased by the treatment, as shown by the tests which he gave in his paper, some of which are quoted below:

Bars in normal state.

Test bars 2 inches parallel by .564 inch diameter.

Steel No.	Carbon	Elastic limit	Maximum stress	Elongation	Reduc. of area
1½	·21	17·08	25·39	42·1	67·8
2	·38	17·95	29·94	34·5	56·3

Bars after annealing.

1½	·21	9·02	21·25	42.3	65·7
2	·38	9·55	25·02	35·5	50·6

It will be seen that the material was very much softened by the treatment; but the ductility was not increased. The per cent of elongation was only .2 and .5 respectively higher, whilst the contraction of areas was reduced 2.1 and 5.7 per cent.

Of course, as it is now five years since the paper was written containing the above-quoted statements, Professor Arnold may have changed his opinions during that time; but, as the author is not aware of any occasion on which the Professor has made a statement to that effect, he thought it as well to point out that he considers it very misleading to make dogmatic statements as to the effect of annealing, on all grades of steel, upon the results obtained by subjecting it to a form of heat treatment which is really a method of "decarbonization."

The author, in his previous paper, showed mechanical tests and grain structure of a steel containing 0.180 per cent of carbon, which had been treated in quicklime at about 1000° C. for only half the time that Professor Arnold's specimens were heated, the results of which did not show the greatest ductility for that particular steel. But in order to make the facts quite clear, a bar of the same steel as that used for experiments Nos. 66–76, 2 feet long, was packed in an iron box with quicklime, and heated to a temperature of about 1000° C. for eighteen hours and allowed to cool in the furnace during twenty-six hours. The bar was then cut in halves. One piece was used for bending test, the result of which was that it bent close double; but after about three blows from the hammer it broke in two places with a crystalline fracture, thus proving its inability to withstand a sudden shock. The other half of the bar was turned to ¾-inch for a length of five inches, and tested for tensile strain and elongation, with the following result:

Size of Sample	Breaking strain. Tons per sq. in.	Elongation per cent., 3 inches	Elastic limit. Tons per sq. in.	Reduction of area, per cent.
·798 diam.	27·7	35·6	13·8	55·8

The microstructure is shown in photographs Nos. 99 and 100 at 50 diameters, etched, in the case of 99, as usual, with nitric acid, and, in the case of 100, etched deeply. No. 101 shows the same section magnified 90 diameters, etched as in No. 99.

It will be seen that, although the bar was heated and cooled during only one-fourth the time that Professor Arnold's were, an enormously coarse and crystalline structure has developed, and the ductility is less than in some of the other experiments upon the same steel, although the breaking strain is lowered considerably; but the elastic limit has been very seriously reduced.

The above results are in perfect agreement with the experimental results obtained by Professor Arnold; but it is calling such a form of heat treatment annealing that the author objects to.

The author, in summing up the results so far obtained, does

so only in a very general way, and refrains from making dogmatic statements as to how any particular grade of steel should be "annealed."

Soft steel, such as is used for constructional purposes, and containing about 0.20 per cent. of carbon, appears to possess improved ductility, with finer grain, after being heated to temperatures between 730° C. and about 850° C. for periods of time varying according to the temperature to which it is raised. If the higher temperature is selected, the pieces must be held there for only a very short time; but the author is strongly of opinion that it is unsafe to heat to a higher temperature than about 800°–820° C. That the piece be heated as rapidly as possible seems to be an essential condition for obtaining a fine-grained microstructure. The author has obtained the most satisfactory results by heating rapidly to about 750° C. for a moderate time, and subsequently slowly cooling through the critical range Ar_1, or the V of Brinell, the point at which the transformation from hardening to cement carbon takes place. The cooling should be pretty rapid until Ar_1 is reached (which is between about 710°–640° C. for soft steel), but should then be as slow as possible, that the transformation of carbon may be complete.

The metal should then possess its maximum amount of ductility, accompanied by fine micro-grain, and be in the most suitable condition to withstand sudden shock. In the case of very soft steel containing, say, 0.10 per cent of carbon, the author has found that a single heating to Ac_1, or the W of Brinell, does not always suffice to completely convert the hardening to cement carbon, and that a second heating to a temperature exceeding V is necessary to give the finest possible grain structure.

Heating to a temperature not exceeding Ar_1, or V, does not produce a fine-grained microstructure; but a prolonged heating at the temperature Ar_1, or V, will affect the complete transformation of the hardening to cement carbon, with practically no alteration in the size of the grains. This is true of all steels, excepting very soft steels and practically carbonless iron, which may (as Mr. Stead has shown), on prolonged heating between 600° C. and 700° C., develop a very coarsely crystalline structure, and produce great brittleness of the metal.

There are some metallurgists who hold the opinion that steel, after heating to a temperature not exceeding V, possesses a finer grain than it would if allowed to cool from a temperature exceeding Ac_1, or W, without having work put upon it. This statement was originally made by Brinnel; but it must be remembered that the historic results of that investigator were obtained by examining the fracture produced by nicking and then breaking, and of only hard steel. It is quite true that the fracture thus produced, when examined by the eye, very often appears to be finer after heating to V; but the microstructure remains practically unchanged, in ordinary carbon steels, as regards size of grains. The author believes that attention was first directed to this fact by Mr. Sauveur at the Manchester meeting of the Iron and Steel Institute, in 1899.* During that meeting, our late President, Mr. Paul, experimented with some steels, with a view of deciding the point. Mr. Stead examined them and reported that the microscopical examination confirmed Mr. Sauveur's statement.

In the case of steel containing about 0.35 per cent carbon, the limits of temperature to which the metal may be heated are less wide, and, apparently, the maximum is 800° C., and that for only a very short heating; but a moderate time at about 700°–720° C. seems to be the most suitable.

The steel with .44 per cent carbon has given good results between 620° C. and 720° C. Probably the most suitable temperature for this grade of steel is about 660° C.; but much more work must be done before saying anything further on this point.

The results of experiments 95 and 97 seem to point to the fact that a prolonged heating below the critical range Ar_1 does not produce the detrimental effects which it does in the case of soft steel, and thus appears to confirm Mr. Sauveur's explanation of the cause of the great brittleness which Mr. Stead obtained with nearly carbonless irons. Mr. Sauveur considers that, as very soft steels and nearly carbonless irons are made up of practically 100 per cent ferrite, the crystallization of the ferrite increases the dimensions of the grains of the metal itself; but, in the case of high carbon steels, the structure consists of pearlite areas surrounded by ferrite; and, assuming that this ferrite does crystallize in the same way, it could not alter the size of grain in

---

* "The Relation between the Structure of Steel and its Thermal and Mechanical Treatment." *Journal Iron and Steel Institute*, Vol. II, 1899. Also *The Metallographist*, 1899, Vol. II, page 264.

the steel, and therefore is unable to exert any great effect on its physical properties.

Heating carbon steels for a lengthened period does, however, in some cases cause segregation of the carbon towards the center of the mass.

The investigation is being continued by the author, who will present the results to this Society at a future time. But a great deal yet remains to be done; it is only the very fringe of the subject which has so far been approached, but what little has been accomplished points to the fact that the process of annealing requires to be carried out with much more care and intelligence than has hitherto been considered necessary. It is hoped that, incomplete as the present paper is, it may be the means of inducing some of the other members to study the subject from various aspects, and to give the results of their investigations before the Institute.

In conclusion, the author must acknowledge the assistance which he has received from several gentlemen in carrying out the investigation, especially his colleague, Mr. J. D. Evans, who has superintended the mechanical testing of a large number of the samples.

The Kennedy-Morrison Rail Finishing Process.* — For several years past there has been considerable agitation among railroad and inspecting engineers, who claim that the wearing qualities of the rails furnished them 12 or 15 years ago were very much superior to those of the rails they are getting at the present time, and they have been insisting upon the manufacturers giving them a better wearing rail. In the attempt to accomplish this the makers have increased the contents of carbon and other hardening elements to as great an extent as they have thought could be done with safety. It is well known that more attention is being given to the chemical and metallurgical branches of rail making than ever before, and the analyses of rails made to-day are more uniform and better than they were a number of years ago. Notwithstanding the improvements in this direction, the

* *The Iron Age*, Dec. 20, 1900.

The photomicrographs illustrating this article are reproduced here through the courtesy of *The Iron Age*.

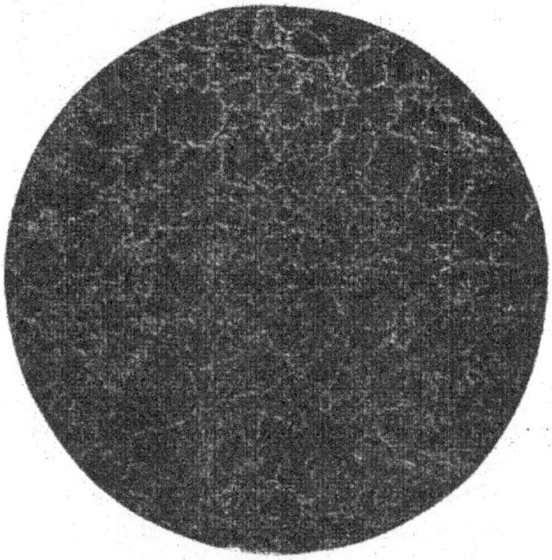

Fig. 1. Section from head of treated rail.
Seventy-five-pound Steel Rail.

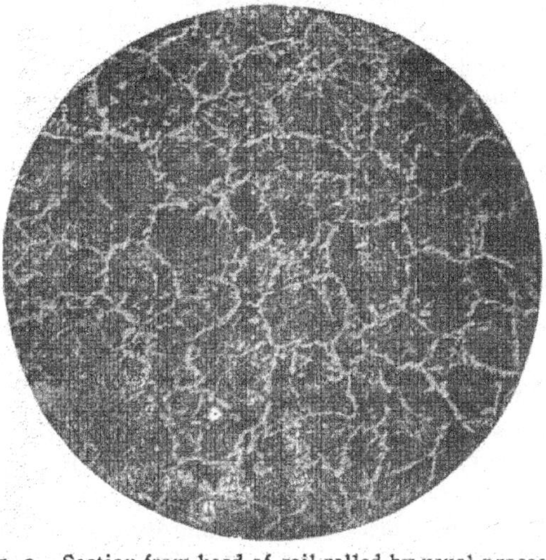

Fig. 2. Section from head of rail rolled by usual process.
Seventy-five-pound Steel Rail.

rails are not giving as good results in the track as they did 15 years ago; there can be no doubt that to a great extent the trouble with the rails being made at present is due to their physical treatment and not to their chemical composition. Owing to the fact that the weight of rails has been increased, and also that the modern rail sections have a large proportion of the metal concentrated in the head, as well as to the fact that the flanges are wide and thin, the heavy rails at present rolled in modern mills are finished while the head is extremely hot, whereas the older rails of smaller section, with a less proportion of metal in the head, and rolled on slower mills, were finished at a much lower temperature. In order to improve the quality of the heavy rails there is no doubt that the head should be finished at a lower heat, but in striving to accomplish this it is necessary not to reduce the initial heat of the bloom, which is none too high for the first forming passes of the mill, and also not to decrease to too great an extent the heat of the web and flanges.

Julian Kennedy, the well-known Pittsburgh engineer, and Thomas Morrison, general superintendent of the Edgar Thomson Steel Works of Carnegie Steel Company, some time ago invented a method of rolling steel rails at a much lower temperature than has hitherto been accomplished. In order to put this process in operation and give the railroads the benefits of improved quality rails as early as possible, the Edgar Thomson Steel Works were entirely closed down for about three weeks, and resumed operations on Wednesday morning, December 5th, 1900, during which time the new machinery necessary for re-arranging their rail mill to adapt it for making rails by this process was installed at a heavy outlay of money. In order to put in this cooling arrangement it was necessary to tear out the old finishing rolls and the large engine which drives them, and place them 56 feet further down the mill, and 8 feet out of line of the intermediate rolls. It was also necessary to move the hot saws, saw run, curver, hot bed run and hot beds 56 feet further down the mill, which, of course, made it necessary to abandon the old foundations and put in new ones for these improvements. A steel structure 78 × 71 feet was erected for covering the new hot beds, etc., which replaced those torn out. In addition to moving the mills, hot beds, etc., there had, of course, to be a good deal of new apparatus installed.

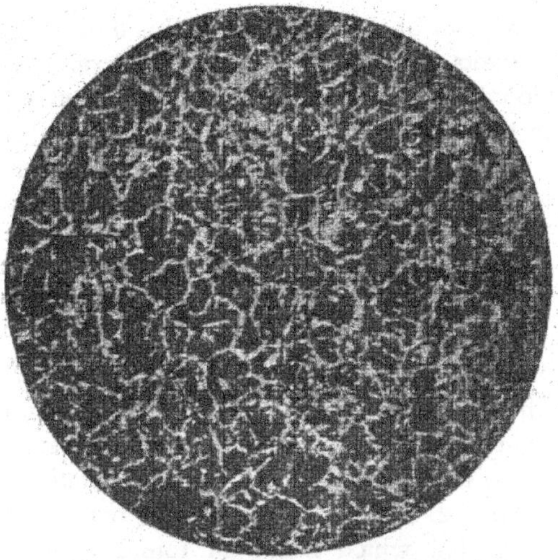

Fig. 3. Section from head of treated rail.
Eighty-five-pound Steel Rail.

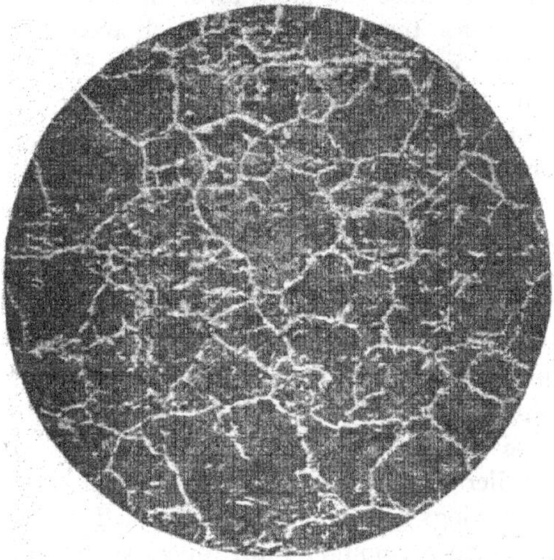

Fig. 4. Section from head of rail rolled by usual process.
Eighty-five-pound Steel Rail.

A brief description of the operation at this mill is as follows: After the steel ingots have been rolled in the blooming mill and sheared into blooms, they are conveyed to the Siemens heating furnaces in the rail mill and allowed to remain there until they are sufficiently hot to be rolled into rails. The rail mill, as it is now equipped, consists of three trains of rolls; the roughing train, where the bloom is passed forward and backward through five passes; it is then run to the intermediate or "short" rolls, where it is given five passes in the same manner. The partially rolled rail, which has been elongated until it is now about 90 feet in length, is then run to the special cooling table, which is but a few feet to the right of the main run, on which the rails are passed and which affords an interval of time of passage from the intermediate to the finishing rolls, during which time every individual rail can be brought to the same temperature, this temperature being that which has been found to produce the best results. It can be seen that on account of the greater amount of metal and heat contained in the head of the rail, it would, when cooling, cause the piece to be drawn into a curved shape, were it not for the precautions which have been taken, which, while being very simple, are still quite effective. The rail is laid on the cooling bed on its side with the head close up to the flange (or bottom) of the rail in front of it, so that the metal in the combination of two rails is about equally distributed, which allows the rails to cool gradually without causing them to become distorted. A great deal of the heat which leaves the head of the rail is absorbed in the flange of the rail adjoining it, thus keeping the temperature of the flange from being unduly lowered while the head is cooling. This is done without interfering with the rapidity or continuity of action of the mill, and it results in improving greatly the quality of the product.

After the rail leaves the intermediate rolls it is carried by a table of live rollers and delivered to the special cooling table by means of six "dogs" or "latches" attached to ropes connected with drums on a common shaft, operated by means of a hydraulic cylinder and rack. Each of these ropes carries a second dog, the first set of dogs being adapted to drawing the rails onto the cooling table and to push over all the rails then on the table, which may amount to six or more, while the second set of dogs is arranged to draw one piece at a time onto the feed

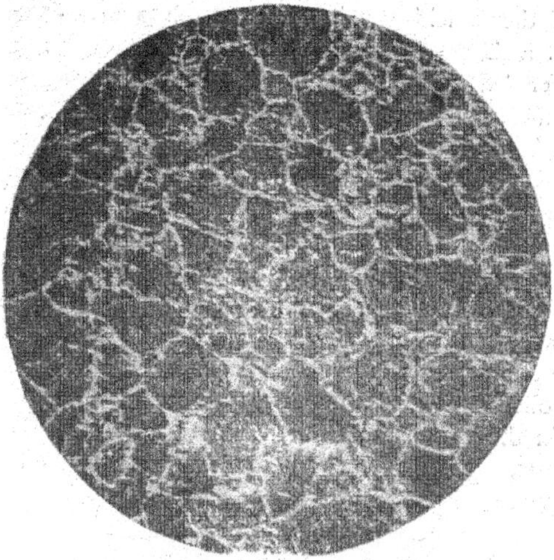

Fig. 5. Section from head of treated rail.
One-hundred-pound Steel Rail.

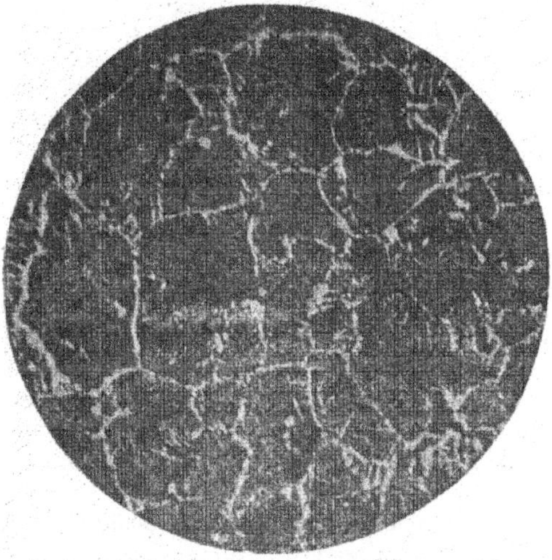

Fig. 6. Section from head of rail rolled by usual process.
One-hundred-pound Steel rail.

table of the finishing rolls, thus avoiding any danger of two pieces at a time being carried forward to the rolls.

After leaving the finishing rolls the usual processes are carried through, except that the rails do not need quite so much curving in the cambering rolls, on account of having less difference of temperature between head and flange than would be the case in the ordinary method of rolling. A great many experimental tests of various descriptions were made with this method of rolling, before it was decided to make such large and expensive changes at the Edgar Thomson Steel Works. Test pieces were taken from the heads of rails of different sections, ranging from 75 to 100 pounds per yard — one rail being from a bloom rolled in the usual manner, and another from the adjoining bloom in the same ingot rolled at a lower temperature. These test pieces, after being highly polished and etched, were very critically examined, microscopically, and photographed. We are able to show photographs taken from a section in the center of the head in each case, and all magnified to 46 diameters; the photographs marked (1) are those from the rails rolled at the lower temperature, and those marked (2) are from rails rolled in the usual manner. The rails from which the sections were photographed were, in the case of each weight, rolled from blooms made from the same ingot, the only change in the manufacture being in the variation in treatment at the rolls, as described, and it can be seen at a glance that, without exception, the steel rolled at the lower temperature is very much finer in structure, indicating a tougher and better wearing quality of steel. The tensile, drop and other tests also showed very much better results than in the same section of rails rolled in the regular way.

Several railroad engineers and engineers of tests, in endeavoring to arrive at the proper method of controlling the heat treatment of rails during rolling, have thought that it would be advisable for them to stipulate in their specifications that the hot saws should be set a certain distance apart, sufficiently greater than the length of rails required to provide for the contraction of the steel; the rails, of course, to be cut within a certain time, say 12 seconds after leaving the finishing pass. It has been found that the necessary amount allowed for shrinkage will be reduced at least from 15 to 20 per cent. Such well-known men

as R. W. Hunt of Chicago, Col. Thomas Johnson, chief engineer Philadelphia, Cincinnati, Chicago & St. Louis Railway, and a great many other well-known engineers consider this action on the part of the Carnegie Steel Company as being a long stride in the direction of furnishing the railroads with rails that will give them as good service as the small rails did years ago.

Steel Rails.* — The Committee appointed by the Board of Trade to inquire into the loss of strength in steel rails through use on railways have issued their report. The Committee, in the first instance, addressed a communication to the principal railway companies, asking them to assist the Committee by supplying the results of any experiments made on worn or broken rails, and by furnishing specimens of iron and broken rails, should the Committee decide to carry out any experiments.

The question of the loss of strength in steel rails is one which has excited much attention abroad. In Austria-Hungary and also in Germany "trial stretches" (*Versuchstrecken*) have been established on the lines of companies belonging to the Association of German Railway Boards, to which all railway companies in those two countries, and some in Belgium and Holland belong. On each of these trial stretches rails from one cast of steel, of which the chemical composition and physical qualities have been ascertained, are laid, and are carefully observed. These stretches were established in 1891, but so far the time has not been sufficient to enable the Association to draw practical conclusions from the result of the observations. In Austria-Hungary, on the State railways, a regulation has been made requiring the renewal of rails when through wear they have suffered a diminution in height of 0.24 inch, the ordinary height of a rail on these railways being 5 inches, but in Germany no precise limit has been fixed. Two reports, dealing mainly with the effects of corrosion, were received from the Italian Government, and the further information was received from the Belgian, Russian, and United States Governments.

The Committee decided to undertake a series of experiments on worn and broken rails, of which a number should be subjected

* *Journal Iron and Steel Institute*, 1900, II.

to the ordinary falling-weight tests, and selected specimens to tensile and bending tests, the latter being also chemically analysed and examined by the recent methods of photomicrography. A considerable number of rails were tested in the presence of members of the Committee at the North-Eastern Railway Company's works at Gateshead and at the Lancashire and Yorkshire Railway Company's works at Castleton, near Rochdale, and from them a selection of sixteen rails for further investigation was made by W. C. Unwin and E. Windsor Richards. The Lancashire and Yorkshire and North-Eastern Railway Companies supplied specimens of the chosen rails cut in a suitable manner for the further investigations proposed, which were apportioned amongst members of the Committee as follows: Mechanical tests, W. C. Unwin; chemical analysis, T. E. Thorpe; photomicrographic examination, Sir W. Roberts-Austen; atmospheric corrosion tests, W. R. Dunstan. W. C. Unwin gives the result of transverse, hardness, and tension tests on each of the selected rails, and also Tetmajer's and Dormus's coefficients of quality for each rail. T. E. Thorpe gives an analysis of each of the sixteen selected rails, and also of the rail, the fracture of which on the Great Northern Railway at St. Neots in December, 1895, led to the appointment of the Committee, together with a summary of methods employed in his analysis. Sir William Roberts-Austen describes the method of examining the rails by photomicrography, the effect of the various constituent elements, and the manner in which they may combine; he also compares the mechanical, chemical, and photomicrographic results of the experiments on the selected rails. His examination of the St. Neots rail disclosed the presence of patches of martensite in portions of the rail, and in order to ascertain whether this, the characteristic constituent of hardened steel, can be produced by other means than by quenching the steel from a high temperature, Sir William Roberts-Austen has examined a portion of the tube of a 4.7 in. quick-firing gun, making a study of the nature of the change produced by the action of the explosive. He is satisfied that the conditions which prevail in the gun do produce changes in the steel which resemble the local changes in structure of the St. Neots rail, but do not enable the local patches of martensite occurring in this rail to be attributed to any other known cause than the quenching of the steel from a high temperature. The structure of this rail is evidently abnormal. W. E. Dunstan gives an account of the manner in which atmospheric corrosion affects steel, having regard to its chemical composition and the nature of its structure. The results of his experiments are illustrated by selected photographs of the specially corroded surfaces of certain of the rails chosen by the Committee. These results show how important it is that the condition and mode of distribution of the constituents of steel, and not merely their nature and quantity as determined by chemical analysis, should be taken into account in the judging of the suitability of the metal for the manufacture of rails. Local atmospheric corrosion is often determined by the presence of patches of irregularly distributed impurity. E. Windsor Richards, in addition to preparing an account of the following weight tests, has drawn certain deductions from them, and in his conclusions refers to the effects on rails of incipient small cracks forming "nicks." E. P. Martin has conducted some experiments on rails in which fissures or nicks have been purposely made by a cutting tool. In addition to these experiments conducted by the Committee, Sir Lowthian Bell has tested a very large number of rails, and he has favored the Committee with a memorandum thereon, in which he calls attention to the small proportion of rails broken annually, which he estimates as only one to every 25,000 laid down, and he also discusses the effect on a rail of the rolling weight of a train, and gives the vertical and lateral deflections due to the passage of a train at different rates of speed.

The Committee requested the chemical and engineering members to prepare criticisms on the evidence in possession of the Committee from the chemical and mechanical points of view respectively. Sir William Roberts-Austen and T. E. Thorpe accordingly prepared a memorandum on the chemical aspect of the evidence before the Committee, and their own experimental work. In this memorandum they came to the following conclusions: *a.* The evidence before the Committee indicates what limiting proportions of carbon, sulphur, phosphorus, manganese, and silicon should be. As regards the influence of phosphorus, it is pointed out that, in a broad sense, brittleness of steel does not depend on the total amount of phosphorus present, as that element may exist in steel in at least two different forms, one of which is comparatively innocuous. *b.* It is very important that all who are responsible for the manufacture or use of steel rails should realise that steel is not the homogeneous mass it is often supposed to be, but possesses a complex structure. The nature of this structure will vary greatly with the mechanical and thermal treatment to which the metal has been subjected. The durability of the rail depends in no small measure on its structure, which may, if the specimens of steel have been suitably prepared, be revealed by the microscope. The peculiar structure of the St. Neots rail, for instance, can be exactly imitated.

W. C. Unwin, with the assistance of Sir Benjamin Baker and W. Kennedy, prepared a memorandum reviewing from the mechanical point of view the information before the Committee. In this memorandum the following conclusions are arrived at: *a.* The preponderance of fractures near the ends of rails seems to show that the greater straining action due to discontinuity at the joint is a contributing cause of fracture, and this can be remedied by adopting rails of sufficient strength with webs of ample thickness and secure types of fastening, and by care that no looseness arises in service. The fact that worn rails are improved in strength and ductility by annealing proves that part of the deterioration of rails in service is of the nature of what is sometimes termed "fatigue." It appears certain that in some cases fractures of rails have been due to fissures formed during service. How far the very minute transverse fissures, very often noticed in the running surface of old rails, give rise to these larger induced flaws, requires further investigation. It is not likely that they usually spread into the substance of the rail, because they are common in old rails, and fractures would be more frequent than they are if that were the case. Also, the evidence as to the existence of visible flaws or defects in the fractured surface of rails is very conflicting. In some cases, undoubtedly, the combined effects of the weakening of a rail by wear and corrosion, possibly also increased straining action from defective packing of sleepers, and the presence of a flaw or fissure of not inconsiderable size, have led to fracture of the rail. That such defects appear most commonly in the head of the rail is evidence, to a certain extent, that they are induced by the hammer-hardening of the top surface; *b.* it is very desirable that the mechanical tests to which rails are subjected should be, as far as possible, standardized in connection with (1) the weight, (2) the section, and (3) the chemical composition of the rail.

With these conclusions the Committee agree, and as regards chemical composition they do not think it desirable to insist upon too high a proportion of carbon, manganese, or silicon in the steel, having regard to the ordinary contingencies of manufacture, and the greater susceptibility of high carbon steel to thermal influences.

The Phase Theory of Iron and Steel.* — Until within the past few years the knowledge of the real conditions which exist in the various forms of iron and steel was very vague, and the theories contradictory, but in very recent years the researches of a few scientists have brought some measure of order out of the chaos, and the study of iron and its combinations may fairly be said to have entered upon a new period. The whole period of investigation and research upon which the modern ideas rest is well covered by an important paper translated from the German of Prof. H. W. Bakhuis Roozeboom by M. F. Osmond, and published, with full comments, by Messrs. Osmond and Le Chatelier, in the *Bulletin de la Société d'Encouragement pour l'Industrie Nationale.**

In 1896, Baron Jüptner von Jonstorff called attention to the necessity for gathering the observed and hitherto contradictory facts into some sort of orderly system, and early in the following year appeared the masterly paper of Mr. Le Chatelier upon the existing condition of the theories of the tempering of steel. Le Chatelier then enunciated the theory that the alloys of iron and carbon, known as steel, were in reality solid solutions, which during cooling were capable, in a manner similar to liquid solutions, of depositing either simple or compound bodies, or both. Le Chatelier also showed that these transformations were closely connected with the allotropic transformation of the iron itself, a fact which had already been suspected by Osmond. The fourth report of Professor Roberts-Austen to the Alloys Committee of the Institution of Mechanical Engineers appeared about the same time, treating the subject in a similar manner, and gathering the observations in the form of a graphical diagram representing the phenomena of solidification and those of the

* *The Engineering Magazine*, February 1901.

† Messrs. Osmond and Le Chatelier's comments will be published in an early number of the *Metallographist*. ED.

transformation into a solid mass. This was the first attempt to give a complete view of the relations between iron and carbon.

In 1898 appeared further communications from Baron Jüptner, discussing the "solution theory" of iron and steel, followed by the researches of the author, Professor Bakhuis Roozeboom, the fifth report of Sir Roberts-Austen, and the work of Stansfield. These later researches have furnished data enabling more complete diagrams to be formed, and in discussing these Prof. Bakhuis Roozeboom enunciated the so-called "phase theory" as applied especially to the phenomena of solidification.

The very full discussion of Professor Bakhuis Roozeboom takes up the behavior of mixtures containing various percentages of carbon, showing in each case how the proportion of carbon present affects the behavior at various temperatures. For the details the reader must be referred to the original paper, but some of the conclusions may be given here, and the practical value of the researches shown.

The general conclusions may be summed up as follows:

Cast alloys containing from 0 to 2 per cent of carbon, solidify in crystals of a homogeneous mixture in which the iron is in the so-called *Gamma* state, that is, stable above 890°; this combination being called martensite. Cast alloys containing from 2 to 4.3 per cent of carbon form, in solidifying, a skeleton of mixed crystals of which the intervals are filled with the eutectic alloy consisting of a conglomerate of mixed crystals and graphite containing 4.3 per cent of carbon. By the eutectic alloy is meant the alloy which solidifies at the lowest possible temperature for the series under consideration. Cast alloys containing more than 4.3 per cent of carbon form, in solidifying, a skeleton of crystals of graphite, of which the interstices are filled with the eutectic alloy.

Alloys containing more than 2 per cent of carbon deposit their carbon at temperatures between 1130° and 1000°. At 1000° there occurs a transformation which produces the carbide of iron known as cementite, at the expense of the martensite and the graphite. When the total carbon content is less than 6.6 per cent there is thus obtained a conglomerate of martensite and cementite. This transformation is the first example in which a chemical combination (cementite) is formed of a solid solution (martensite), and one of its elements (graphite), and that at a temperature quite comparable with the temperature of the transformation at which combinations are produced in the liquid solution and one of its elements.

The formation of cementite advances progressively between 1000° and 690°, the crystals of martensite in the conglomerate being impoverished in carbon down to 0.85 per cent. These crystals disappear altogether at 690°, decomposing into *Alpha* iron, or ferrite (stable below 770°), and cementite.

When the normal order of events is not followed it is due to sudden cooling. The action in this case is difficult to determine, but it appears that cementite is immediately formed, and that, from the instant that cementite is formed from the martensite and the graphite, a portion of the latter escapes entirely the transformation.

Alloys containing from 2 to 0.85 per cent of carbon may, during the period between 1000° and 690°, allow the cementite to separate, and, at 690°, to be decomposed into cementite and ferrite; but these two transformations may be checked by rapid cooling, the mixed crystals then being retained as martensite.

Alloys containing from 0 to 0.35 per cent of carbon show a gradual separation of *Beta* iron between 890° and 770°, and then, at 770° a total transformation of the *Beta* into *Alpha* iron, followed by a gradual separation of the *Alpha* iron, the mixed crystals remaining at 770°. Alloys containing 0 to 0.85 per cent of carbon only show a separation of *Alpha* iron between 770° and 690°. In all the alloys from 0 to 0.85 per cent of carbon, the transformation of the mixed crystals of martensite existing in ferrite and cementite is produced at 690°. The conglomerate of these last two (pearlite) contains 0.85 per cent of carbon. The transformation of the mixed crystals containing from 0 to 0.85 per cent of carbon like those of the alloys containing from 0.85 to 2 per cent, may be suppressed by a sufficiently rapid cooling from the higher temperature to the point of transformation, this constituting what is commonly known as tempering.

The description of the transformations of the mixed crystals containing from 0 to 0.85 per cent of carbon may be slightly modified if it is admitted that both the *Alpha* and *Beta* forms of iron are capable of dissolving small quantities of carbon, and this supposition is not contradictory to any observed facts.

The alloy known as austenite, like martensite, may be compared to mixed crystals separating from a concentrated liquid solution, and may exist by the side of martensite and cementite, being transformed by slow cooling into a conglomerate of the two latter.

The comments of Messrs. Osmond and Le Chatelier upon the exhaustive paper of Professor Roozeboom confirm in nearly every point the correctness of his judgment in the readings drawn fiom experiment.

The Metallographist

A QUARTERLY PUBLICATION DEVOTED TO THE STUDY OF METALS, WITH SPECIAL REFERENCE TO THEIR PHYSICS AND MICROSTRUCTURE, THEIR INDUSTRIAL TREATMENT AND APPLICATIONS.

Vol. IV APRIL 1901 No. 2

IRON AND PHOSPHORUS *

By J. E. STEAD

Part I

ACCORDING to ([2]) Dr. Percy,† no less than seven definite chemical compounds of phosphorus and iron, or at least compounds containing definite atomic proportions of these elements, have been isolated. They are as follows: —

	Fe_6P	Fe_3P	Fe_2P	Fe_4P_3	Fe_5P_4	FeP	Fe_3P_4
	Per Cent	Per Cent	Per Cent	Per Cent	Per Cent	Per Cent	Per Cent
Iron . . .	91.54	84.42	78.32	70.67	69.31	64.37	57.54
Phosphorus .	8.46	15.58	21.68	29.33	30.69	35.63	42.46

Since Percy's time no very elaborate research has been made, but various workers have added to our knowledge from time to time. What has been done in recent years, however, has enabled metallurgists to see that they had yet much to learn, and that the last word has not been written about the relation which exists between iron and phosphorus.

Excepting in a short note by ([8]) Professor Arnold, there has been little published respecting the microstructure of iron

* *Journal Iron and Steel Institute*, 1900, No. II, page 60.

† The small figures coincide with the index numbers of the Bibliography in Appendix V.

and steel containing phosphorus; and as this branch of the research has practically remained untouched, I have, with the coöperation of a large number of metallurgical friends and assistants, during the last few years devoted much time in endeavoring to clear away the fog which surrounds the subject by studying with the aid of the microscope and chemical analysis. The results have far exceeded anticipation, but there still remains much to be done; and it is sincerely wished that metallurgical students will not only repeat the original experimental work recorded, but will follow up the research from the point where it has been left.

Preparation of Specimens

In the prosecution of the research it was necessary to prepare many hundreds of specimens containing from traces up to a very large amount of phosphorus. The class of steels, or rather irons, containing under 1 per cent of phosphorus were prepared for me by Mr. C. H. Ridsdale, through the kind permission of Mr. Arthur Cooper, and by Mr. Arthur Richards, of the Cleveland Steelworks. They consisted of samples taken from the Bessemer converter during the after-blow, when practically all elements other than phosphorus and iron had been removed. For the samples containing from 1 per cent to 2 per cent phosphorus I am indebted to Messrs. J. H. Andrew & Co., Ltd., of Sheffield, who made several crucible melts of the purest Swedish bar iron and phosphorus, which latter was simply dropped into the crucibles containing the pieces of bar in a white hot state.

The metals with higher percentages of phosphorus were prepared in my own laboratory by melting either Swedish iron or ordinary puddled iron in crucibles with phosphorus.

Other samples of a curious and interesting nature were produced incidentally in iron manfacturing processes.

Classification of Results

The results of the examination of iron and phosphorus may be conveniently divided into the following classes:—

1*st Class.* In this class are included all compounds containing between 0 per cent and about 1.70 per cent phosphorus.

2d Class. Compounds containing between about 1.70 per cent and about 10.2 per cent phosphorus.

3d Class. Compounds containing between about 10.2 per cent and 15.58 per cent phosphorus.

4th Class. Compounds with from 15.58 per cent to 21.68 per cent phosphorus.

The metals containing much above 21 per cent phosphorus have not been investigated, as they are of little metallurgical interest.

Class I

Phosphorus from traces to about 1.70 per cent

The preliminary examination of a few samples containing between 1 per cent and 2 per cent phosphorus clearly showed in the latter two distinct constituents, one consisting of crystalline grains, the other of a hard constituent enveloping the grains, closely resembling carbide of iron or cementite.

The samples containing between 1 per cent and 1.4 per cent were quite free from the hard constituent. It thus became clear that in iron containing between 1.4 per cent and 2 per cent phosphorus there must exist a critical point.

After making many "melts" and careful micro-examinations of the specimens prepared, one of the small masses of metal, which had been allowed to cool *very slowly* in the crucible, on sectioning and inspection, was found to be free from the hard constituent near the outside, but contained much of it in the central portions. The exterior part was cut off and carefully analyzed, with the following result:—

	Per Cent
Carbon	0.02
Silicon	trace
Sulphur	0.03
Phosphorus	1.63
Manganese	trace

Valuable data was also obtained from a very crystalline piece of metal which was found lodged between the bricks in a puddling furnace, and was taken out when the furnace was off for repairs.

The crystalline particles were arranged in columnar form, which averaged about one inch in length. The macrostructure had much the same appearance, on a small scale, of the basaltic formation at Staffa. Fracture was readily effected along the columnar junctions. Each complete column constituted a single allotriomorphic crystal, and its external form had no relation to the orientation or cleavages, which were different in each parallel column. They were quite free from carbon, silicon, and manganese.

On examination of sections made vertically and horizontally through the columnar masses, the columns were found to be enveloped by the hard constituent above referred to.

The following is an analysis of this interesting specimen: —

	Per Cent
Iron	97.30
Carbon	Nil.
Silicon	Nil.
Manganese	Nil.
Sulphur	0.33
Phosphorus	1.86
	99.49

By crushing and sieving off the finer powder, most of the hard constituent and free sulphide of iron was got rid of, and by treating the metallic powder with dilute acid, practically all the sulphides were dissolved away. The residue was washed with water, alcohol, and ether, dried and tested. It still contained some of the hard constituent. By analysis it was found to contain —

	Per Cent
Iron	98.00
Carbon	Nil.
Manganese	Nil.
Silicon	Nil.
Sulphur	0.07
Phosphorus	1.77
	99.84

By a method presently to be described, the hard constituent was separated and independently examined. It had almost exactly the composition demanded by the chemical formula Fe_3P. The actual amount of free phosphide present was equal to 0.103 per cent phosphorus, or 0.766 per cent Fe_3P.

The constitutional analysis of this specimen is as follows: —

Crystalline columns	Iron	97.217	98.884
	Phosphorous	1.667	
Free constituent	Phosphide of iron (free)		0.766
	Sulphide of iron		0.190
			99.840

A further trial was made on material practically free from carbon, silicon, sulphur, and manganese, but which contained more than 1.7 per cent phosphorus. Instead of cooling the melted mass by allowing the crucible to remain in the fire, as soon as the metal was perfectly fused the crucible with contents was removed, and the solidification allowed to proceed rapidly. The object of this more rapid cooling was to ascertain whether or not more phosphide, under such conditions, would be left in the crystalline mass. The analysis of the metal was as follows: —

	Per Cent
Iron	97.35
Carbon	0.02
Manganese	Nil.
Silicon	Nil.
Phosphorus	1.94
Oxygen, etc.	0.69
	100.00
Phosphorus in free phosphide.	0.19
Phosphorus in the crystalline mass.	1.75

It is apparent, then, that iron will retain, at its solidifying point, between 1.63 per cent and 1.75 per cent phosphorus, the amount varying slightly according as the passage from the liquid to the solid state is rapid or retarded.

Under normal conditions of cooling, in which the molten metal is withdrawn from the fire and allowed to solidify in the crucible, it may be accepted that 1.70 per cent phosphorus will be retained in the crystalline mass, and that any of the element exceeding 1.70 per cent will fall out of solution in the definite compound Fe_3P, which will then form cell walls to the crystalline grains or irregular-shaped eutectic * masses distributed throughout the metal.

Judging from the variable amount of phosphide retained by the iron at its solidifying point, and the ease with which some of

* Notes on eutectics will be found in Appendix I.

what is so retained can be thrown out of solid solution,* proof of which will be given later, I am justified in concluding that all the crystalline compounds containing no free phosphide of iron (triferric phosphide) are solid solutions, in iron, not of phosphorus, but of the definite chemical compound Fe_3P.

On the Action of Reagents on the Metals of Class I

Metals containing between 0.10 per cent and 1.70 per cent phosphorus leave a black residue on treatment with dilute hydrochloric acid or sulphuric acid. The amount of the insoluble residue increases with the phosphorus. The less the phosphorus the slower is the action of the acid.

Metal with 1 per cent of phosphorus dissolves more than twice as rapidly as one with 0.05 per cent.

The black residue is completely oxidized by a ferric salt, and by that means is distinguished from a carbon residue. When washed rapidly and dried at 60° over sulphuric acid, and afterwards heated to 80° or 90° C., it spontaneously ignites and burns with a phosphoretic flame. If it is washed with alcohol and ether this peculiarity does not appear.

After dissolving in dilute hydrochloric acid an iron containing about 1 per cent phosphorus, on long-continued boiling of the solution some of the black decomposition product of Fe_3P suspended in it is dissolved, yielding a dark brown solution, the color of which is destroyed by ferric chloride.

([14]) Messrs. Osmond and Werth have made a preliminary examination of the black residue, and they state that it contains phosphorus, iron, oxygen, hydrogen, and water.

It is of very complicated composition, and, beyond confirming the observations of the gentlemen just named and determining the ratio of the iron to the phosphorus present, I have not more thoroughly examined it.

Two samples containing about 1.7 per cent phosphorus were dissolved in cold dilute hydrochloric acid. The residues were filtered off, washed with water, and dissolved off the filter in acidulated bromine water. The solutions were examined for iron and phosphorus. The results showed an atomic relation, viz.: —

---

* Notes on solid solutions will be found in Appendix II.

In 1st sample	$FeP_{1.66}$
In 2d sample	$FeP_{1.85}$

It was observed that the ratio of iron to phosphorus was materially influenced by boiling the solution of the dissolved iron and the length of time the boiling was continued.

The black residues were not attracted by the magnet.

When pure triferric phosphide, Fe_3P, was digested in cold dilute hydrochloric acid in an open vessel for several days, the metallic lustre was removed and the powder changed to a dense black substance. During this treatment streams of gas were evolved. The residue was not attracted by the magnet. It was filtered, washed, and tested for iron and phosphorus. In two separate specimens of Fe_3P, one was kept in contact with the acid for two weeks, the other for four days. The atomic ratio of the iron and phosphorus in the black residue was as follows: —

1st sample, treated with acid for 14 days $= FeP_{1.75}$
2d sample, treated with acid for 4 days $= FeP_{1.30}$

The fact that the long-continued action of hydrochloric acid on Fe_3P leaves a black non-magnetic residue similar, as regards the ratio of the iron to phosphorus, to the residues from iron and phosphorus compounds containing no free phosphide, is strong evidence in proof that such compounds are solidified solutions of phosphide of iron in iron.

There can be no doubt that the residues in both cases are decomposition products, and are not definite chemical compounds.

The property of spontaneous combustion at 80° C. may possibly be due to the presence of one of the higher hydrides of phosphorus.

The study of these residues is a research in itself, and one which must be carefully followed before we can feel justified in forming definite conclusions.

On etching together in dilute acids polished pieces of iron or steel containing varying proportions of phosphorus, they darken relatively with the proportion of phosphorus present. Taking advantage of this observation, I have found it valuable in detecting the segregation of phosphorus in soft steel bars; for where there was most phosphorus the ferrite darkened relatively more rapidly (on the continued acid of dilute acid) than did the ferrite in the portion external to the segregate.

An instance of this has quite recently come under my notice. A test-piece cut from a gun-forging, on etching, darkened relatively more rapidly in one part than in the greater mass of the section. On drilling out the dark and light portions and subjecting them to chemical analysis, the phosphorus was found to be 0.12 per cent in the dark portion, while the light part contained only 0.03 per cent. As the bar failed to stand the regulation test entirely owing to the segregation, it is reasonable to conclude that if a steel shaft, or other structure which has been turned up at one end in a lathe and roughly polished on ordinary fine emery-paper, were to be placed in, or washed by, dilute acid for a minute or two, such segregation would be at once detected. I have actually tested many samples in this way, and have proved beyond doubt that it is a safe and reliable method. The advantage is that it can be applied to the steel in the machine-shop, and does not require the assistance of an experienced metallographist.

It should be remarked in passing that it frequently and nearly always happens that where the phosphorus is segregated, both carbon and sulphur are also concentrated, so that in carbon steels the darkening produced is caused by an increased proportion of pearlite and sulphide of manganese in addition to the phosphorus. The test, however, is the more valuable on that account.

([14]) Osmond and Werth have shown that on treating steels, etc., containing phosphorus, with dilute hydrochloric acid, a portion of the phosphorus passes off as PH_3, and that generally the freer from carbon the greater is the proportion so evolved. The actual proportion of the whole amount given off as PH_3 on dissolving blown Bessemer steel, before the spiegel was added to it, amounted to about two-thirds.

In a soft steel containing only 0.18 per cent carbon, 71 per cent of the phosphorus was gasified.

In the case of a sample of burnt Moselle iron, however, which contained 0.81 per cent phosphorus and 0.11 per cent carbon, only 18 per cent was evolved as PH_3.

Messrs. Osmond and Werth remark that the quantity evolved depends also upon the strength of the acid.

([11]) Baron Jüptner's results, obtained by dissolving the metals in dilute sulphuric acid and absorbing the PH_3 in neutral nitrate of silver solution, are somewhat different from those quoted. In the case of certain steels with under 0.1 per cent phosphorus and about 0.16 per cent carbon, only about 30 per cent of the total passed off as PH_3 instead of 71 per cent. When the phosphorus was 0.6 per cent with carbon 0.12 per cent, only 3 per cent was evolved as gas.

Both researches, however, show that with high phosphorus there is very little of it converted into PH_3.

In order to ascertain if my own experiments would confirm these results, I examined a set of specimens containing between 0.07 and 1.70 per cent phosphorus, obtained for me by Mr. Arthur Richards, director of steelmaking at the Cleveland Works. They were taken from a charge of metal in a basic-lined converter during the after-blow. The small ingots were much honeycombed, but, for all that, a sufficient quantity of clean drillings was obtained from each upon which to experiment. In each case one gramme of drillings was used. They were dissolved under different conditions. One set was dissolved in strong concentrated hydrochloric acid, using 20 c.c. of acid. Solution was accelerated by heating. When dissolved, the solutions were just brought to the boiling point. They were then oxidized with nitric acid, well boiled, and finally the phosphorus remaining in the solution was gravimetrically determined.

In a second series the same treatment was followed, with the exception that the acid was diluted with an equal volume of water.

In a third series the treatment was the same as in the second, excepting that solution was allowed to take place in the cold.

The total phosphorus having been determined in each case, the difference between that obtained by the special treatment and the total was accepted as that portion which had passed off as gas. This method was much easier to work than that of Osmond and Jüptner, and had the advantage that a large number of assays could be made simultaneously.

For results obtained see table on following page.

These results, although varying with the way the metal is dissolved, show clearly enough that the more dilute the solid solution of the phosphorus in iron, the greater the proportion of phosphine, PH_3, gas which is formed.

They also show that, if any comparison has to be made, the

Analyses of Samples

	1	2	3	4	5
	Per Cent	Per Cent	Per Cent	Per Cent	Per Cent
Iron	97.80	98.20	98.70	98.90	99.60
Carbon	0.06	0.05	0.05	0.04	0.04
Phosphorus	1.69	1.32	0.62	0.23	0.065
Not determined	0.45	0.43	0.63	0.83	0.295
	100.00	100.00	100.00	100.00	100.00
Dissolved in strong hot acid, 1 grm. iron to 20 c.c. HCl					
Phosphorus evolved as PH_3	0.26	0.42	0.30	0.17	0.052
Phosphorus in residue	1.33	0.91	0.32	0.06	0.016
Per cent gasified	15.00	33.00	48.00	74.00	80.000
Dissolved in weak hot acid					
Phosphorus evolved as PH_3	0.28	0.26	0.14	0.11	0.028
Phosphorus in residue	1.41	1.06	0.48	0.12	0.037
Per cent gasified	16.00	20.00	23.5	48.00	74.000
Dissolved in weak cold acid					
Phosphorus evolved as PH_3	0.06	0.09	0.07	0.11	0.035
Phosphorus in residue	1.63	1.23	0.55	0.12	0.030
Per cent gasified	3.5	6.5	11.00	48.00	54.000

chemical treatment must be identical. The strength of acid, the temperature, and the time the solutions are boiled after the irons are dissolved, all have an influence on the proportion of phosphine gas formed.

The following results were obtained by dissolving one gramme of each sample in 10 c.c. strong hydrochloric acid in the cold, and boiling afterwards for one minute.

The samples were obtained from different charges during the after-blow period.

Nos.	Phosphorus			
	Gasified	Not Gasified	Total	Parts per 100 Gasified
	Per Cent	Per Cent	Per Cent	Per Cent
1	0.260	0.63	0.89	30.00
2	0.18	0.34	0.52	34.00
3	0.12	0.18	0.30	40.00
4	0.09	0.09	0.18	50.00
5	0.03	0.03	0.06	60.00

Many other cases might be brought forward, but they go to prove that the higher the concentration of the solid solution, the less proportion of the total phosphorus which is given off as PH_3 — a most important general law, as will be seen in Part II of this paper.

Cupric chloride, although it has little action on free Fe_3P, behaves like sulphuric and hydrochloric acid in one respect, that when acting on solid phosphide solutions it leaves a black decomposition product.

([14]) Osmond and Werth have found that the residue left on dissolving these metals electrically by Weyl's method resembles that left by dilute acids.

There is much room for hypothetical and inductive reasoning to account for the facts, and it is hoped that members of this Institute will endeavor to find a satisfactory solution to account for the phenomena. A probable explanation is that the more attenuated the phosphide, and the greater the proportion of iron, and the more nascent hydrogen available, the more readily the phosphide will be decomposed and hydrogenized.

Certain results of Baron Jüptner, referred to subsequently, do not appear to confirm such a view; it is therefore necessary to await the results of further research before we can form other than tentative conclusions, and seems to point in the direction that the more diffused the phosphorus, and the greater the proportion of phosphine formed, the safer the steel.

This question, however, will be further discussed in Part II, when the influence of carbon is considered.

When nitric acid of specific gravity 1.42 is poured upon iron containing little phosphorus, the iron becomes passive, and will not dissolve until the solution is heated; but when the percentage of phosphorus is high, say between 0.6 per cent and 1.7 per cent, the passivity disappears and the acid attacks the metal violently.

In very dilute solutions its action resembles that of sulphuric and hydrochloric acids, ferrous nitrate and the black decomposition product being formed.

With nitric acid of specific gravity 1.20, and at a temperature of about 20° C., a perfectly clear solution results, free from any black decomposition product, whilst under like conditions free phosphide of iron, Fe_3P, is but little acted upon. This reagent then is a very good one for separating free phosphide from phosphide of iron in solid solution. Practically all the results showing the amount of free phosphide of iron given in this paper were obtained by a method based upon this behavior of nitric acid. Details of the method are given in Appendix III.

The Effect of Annealing Iron Saturated with Phosphide of Iron

The next question which arose was, Will phosphide of iron separate from the solid saturated compound containing above 1.8 per cent of phosphorus by annealing and slow cooling?

To answer this, a large piece of metal, containing about 1.8 per cent phosphorus and 0.18 per cent carbon, was passed repeatedly through a furnace used for the manufacture of malleable castings.

The sample was placed in the centre of the furnace near the top. It was packed in an iron box with a mixture of five parts of old ore and one part of new.

After closing the oven, the fire was started and it was fired continuously for thirty-six hours to get up the full heat. This was maintained for ninety-six hours. It was cooled down in about forty hours, and was then removed. The temperature was estimated to be 900° C.

The same treatment was repeated a second time.

The object of using this furnace was twofold: first, to remove the carbon; second, to maintain the metal at a temperature of near 900° C. for a long period, and to cool down from that temperature very slowly.

The metal, although 20 cm. in thickness, was completely decarburized. Drillings were taken from different parts of it and separately analyzed, as shown on following page.

The outside layer was very spongy and porous, and contained oxide of iron in considerable quantity.

In this layer it was noticed, on micro-examination, that the areas which had been originally filled with phosphide eutectic were quite empty. It had evidently run out, and had been absorbed in the spongy oxide surrounding it, leaving cavities of the characteristic shape shown in Photo No. 8.* In no other parts excepting these could the cavities be detected.

---

* All photomicrographs bearing a higher number than 6 will be published in subsequent installments of this paper. — ED.

	Before Treatment	Outside ¼ In.	½ In. below Top	¼ In. from Bottom	½ In. from Bottom
	Per Cent	Per Cent	Per Cent	Per Cent	Per Cent
Iron	97.20	96.96	97.60	97.40	97.80
Carbon	0.18	—	trace	trace	trace
Manganese . . .	trace	trace	trace	trace	trace
Silicon	trace	trace	trace	trace	trace
Sulphur	0.02	0.02	0.02	0.02	0.02
Phosphorus . . .	1.82	1.63	1.85	1.94	1.78
Oxygen, etc. . . .	0.78	1.39	0.53	0.64	0.40
	100.00	100.00	100.00	100.00	100.00
Phosphorus in free Fe_3P .	0.51	0.28	0.92	0.88	0.80
Phosphorus in solid solution . . .	1.31	1.35	0.93	1.06	0.98
Total*	1.82	1.63	1.85	1.94	1.78
Phosphorus thrown out of solid solution .	—	?	0.41	0.37	0.29

It is probable that the figure 1.35 per cent does not correctly represent the phosphide held in the iron, as some of it was no doubt in the state of phosphoric acid. The other analyses made on the solid and non-oxidized material indicate beyond doubt that, on long annealing the metal saturated with phosphide at its solidifying point, a portion of that substance separates out, leaving the mass with only about 1 per cent phosphorus.

This most interesting experiment disposes of the conception that the solid solution at the solidifying point, containing about 1.70 per cent phosphorus, is a definite chemical compound.

In order to ascertain whether a portion of what separated on annealing at about 900° C. would recombine at 1100° and 1300° and after melting, portions of the annealed metals were heated for half-an-hour at these temperatures, and a third portion was melted in a magnesia crucible. They were then cooled and analyzed. The following were the results obtained:—

* It will be noticed that the amount of phosphorus is very variable in the same piece, a peculiarity of all rather large pieces of iron containing much phosphorus. The ingot from which this was obtained weighed only about 14 kilos., yet the phosphorus varied between 1.6 per cent and 2.4 per cent in drillings taken from different parts. As a rule, in slowly cooled ingots my experience is that the highest phosphorus was generally found near the top.

	Heated to 1100° C.	Heated to 1300° C.	Melted
	Per Cent	Per Cent	Per Cent
Phosphorus in free phosphide . .	0.95	0.91	0.20
Phosphorus in solid solution . .	1.01	1.06	1.74
	1.96	1.97	1.94

They show clearly enough that even at 1300° C., a point not far removed from the point of fusion, practically no phosphide was reabsorbed.

Microscopic Examination of the Annealed Metal

A section of the metal before annealing and decarburizing is represented in Photo No. 8, in which it will be seen the phosphide eutectic is present in irregular masses and is surrounded by dark fringes of pearlite. After decarburization the pearlite was absent, and what is most remarkable, the phosphide of the phosphide eutectic was found not in irregular masses, but in what appeared to be rectilinear prisms, many terminals of which had true crystalline angles.

They were not always continuous, but were generally much broken up. Photo No. 20 illustrates this.

In the specimen heated for half-an-hour to 1100° C. the change was also remarkable. Most of the prismatic crystals retained their position, but the ends were quite rounded. Many of them had broken up and contracted, forming chains of globules or lenticular-shaped bodies, the longer axes of which invariably corresponded to the longer axes of the original prisms. There were also present irregular-shaped masses of eutectic somewhat resembling those of the sample before it was annealed.

In the specimen heated to 1300° C. the prismatic formation and the globular particles of phosphide were entirely absent, and had been replaced by the irregular-shaped eutectic masses.

Photo No. 21 illustrates this.

On remelting a portion of the annealed material, the microstructure was similar to No. 8 if the pearlite is excluded.

On looking at the three photos, Nos. 8, 20, 21, placed side by side, one is impressed with the facts: 1st, That by annealing, the phosphide illustrated in Photo No. 8 has entirely rearranged its position, and must either have diffused and crystallized in the new order, or grown out of the old position, the liquid eutectic being the reservoir or base, supplemented by what fell out of solution during the long-continued heating. 2d, That on heating to above the melting-point of the eutectic and phosphide of iron, the latter, becoming liquid, separated into oblong and globular drops, and combined with a portion of the surrounding metal, formed a fusible eutectic the separate portions of which coalesced to form the irregular masses shown in Photo No. 8. I cannot offer any explanation as to why it should take this particular form. 3d, That the molecular movement of the ground mass must have been very great, behaving almost as if it was a viscous fluid, in rearranging its position to suit the altered areas occupied by the eutectic.

Physical Properties

The peculiar property of brittleness which phosphorus gives iron is well known, and it is the universal opinion that that element is the most objectionable of all the metalloids which enter into the composition of steel and iron.

Various observers do not agree exactly as to the effect of phosphorus on the physical properties. ([19]) Professor Howe, in his work on "The Metallurgy of Steel," gives a condensed account of the results obtained by many observers, and forms the following conclusions: —

"1st. We may infer that when the phosphorus is above, say, 0.12 per cent, it probably has no important constant effect, for if it had, the analysis of statistics should yield concordant results.

"2d. Phosphorus usually raises the elastic limit, and thus the elastic ratio, an index of brittleness; indeed, the elastic limit and breaking strength of steel occasionally coincide."

The whole question as to the effect of phosphorus requires to be reinvestigated, and such investigations must not be confined to chemical analyses and mechanical behavior in the testing-machine, but should include the character of the microstructure, and more particularly the orientation and size of the crystalline grains.

The great variation in the mechanical properties of phosphoretic steels containing approximately the same amount of phosphorus most probably is caused by difference in structure. I have myself obtained widely different results from the same high phosphorus steel by treating it so as to obtain such variations.

([8]) Professor Arnold has given the results on examining iron containing 1.37 per cent phosphorus and 0.07 per cent carbon.

Such iron may be considered as nearly saturated with phosphide of iron, and to have the following constitutional analysis:—

	Per Cent
Iron	90.16
Phosphide of iron in solid solution . .	8.79
Carbide of Iron	1.05
	100.00

It was drilled and turned with difficulty. On etching, the crystalline joints appeared to be readily penetrated by the etching acid. The mechanical tests, compared with carbon steel and nearly pure iron, were as follows:—

Test Pieces 0.564″ × 2″	Breaking Weight	Elastic Limit	Elongation	Contraction of Area
	Tons	Tons	Per Cent	Per Cent
Iron with 0.04 per cent carbon .	21.77	14.39	47	76.5
Steel with 1.35 per cent carbon .	57.50	46.53	5	5.6
Iron with 1.37 per cent phosphorus	28.98	28.98	0	0.0

Crushing Tests

	20 Tons	40 Tons	60 Tons	80 Tons	100 Tons
Iron with 0.04 per cent carbon . . .	5.80–1.30	25.2–12.8	43.8–30.5	55.3–43.8	62.4–52.2
Iron with 1.35 per cent carbon . . .	1.30–0.00	3.0–0.0	8.4–0.0	20.0–0.0	33.0–0.0
Iron with 1.37 per cent phosphorus . .	0.70–0.00	3.0–0.9	8.9–5.8	20.8–13.6	26.1–23.0

It is a well-known fact that in puddled iron bars phosphorus increases the tenacity, reduces the property of drawing out under tension, and increases the liability to break under shock. It, however, has one good property, as it makes the iron weld easily, iron free from phosphorus being much more difficult to weld than phosphoretic material; but this good property is greatly discounted by the iron at the weld being left in a more or less coarsely granular state and being easily fractured.

Iron containing little or no phosphorus can always be obtained in a fine granular state by heating to about 900° C.,* when, no matter how brittle and coarse grained it was previously, its structure will be refined. With phosphoretic iron no such refining follows heating, and the only method of making such material fine in structure is by forging or rolling at a comparatively low temperature; and if the phosphorus is very high, it is very difficult to obtain the fine condition even by forging.

The importance of having a fine granular structure in phosphoretic iron is much greater than in irons and steels containing little of that element.

Crystalline Form

The crystalline structure of irons containing much phosphorus is apparently the same as pure iron. In the drusy cavities of ingots of each there are to be found the fir-tree growths of crystallites of the cubic system. It does not appear that the dissolved phosphide modifies the crystalline form, but possibly when perfectly measurable crystals are isolated, if they ever are, it will be found that the phosphide does cause some slight variation. It is easy to obtain rectangular forms by cleavage from large crystalline grains of iron saturated with phosphide.

There is an increasing tendency, under like conditions of treatment with increasing phosphorus, for iron to crystallize in larger and larger crystalline grains, and I have frequently obtained in fractured metals containing above 1 per cent phosphorus cleavage faces measuring more than an inch across.

Mr. T. W. Sorby, of Sheffield, very kindly annealed for me, in one of his cementation furnaces, a portion of a metal containing 1.15 per cent phosphorus. The coarsely crystalline metal was inserted and packed with sand in an iron tube which was closed at both ends with fine clay.

The temperature to which it was subjected must have been about 1000° C., and it was maintained at that temperature for a

---

* It has been shown by ([8]) Arnold that 1.37 per cent phosphorus prevents the internal critical change Ar_3, about 860° C.

considerable period. When cold, the appearance was found not to have altered, and, on breaking it to pieces and selecting such external portions as had, in fracturing the metal previous to annealing, broken along three cleavages, leaving corners in projection bounded by three cleavage faces, it was found that, on breaking these, cleavage followed the planes parallel to the faces, proving that the crystalline organization had not been altered by the prolonged heating, and that there had therefore been no allotropic change to the gamma state.

Hardness

The hardness of iron is steadily increased with each addition of phosphorus, until, when it is saturated, and contains 1.7 per cent phosphorus as phosphide, it takes a well-hardened steel drill to drill it properly, and it apparently has a hardness on Mohr's scale of just under 6, probably 5.50; a sharp point will easily scratch apatite, but it is soft enough to be easily scratched by feldspar.

On heating polished semi-saturated metals until they take characteristic oxidation tints, no gradation in color could be detected between the interior and exterior portions of the grain, from which one would conclude that they are homogeneous, and that the first portions which commence to crystallize in the centres do not contain more phosphide than the parts last to solidify at the external portions.

The hardness of cementite is between that of feldspar and quartz, or about 6.50°. Massive phosphide of iron appears to have a hardness equal to feldspar, or 6° on Mohr's scale. It is difficult, however, to determine the degrees of hardness, with any accuracy, by means of the existing methods; therefore the figures given must be considered as only approximate.

General Microstructure of Metals of Class I

The microstructures of all metals of this class are similar, and resemble in appearance similar sections of pure iron, but, generally speaking, the crystalline grains are of greater size and increase with increasing phosphorus. ([8]) Professor Arnold found that on etching with nitric acid the fluid acted more severely on the junctions of the grains than on their mass, an observation which I have confirmed in many cases, but I have also found that it did not occur invariably. Photo No. 1 is a type of all metals of this class.

Although the microstructure of pure iron and of the metals of Class I were similar, there were certain characteristic differences.

The polished surfaces of the phosphoretic metals, on being etched with exceedingly dilute nitric acid, were transiently pris-

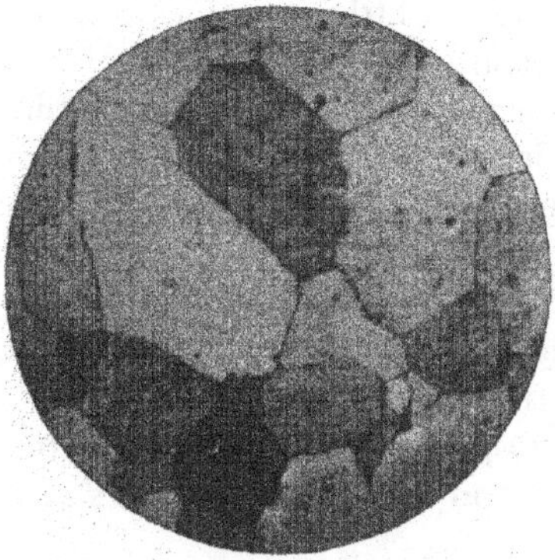

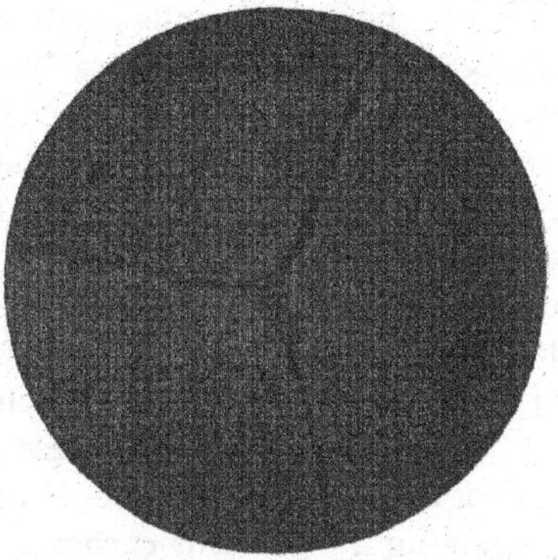

Fig. 1. Type of all the metals in Class I. Polygonal crystalline grains. Solid solutions of Fe_3P in iron. The grains, under like conditions of heating and cooling, increase in size with the increase of phosphorus. V × 50.

Fig. 2. Metal ingot with 1.8 per cent phosphorus, showing portions of three contiguous grains surrounded by a cellular envelope of Fe_3P with a little of the eutectic at the part where the three grains meet. V × 350.

matically colored. The different crystalline grains at the same time assumed different colors, and on a single specimen were to be seen grains colored yellow, orange, red, purple, green, and blue. They changed rapidly from one color to the others, and finally the coloring vanished with the continued action of the acid, and a brown stain or deposit was left on the surface, which eventually became quite black.

The black matter could be removed from the surface by slight friction, and acid tincture of iodine partially destroyed it, a reaction which did not follow when the black deposit was formed by etching hardened pure carbon steels. It evidently consisted of the same substance which is left insoluble on dissolving the metals in acid.

Class II

Phosphorus 1.70 per cent to 10.2 per cent

In the compounds containing a little over 1.70 phosphorus, the eutectic or pearly constituent, being last to freeze or solidify, is thrown out or expelled from the crystalline grains, and forms fluid boundaries or cells round about them. As the temperature falls these finally solidify, breaking up into two constituents. When cold the polished and etched structure, under low power magnification, has the appearance of an irregular meshed net, the spaces being filled with that part of the crystalline grains of metal which first solidified.

The metal with 1.8 per cent phosphorus (represented in No. 2 photograph) is an example of this, and somewhat resembles steel containing 1½ per cent carbon.

When the phosphorus approaches 8 per cent a pearly constituent occupies the greater area, and on this ground mass skeleton arboriform crystallites abound, consisting of the saturated solid solution of phosphide of iron in iron. (No. 3 photograph represents this compound.)

As 10.2 per cent is approached these crystallites diminish, and when it is reached the whole mass or area is pearl-like in structure and no crystals are visible. (Photo No. 4.)

All the eutectics of the metallic alloys have not the same appearance or form, but most of those I have examined (and I have examined a very large number) had one common property — that of giving the pearly structure, first discovered by Dr. Sorby in carbon steels after they had been polished and etched by suitable reagents.

The pearly mass with about 10 per cent phosphorus and 90 per cent iron is the true eutectic, or the most fusible of the compounds of iron and phosphorus. It has only one melting-point, which is about 980° C.

The melting-points in the metals of Class I steadily fall with each increment of phosphorus, but each of them has only one solidifying point.

Fig. 3. Metal ingot containing about 8 per cent phosphorus. The broad light parts are crystallites of the metal containing about 1.7 per cent phosphorus. The white mottled ground mass is the eutectic containing 10.2 per cent phosphorus. V × 250.

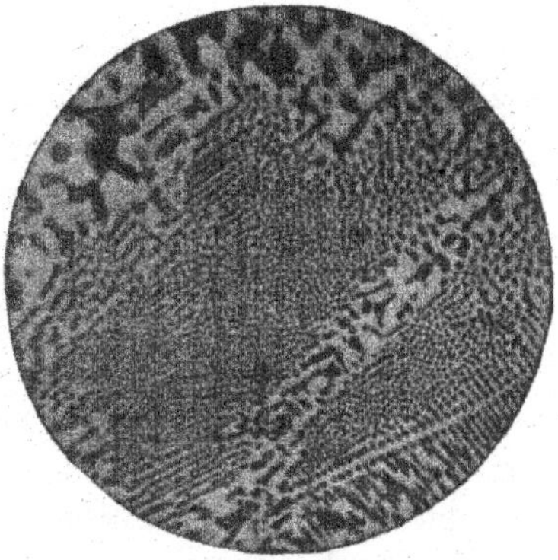

Fig. 4. Ingot containing 10.2 per cent phosphorus and 89.8 per cent iron. It is the eutectic of phosphorus and iron. It has only one critical point at about 950° C. Etched with nitric acid. V × 350.

Fig. 5. Ingot containing 11.07 per cent. phosphorus and 86.9 per cent iron, showing sections of rhombic or oblique idiomorphic crystals of Fe_3P embedded in a ground mass of the eutectic. V × 60.

In Class II the metals intermediate between 1.7 per cent and 10.2 per cent show two arrests during their passage from the perfectly liquid to the completely solid condition; the first, when the crystallites fall out of solution; the second, when the eutectic solidifies.

The first arrest is lowered with each addition of phosphorus, the second remains constant throughout the series.

The composition of the hard element of the eutectic was determined by treating the powdered metal with dilute nitric acid. The softer constituent was in this way dissolved away, leaving insoluble the harder constituent.

Prepared in this way, on analysis it was found to contain:—

	1	2	Calculated
	Per Cent	Per Cent	Per Cent
Iron	83.20	83.20	84.42
Phosphorus	15.32	15.48	15.58
Oxidation products, etc.	1.48	1.32	—
	100.00	100.00	100.00

It is therefore proved that the chemical formula is Fe_3P.

Class III

Phosphorus 10.2 per cent to 15.58 per cent

The slightest increase of phosphorus above 10.2 per cent results in the formation of well-formed crystals and skeletons of crystals of the same chemical composition as the phosphide in the eutectic, and these increase with the phosphorus until, when 15.58 per cent is reached, the whole mass is composed of them and is homogeneous. (Photos Nos. 5, 6.)

In the metals of this class, as in Class II, there are two arrest-points during solidification.

Each addition of phosphorus raises the initial point of solidification; the final arrest coincides with that of the eutectic, but when about 15.6 per cent phosphorus is reached, only one arrest-point can be observed, which is about 1060° C.

The microscopic, chemical, and thermal results all tend to prove that the crystals are a definite chemical compound, of the formula Fe_3P.

The metal sections containing the well-formed crystals of Fe_3P have been kindly examined for me by Professor Bauerman, Professor Arnold, and Mr. MacWilliam, who consider that they will most likely prove to be rhombic or oblique.

So far only the *sections of the crystals* have been available for examination, but it is hoped that before long they will be isolated so as to enable the angles to be correctly measured.

In my paper on "The Crystalline Structure of Iron," I gave

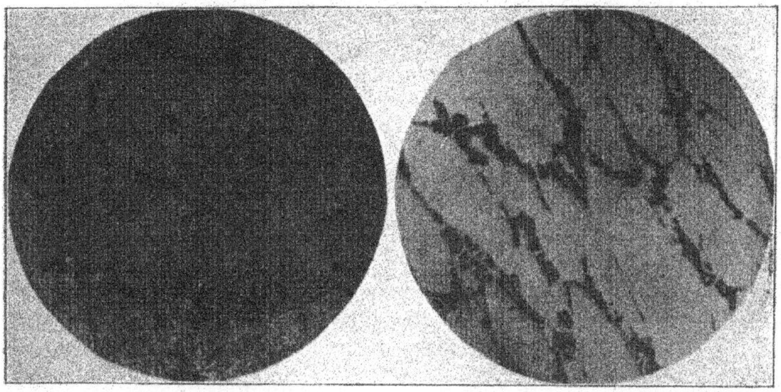

Fig. 1a. Metal ingot containing 0.8 per cent phosphorus and no carbon, and some oxide of iron in globules and in incomplete cells surrounding the grains. Strongly etched with nitric acid (20 per cent). The dark bands in the illustration are spaces dissolved away by the acid. V × 50.

Fig. 6. Ingot containing 15.15 per cent phosphorus, consisting of crystalline grains or allotriomorphic crystals of Fe_3P surrounded by the eutectic. V × 100.

a photograph of what was believed to be an idiomorphic crystal of carbide of iron.

The metal section photographed contained phosphorus. On reëxamination by the heat-tinting process, I found that the crystals believed to be carbide were not so, but were phosphide of iron. I am glad of this opportunity of correcting an erroneous conclusion which any one might have made previous to the discoveries of the method of distinguishing one from the other.

Other metallographists would probably find, on reëxamining their specimens by the new method, that what they had taken to be cementite was in reality phosphide of iron.

Class IV

Phosphorus between 15.58 per cent and 21.68 per cent

When the phosphorus is increased above 15.58 per cent somewhat porous ingots result, and until 21.68 per cent is present it is easy to separate the powdered metal into two distinct parts.

On endeavoring to color or stain the constituents of polished sections so as to identify them by the usual etching methods, no success followed the effort; but on heating them till a blue oxidation temper tint was just visible to the eye, on examination under the microscope two differently colored parts became evident, one blue, more readily oxidized than the other, and a second which appeared yellow.

Both constituents were practically homogeneous, and were evidently definite chemical compounds. One of these compounds was less soluble in nitro-hydrochloric acid than the other. The more insoluble part was also least readily attracted by the magnet.

The constituent, readily colored blue by heating, was the more soluble in acid, and was strongly attracted by the magnet. It consisted of Fe_3P.

The less readily colored constituent, very slowly acted upon by acids and slightly attracted by the magnet, consisted of Fe_2P.

The microstructure leads one to conclude that the Fe_2P crystallizes at a higher temperature than the Fe_3P, for the crystal faces of Fe_2P in the drusy cavities were generally more perfect than those of Fe_3P.

The more insoluble compound, Fe_2P, separated by acid, contained —

	Per Cent
Iron	78.30
Phosphorus	21.50
Not determined	0.20
	100.00

The compound of iron and phosphorus containing about 18 per cent phosphorus, which had about equal parts of Fe_3P and

Fe_2P, was most fragile and very porous, and the crystals loosely attached to each other were easily separated by slight pressure. On placing the coarse crystalline matter on a steel plate and heating it till some of the particles took a "temper blue color," and then rapidly cooling the whole, it was easy to pick out the blue particles from those of a pale yellow color. On testing them chemically side by side, the blue particles were found to approximate to the composition of Fe_3P, the yellow crystals to Fe_2P.

When the mixed mass of blue and yellow particles was approached by a magnet to within a distance of 2 mm., the blue particles were attracted and adhered to the magnet, the yellow ones remaining behind.

The analysis of the magnetically separated compounds were as follows: —

	Magnetic	Slightly Magnetic
	Per Cent	Per Cent
Iron	84.00	78.40
Phosphorus	15.82	21.50
Not determined	0.18	0.10
	100.00	100.00
Corresponding to the formula of =	Fe_3P.	Fe_2P.

It is not often that compounds can be separated by such simple and diverse methods: —

1st. By chemical reagents.

2d. By the magnet.

3d. By heat-coloring and by hand-separation.

Of the three, the magnetic method is by far the most perfect, for whereas Fe_3P is strongly attracted, Fe_2P will not rise to the magnet, and only slightly adheres to it when placed in contact with the finest powder.

Class V

Phosphorus 21.68 per cent to 24 per cent

By the expenditure of a very excessive amount of phosphorus and the use of a very high temperature, iron may be made to combine directly with sufficient phosphorus to give a compound containing 24 per cent of that element. This is contrary to what was found in Dr. Percy's laboratory, where a maximum of 8.46 per cent was obtained by that method. It is nevertheless an undoubted fact that the higher phosphide can be obtained by direct union.

On polishing the compound so obtained and tinting it by heating, the microscope revealed the presence of two differently colored constituents. The more easily oxidized was undoubtedly the definite compound Fe_2P, but what the second was has not yet been determined.

Summary of Part I

Class I. — Phosphorus, traces to 1.7 per cent. — The metals of this class consist of iron containing phosphide of iron, Fe_3P, in solidified solution.

All the metals used commercially, such as wrought iron and steels containing practically no carbon, may be included in this class.

In proportion as the phosphorus is increased, the polished and etched specimens under like acid etching darken relatively with the quantity of phosphorus present.

The dimensions of the grains in solidified ingots, *ceteris paribus,* tend to become larger as the phosphorus is increased, and when about 1 per cent is present, it is easily possible to get cleavage faces more than an inch in diameter on the fractured material.

Under certain conditions, on annealing metals containing a very large quantity of phosphorus, some of what is in solution will fall out, and separate in the free state as phosphide of iron.

The hardness increases steadily with the proportion of phosphide in solid solution, until, when it reaches a maximum of about 11 per cent, equal to 1.75 per cent phosphorus, it requires a well-hardened tool to drill it properly.

Probably the great variations in the mechanical properties of phosphoretic material can be explained by the difference in structure, for it has been proved that the same steel or iron can be made very brittle or comparatively tough according to whether the structure has been made fine or coarse.

It is much more important to have a fine structure in phosphoretic material than it is in ordinary steel low in phosphorus.

It is suggested that the study of the mechanical properties

of phosphorus irons and steels should be carefully continued, and that observations should be made not only of the chemical analyses and mechanical properties, but of the coincident structure of the samples operated upon, and the best treatment be ascertained which will give the best mechanical properties.

Class II. — *Phosphorus* 1.7 *per cent to* 10.2 *per cent* consists of varying quantities of a saturated solidified solution of Fe_3P in iron and a eutectic containing about 10.2 per cent phosphorus, which is composed of about 61 per cent of Fe_3P and 39 per cent of saturated solution of Fe_3P in iron. (Photos Nos. 2, 3, and 4.)

Class III. — *Phosphorus* 10.2 *per cent to* 15.58 *per cent* contains idiomorphic crystals of Fe_3P, very perfectly formed in compounds containing between 10.5 per cent and 12 per cent phosphorus and allotriomorphic crystals, or crystalline grains of the same body, in compounds containing 15 per cent phosphorus, surrounded the eutectic containing about 10.2 per cent phosphorus. (Photos Nos. 5 and 6.)

Class IV. — *Phosphorus* 15.58 *per cent to* 21.68 *per cent.* — In the slowly cooled compounds of this class we commence with pure Fe_3P and end with Fe_2P. The intermediate compounds consist of mixtures of the two bodies, easily separated from each other, when in powder, by a magnet. Fe_3P is readily attracted, Fe_2P very slightly so.

Class V. — *Phosphorus* 21.68 *per cent to* 24 *per cent.* — This class has not been sufficiently studied, but it contains two constituents, one of which is Fe_2P.

(To be continued)

SEGREGATION OF PHOSPHORUS IN A PIECE OF COLD ROLLED SHAFTING*

By HENRY FAY

THROUGH the kindness of Mr. P. Kreuzpointner, of the Pennsylvania Railroad testing laboratory, I have come into possession of a piece of cold rolled shafting which shows an interesting case of segregation. The shaft, 6 cm. in diameter, was

* Received March 30, 1901

in service about one week when it fractured, and on examination it was found that there was a spiral crack running the entire length of the shaft from the point of fracture. So far as known, the material was of a high-grade steel, and the process of manufacture was identically the same as used in making other shafts, which had stood the severest tests. This case helps to prove how

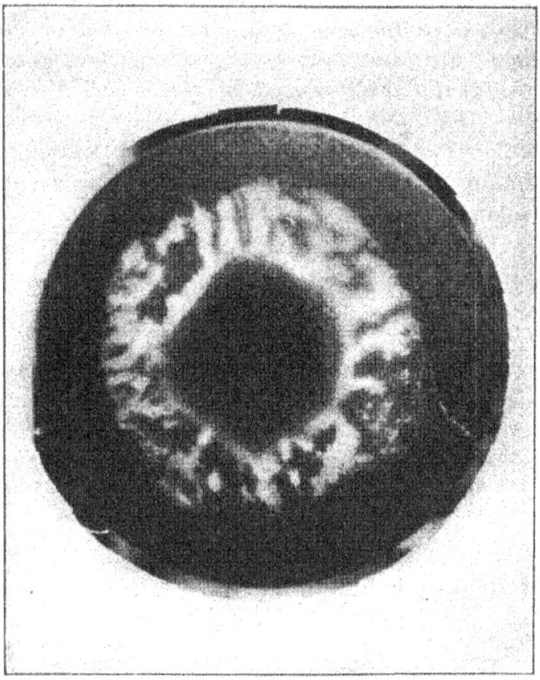

Fig. 1

simple is the explanation of some of the so-called mysterious fractures.

It was thought of sufficient interest to examine the broken shaft microscopically and chemically, to see what light could be thrown upon it. An entire section of the shaft was polished. In the final polishing with rouge, licorice extract was used to keep the chamois leather moist. The surface etched rapidly and showed a very decided segregation to the naked eye.

Fig. 1 shows the general appearance of the polished section. The central and outer portions showed the usual appearance of a polished iron surface, but the central ring was very much lighter and brighter in appearance, and the crack was found to pass from the outer edge into, and ending in, the segregated part.

Borings for analysis were taken from the inner and outer parts and from the most thickly segregated portion of the ring, and analyzed, with the following results: —

	Centre	Ring	Outside
Carbon	0.026	0.079	0.05
Manganese	0.24	0.32	0.30
Phosphorus	0.084	0.214	0.094

It will be noticed that the segregation of phosphorus and carbon is quite marked. There was not enough material for a determination of sulphur, but under the microscope there was seen a number of iridescent slate-colored patches, mostly round or ovoid, which were very likely the iron sulphide. The outer and inner portions contained very few of these iron sulphide particles.

In the light of the above results and the recent work of Stead, "On the Mutual Relations of Iron, Phosphorus, and Carbon when together in Cast-Iron and Steel," it was thought that the light-colored ring should show the phosphide eutectic, and such was found to be the case.

Fig. 2 shows a photograph of a portion surrounding the crack at its point of origin. There can hardly be any doubt of the fact that this is the eutectic of iron and iron phosphide, as the carbon is so low that it appears only in patches, and no steel of such low carbon content which the author has examined has shown any appearance of well-defined pearlite. The peculiar brilliant-white appearance, characteristic of the ring, has been noticed in many other specimens. Very often, when polishing specimens of low carbon per cent, in relief, the same appearance would manifest itself, and it was thought at the time to be due to the peculiar etching effect of licorice; but, on looking over the results, it is found that all such specimens were rather high in phosphorus. Whether such an appearance would always indicate high phosphorus can only be told by repeated observations on a much larger number of specimens.

That the segregation took place in such a form, and to such a depth, is rather peculiar. On close inspection, it is seen that the inner line of segregation is nearly that of a square. Such a figure would suggest that the original bar was square and that it had been rounded off on the outer edge, while passing through the rolls, leaving the interior more or less in its original condition. That there has been considerable flow of metal, there can be no doubt. It seems more than likely that the temperature of

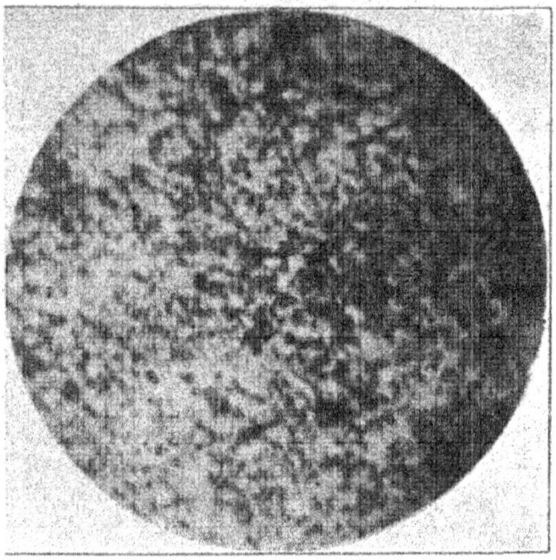

Fig. 2

the outer portion had been raised so high that the metal flowed freely, and the pressure was great enough to force the impurities toward the more cool central portion. Indeed, it seems highly possible that the bar had been heated so rapidly that the heat had not penetrated to the inner portion. This seems quite possible, as the phosphide areas would most naturally be looked for in the portion which was the last to cool. If the centre portion had been heated to the same temperature as the rest of the material, the phosphorus and carbon would have segregated to the centre. That other shafts, made at other times, from the same metal, stood wear successfully would seem to be additional evidence that this specimen was not uniformly heated.

The metal is distinctly harder, as would be expected, in the segregated portion, and, being produced in such a way without subsequent annealing, would cause the mass to act, not as a homogeneous whole, but as three separate portions imperfectly welded, and in which the individual molecules possessed a very low cohesive power. It is not at all surprising that a crack should have been developed, but that the shaft should have accomplished any wörk at all without breaking.

ON THE CRYSTALLOGRAPHY OF IRON *

By F. OSMOND and G. CARTAUD

ONE of us has, in a previous paper,† summed up and discussed the data so far gathered by observation and experiment concerning the crystallography of iron.

The conclusion which we reached, however, could only be a first approximation. The materials which had been studied were industrial products whose history and composition were often imperfectly known. In many cases the external crystalline forms were prevented from developing freely, or there was a possibility of the forms belonging to each molecular condition of the metal having been superposed, and it was difficult to distinguish their characteristics.

In order to acquire a more accurate knowledge of these facts, it was necessary to cause the three allotropic varieties of iron to crystallize:

1. As much as possible in a pure condition.

2. In the exact range of temperature in which every variety is normally stable.

3. Under conditions allowing, at the desired temperature, a free development of the crystals, the external forms acquired under a certain state remaining unaltered by the internal transformations which may occur during subsequent cooling.

Presented in this way, the problem had been solved in prin-

---

* *Annales des Mines,* August, 1900.

† *Annales des Mines,* January, 1900. *The Metallographist,* Vol. III, pages 181 and 275, July and October, 1900.

ciple, many years ago, by the methods of Péligot and of Poumarède.

Péligot* had reduced some ferrous chloride by hydrogen and obtained "some very bright octahedral crystals† and some flexible and malleable plates containing possibly some metallic filaments." It is by this method that Mr. Stanislas Meunier succeeded in his important synthesis of meteoric irons.‡

A little later Poumarède§ used some vapor of zinc instead of hydrogen for the reduction of ferrous chloride, and obtained some "dendritic crystals, with here and there some hollow tetrahedrons." Unfortunately, the transformations of iron were not known fifty years ago. It was necessary, therefore, to repeat these old experiments under well-defined conditions of temperature, i.e.:

Below 700° C., to cause iron to crystallize in the *Alpha* state.

Between 750° and 860°, to cause it to crystallize in the *Beta* state.

Above 860°, to cause it to crystallize in the *Gamma* state.

With one exception, all our experiments were conducted in a porcelain tube, internally varnished, placed in a Mermet furnace. The temperature was ascertained by means of a Le Chatelier couple placed in the centre of the furnace, in contact with the porcelain tube and at the same level with the axis, a second clay tube surrounding it. By properly regulating the amount of gas, the temperature could be maintained constant, within a few degrees, as long as desired. A margin of 50° C., either way, is sufficient to allow in the most unfavorable cases for the irregularities of the calorific field, provided the products formed far away from the couple are not considered.

---

* "Sur un moyen d'obtenir certains métaux parfaitement purs." *Comptes Rendus,* Vol. XIX, p. 670, 1844.

† Not cubes, as stated repeatedly in text-books. Hopfgartner and Hornig in repeating Péligot's experiment in Schrötter's Laboratory (*Jahrbuch k.k. geologischen Reichsanstalt,* 1850, p. 151), did, however, obtain some cubes. There was here a disagreement which it was important to examine.

‡ Mémoires présentés par divers savants à l'Académie des Sciences de l'Institut de France, Vol. XXVII, 1883.

§ "Mémoire sur quelques phénomènes de réduction; nouveau moyen d'isoler le fer de ses combinaisons." *Comptes Rendus,* Vol. XXIX, p. 518, 1849.

Péligot's method answers very well for the preparation of *Alpha* and *Beta* iron, but above 1000° C., a large proportion of the ferrous chloride is volatilized without being reduced. Poumarède's method, on the contrary, works best, as might have been expected, above the boiling point of zinc, and is, therefore, well adapted to the production of *Gamma* iron. It was, however, important, since the surrounding atmosphere may have—and, in reality, has — a marked influence upon the crystalline forms, to produce also some *Gamma* iron by the Péligot method and some *Alpha* and *Beta* iron by the Poumarède method, in order that the results might be comparable. As a matter of fact, there is no difficulty in obtaining a little *Gamma* iron by the hydrogen treatment, especially if the temperature be kept as low as possible, i.e., about 910° C., and the volatility of zinc below its boiling point is sufficient to yield some iron at 800°, and even at a lower temperature. Only a small mass of iron could be produced under these unfavorable conditions, but this is unimportant as long as the crystallographic examination was possible.

The hydrogen used in the application of the Péligot method was prepared either by the electrolysis of alkaline water or by the ordinary process (zinc and hydrochloric acid); it was purified by being made to pass through caustic potash, permanganate of potassium, and concentrated sulphuric acid; it was dried by a long tube filled with pumice-stone, saturated with sulphuric acid and anhydrous phosphoric acid, whose efficiency was carefully ascertained; finally, it was further purified, when necessary, by passing it through a tube filled with copper and heated to a dark red. The end of the operation was indicated by a stop in the evolution of gaseous hydrochloric acid. The speed of the current of hydrogen was generally two or three bubbles per second, but occasionally considerably faster or slower.

The reduction of ferrous chloride by zinc vapor was carried on once in an atmosphere of hydrogen which showed little activity at 1000°, and in subsequent operations a nitrogen atmosphere was used. The nitrogen was generally prepared beforehand, by copper and ammonia, purified by diluted sulphuric acid, caustic potash, and copper heated to a dark red, dried by pumice-stone holding sulphuric acid and anhydrous phosphoric acid. It was ascertained that the resulting iron did not contain any zinc.

Whatever the method used, the ferrous chloride is reduced in part to the vapor state. An iron sponge is generally found in the boat which contained the chloride, and a more or less abundant deposit covers the walls of the boat and the walls of the tube in the vicinity of the boat. In some of the experiments small rough plates of porcelain were placed in front of the boat, and upon these were formed few very small, but often very pure, crystals.

At all events, three kinds of deposits should be noted: the coatings, forming a continuous pellicle; the dendrites, which sometimes separate from these coatings; and the crystals proper, which are generally found near their edges. Each of these products calls for a special study by appropriate methods. None of the crystals were sufficiently large to be detached and measured by a goniometer. They were examined under the microscope by daylight, by the light of a Welsbach lamp reflected by means of

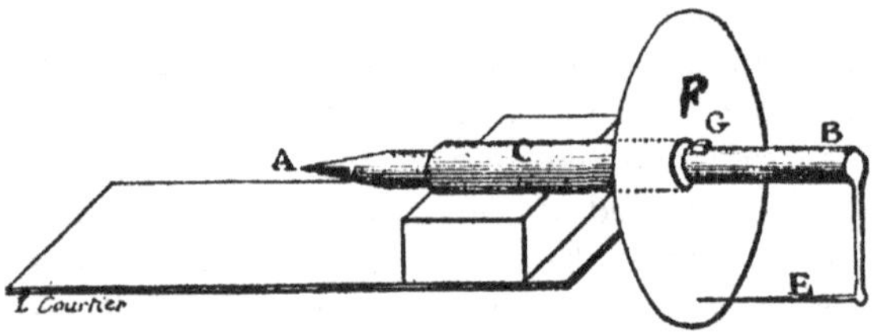

Fig. 1

Sorby's parabolic mirror, and by an incandescent lamp, whose light was reflected in the objective by means of Nachet's totally reflecting prism. In these conditions the facets of the crystals which are lightest appear, respectively, blue, white, or red, according to the illuminating method used. Oblique illumination shows the general forms in perspective, while vertical illumination shows, without much distortion, the true angles of the faces which happen to be in a horizontal plane.

The dendrites are sometimes sufficiently long and isolated to be detached from their support without much injury. It is sometimes possible to separate the largest of them from the surrounding mass by means of an acid which dissolves first the smaller ones. To examine the detached dendrites, they are placed at the extremity *A* of a magnet *B* shaped like a pencil. (Fig 1.) The magnet *B* can be made to rotate in a collar *C*, and at its other extremity there is a crank carrying a needle *E*, which moves opposite the divisions of a graduated circle *F*. A stop *b* prevents the magnet from moving to the left. The apparatus is placed on the stage of the microscope in such a way as to bring the dendrite under the objective. It is possible by these means to make it revolve, to measure the angles of rotation corresponding to characteristic appearances, and, finally, to determine the structure.

The continuous coatings can only be profitably examined by vertical light. They frequently possess a special interest from a metallurgist's standpoint, because they constitute a transition between the artificial crystals obtained by synthesis and the compact metallurgical products.

Crystallography of Alpha Iron and Beta Iron

Three experiments, numbered, respectively, 1, 2, and 3, were conducted by the Péligot method in order to produce *Alpha* iron.

1. In a glass tube heated by a row of gas burners, the maximum temperature, which was not ascertained, corresponding to the softening of the glass.

2. In a porcelain tube placed in a Mermet furnace. The temperature was regulated so as to remain in the neighborhood of 685°, and did not exceed 693° C. The experiment lasted one hour and a quarter after the desired temperature had been reached.

3. In a porcelain tube placed in a Mermet furnace. Hydrochloric acid began to be evolved abundantly at about 600°. The temperature was allowed to increase slowly to 687°, and the formation of acid vapors having then ceased, the experiment was brought to an end, the reduction having lasted three-quarters of an hour.

The ferrous chloride, previously dried by hydrogen, and held or not in a boat, was in every experiment placed in the tube before lighting the furnace.

Four experiments, numbered 4 to 7, were conducted by the Péligot method for the production of *Beta* iron, all of them in a porcelain tube placed in a Mermet furnace.

4. Between 807° and 824°, lasting 45 minutes.
5. Between 812° and 830°, lasting 50 minutes.
6. Between 782° and 825°.
7. Between 802° and 813°, lasting 55 minutes.

In these experiments for the production of *Beta* iron, the boat containing the ferrous chloride was, of course, kept in the cold portion of the tube, outside the furnace, until the desired temperature was reached, when it was pushed in the heated region. By placing at once the ferrous chloride in the heated part, we would have had, during heating, a rather long period of *Alpha*-iron formation, below 750°.

The examination of the products of reduction showed that *Alpha* and *Beta* iron crystallize identically.

It is why we may consider these two varieties under the same heading.

After this fact had been established, three additional experiments were performed (8, 9, and 10) for the production of *Alpha* or *Beta* iron, in a nitrogen atmosphere.

8. By the Poumarède method, below 835°, the boat having been introduced in the tube before heating the furnace.

9. By the Poumarède method, between 706° and 778°, the boat having been pushed in place when the pyrometer registered 736°, the heating lasting one hour and a half.

10. In reducing the ferrous chloride by sodium vapor below 787°, the boat being put in place beforehand, and the experiment lasting one hour from the time the furnace was lighted. The boat with the ferrous chloride was placed upon a larger boat containing the zinc or sodium.

The results obtained are described below.

Isolated Crystals. — They are especially cubes. (See Fig. 2, magnified 250 diameters, vertical illumination.)

In one of the experiments some beautiful pyramidal cubes were produced. (Fig. 3, 400 diameters, vertical illumination.) The pyramid is flattened; the four visible faces may be simultaneously illuminated by the vertical light, and the solid appears to be a tetrahexahedron b^2.* These are probably the crystals which Péligot, who did not have at his disposal the instruments we now possess, took for octahedrons, for, like Hopfgartner and Hornig, we did not obtain any octahedrons, at least by reduction with pure hydrogen below 850°, and Péligot, who used a glass tube, could not have reached as high a temperature.

Few perfect tetrahedrons are found. They are generally combined with cubes. (Fig. 4, 250 diameters, vertical illumina-

---

* These letters refer to Lévy's notations, generally used by French crystallographers. — Ed.

tion.) These same modifications are very likely to be found in the bulky masses of commercial irons, and may be detected in the polished and etched surfaces, where they had previously been attributed to the action of the acid on the edges of the crystals. Mr. Heyn called our attention a few months ago to the appearance of a polished section of iron wire etched with double chloride of cop-

Fig. 2

Fig. 4

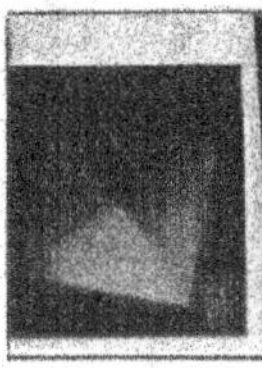

Fig. 3

Fig. 5

per and ammonium, in which some figures could be detected suggesting a perfect tetrahexahedron.

Other modifications along the edges are also possible, especially b^1 modifications, which have already been detected in some metallurgical products and also in meteoric iron exhibiting a cubic crystallization.

Forms of Association. — 1. Two or more cubes may be joined by one face. Fig. 5 (250 diameters, vertical illumination) shows two cubes, incomplete but very large, placed upon each other. When the cubes have the same dimensions, and when the centres of the faces common to all coincide, we have one of the possible methods of generation of lengthened cubes.

2. Two or more parallel cubes with a portion common to them all are disposed as shown in Figs. 6 and 7. In the first instance the axis of the system might be a binary axis, and in the second case, a ternary axis, of the elements.

3. One face of the cube is fastened to a truncation on the edge of another cube, generally to a truncation b^1. This form of

Fig. 6

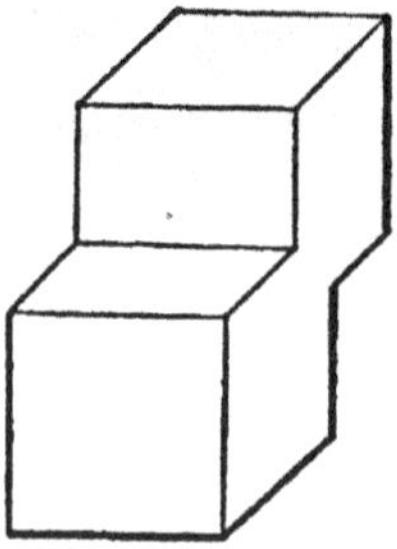

Fig. 7

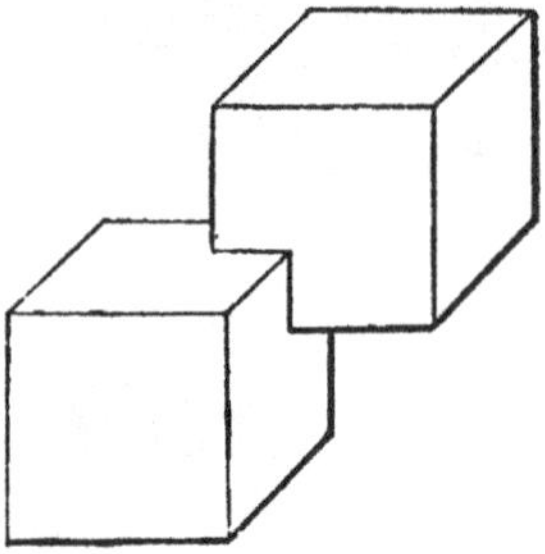

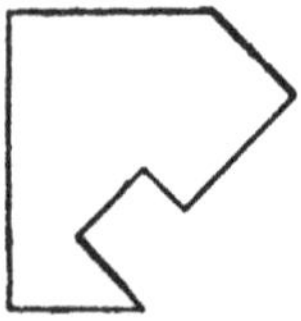

Fig. 8

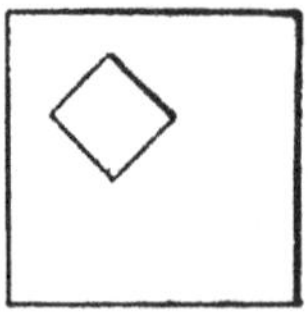

Fig. 9

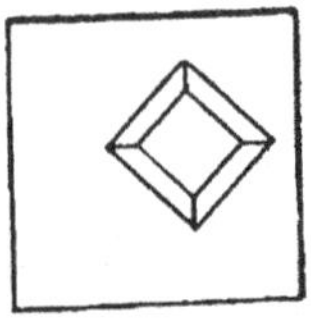

Fig. 10

association is made evident by the presence of many faces crenated in a direction parallel to the diagonals (Fig. 8), and by the inclusion of a cube in another cube, modified or not, and which had turned 45° around the common quaternary axis (Figs. 9 and 10).

4. Fig. 11 (400 diameters, vertical illumination) shows a rather curious association, which at first sight recalls the twin crystals of the diamond, by molecular hemitropy, uniting two tetrahedrons, with their angles modified by the inverse tetrahedrons, but the four visible triangles are here rectangular ones, and in the same plane; they cannot represent a^1 faces.

Dendrites. — The study of the dendrites by means of the rotary apparatus already described (Fig. 1) reveals in them the presence of the three first forms of association just outlined:

1. A row of cubes of same dimensions and united by one face with coinciding centres, yielding a filament with a square section.

2. A series of parallel cubes united as shown in Figs. 6 and 7.

This form of association may exhibit various complications, especially when the reduction of the ferrous chloride is effected by zinc vapor and when the maximum temperature at which *Beta* iron is stable is approached as near as possible. One of the most frequent of these modifications is as follows:

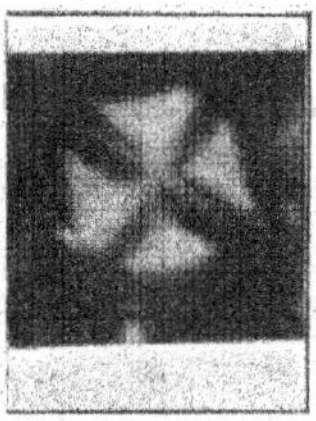

Fig. 11

The unit, instead of being a cube, is made up of a group of five cubes, four of which are placed cruciately against the four faces of the fifth one; these units are then superposed as shown in Fig. 12, which represents the superposition of two such units projected upon a plane parallel to their bases, and Fig. 13 (100 diameters, vertical illumination) exhibits the appearance of such a dendrite to which some parasite crystals are fastened.

This dendrite, and many similar ones, were obtained in experiment No. 8. In experiment No. 9, also by the Poumarède method, in which the temperature had not exceeded 787°, the dendrites were not as well developed, but the form of association was the same.

3. In a dendritic colony some faces may be united to some truncations on the edges, resulting in the formation of curved figures.

These different forms of dendrites may assume a simpler form, at least in appearance, i.e., the units lose their individuality by alteration of their angles and edges and, so to speak, melt into each other. Is this an embryonic state or, on the contrary, a state

Fig. 12

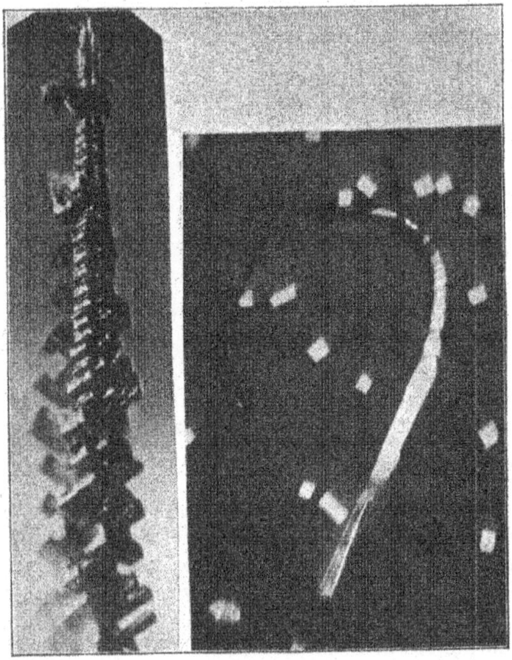

Fig. 13 Fig. 14

of degenerescence? It is difficult to ascertain. At any rate, all the types have as their limit the filiform iron, which had already been observed by Péligot. Between this filiform iron, however, with a smooth surface and a rounded section, and the dendrite with differentiated elements, there exist a series of intermediate forms. The reëntering angles, for instance, of the dendrite represented in Figs. 12 and 13 are progressively reduced to continuous, longitudinal channelings, and the reëntering angles between the successive units are likewise changed into transversal channelings, while the distinct entity of these elements has become undiscernable. These indications of their origin make it possible to infer the true mode of association in the absence of external forms.

The terminals of the dendrites may also throw much light upon their intimate structure, even when this structure has been effaced. The terminals of the dendrites present, generally, two different appearances: the elements become smaller and smaller as they approach the free end, finally forming a sharp point, sometimes crystalline in appearance, and which is in reality the envelope of the decreasing elements; or the terminal may be a solid, representing synthetically the structural units, and whose faces are respectively parallel to the different faces of these units.

In rectilinear associations of *Alpha* and *Beta* iron, the structural elements are always parallel cubes; the terminal solid, if it be present, is itself a cube parallel to all the elementary cubes which make up the colony, and it frequently happens that this terminal cube, obliquely fastened at the end of a rod, is the only evidence of the mode of formation of this rod.

The following illustrations show a number of types of these somewhat simplified dendrites:

Fig. 14 (250 diameters, vertical illumination) shows a rod partly straight and partly curved, twisted and with edges sharp in some places and effaced in others. Some small cubes are also detected in the vicinity.

Fig 15 (250 diameters, vertical illumination) shows a portion of porcelain plate upon which were deposited some small curved rods, subsequently volatilized by hydrochloric acid; one of these rods still exists in part at the lower right angle of the figure. The others have left their traces upon the plate.

Fig. 16 (250 diameters, vertical illumination) exhibits some straight and some curved rods, frequently carrying at one end a cube obliquely fastened upon the axis.

Fig. 17 (250 diameters, vertical illumination) shows some rods with plane faces and sharp edges with pyramidal points.

Fig. 18 (250 diameters, vertical and oblique illumination combined) shows some rods with transversal channelings.

Fig. 19 (250 diameters, vertical and oblique illumination) shows some filiform iron exhibiting traces of spiral figures.

Coatings. — Fig. 20 (250 diameters, vertical illumination) shows the passage between isolated crystals to the continuous

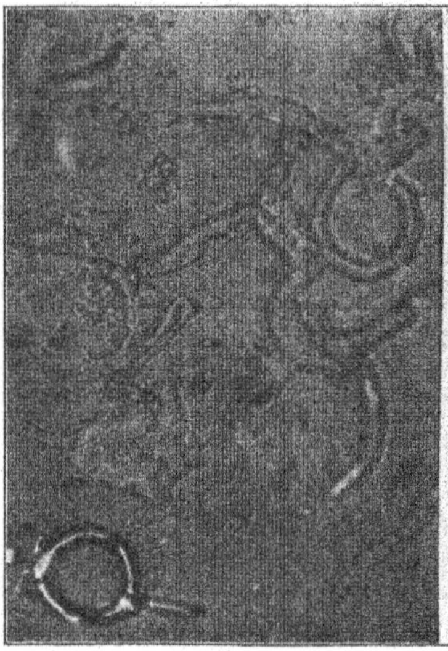

Fig. 15

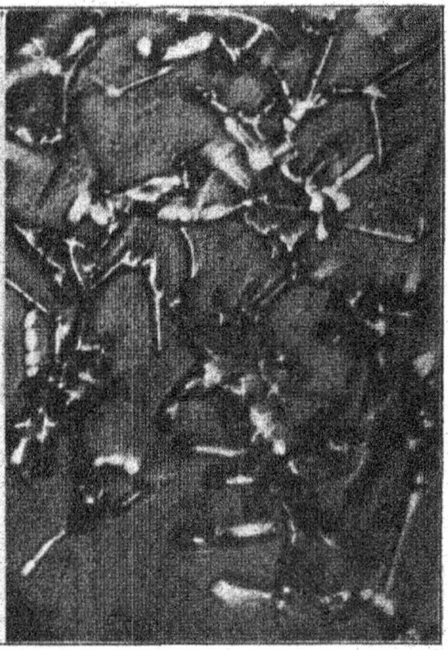

Fig. 16

coating; the crystals, some of the angles of which remain visible, are piled upon each other in a constant direction.

When the coating is thin, well-defined crystals may still be detected, cubes or pyramidal cubes, embedded in a paste having a crystalline appearance. The surface of thicker coatings is sometimes divided in grains by a polygonal network, very similar to that exhibited by a section of iron after etching a few seconds with a 20-per cent solution of nitric acid. The interior of these polygons is not always plane; they quite frequently exhibit the

Fig. 17

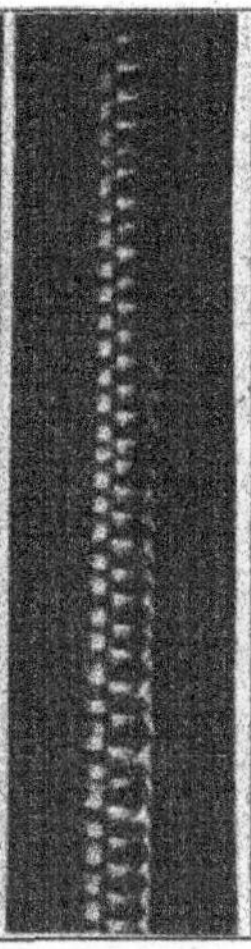

Fig. 18

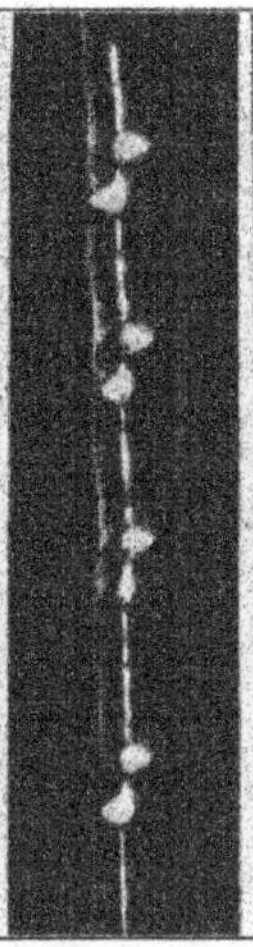

Fig. 19

Fig. 20

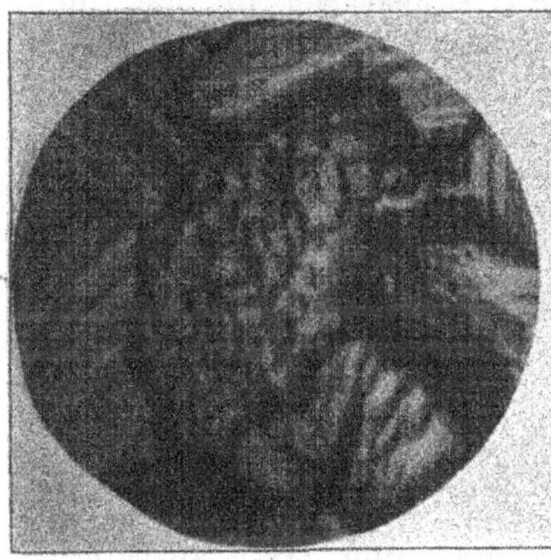

Fig. 21

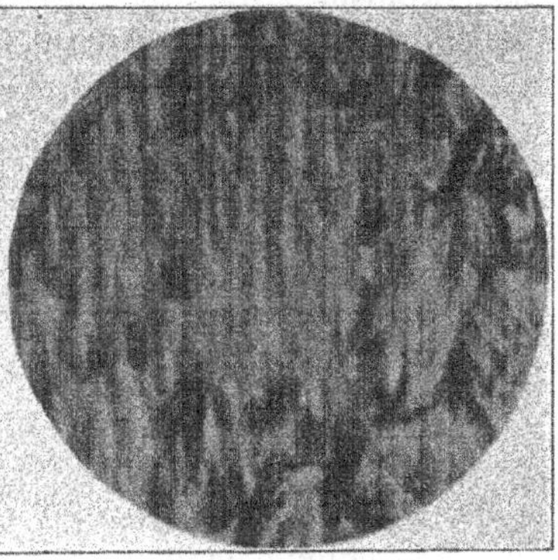

Fig. 22

details of the internal structure of the polyhedric grains. (Figs. 21 and 22, 400 diameters, vertical illumination.) This structure is identical to that assumed by the ferrite of soft steel after a sufficiently long etching with dilute nitric acid (Stead) or with double chloride of copper and ammonium (Heyn). Each grain may be considered as being formed of small cubes, parallel in the same grain, but whose orientation varies from one grain to the next. In spite of the different conditions of their genesis, our synthetic products and industrial products have the same structure, a fact which justifies our experimental method. This identity appears also to prove that the distinction made by Mr. Heyn between the crystalline and the etched figures, a distinction which it was cautious to make, and which still remains accurate, in a general way, is not to be extended, as a matter of fact, to the case which we are studying.

Conclusions. — *Alpha* and *Beta* iron crystallize identically, which is a necessary condition, if not a sufficient one, of their isomorphism; from the progressive character of their mutual transformation their isomorphism was to be expected. The cube is the crystalline form, with frequent modifications on the edges, sometimes b^1, sometimes b^2, as far as can be judged in the absence of accurate measurements. No modification on the angles.

Crystallography of Gamma Iron

Five experiments were made to produce *Gamma* iron, numbered 11 to 15, all in a porcelain tube placed in a Mermet furnace.

11. Péligot method at the maximum temperature of the furnace, i.e., about 1000° C.

12. Péligot method, between 924° and 965°. In both cases the boat containing the ferrous chloride was quickly placed in the middle of the furnace when the desired temperature had been reached.

13. Poumarède method at the maximum temperature of the furnace, i.e., about 1000°, in a hydrogen atmosphere. The boat containing the chloride, placed in a larger boat containing the zinc, was introduced progressively in the hot portion of the tube after the temperature had been regulated.

14. Same conditions as in experiment 13, but the hydrogen was replaced by nitrogen.

In experiments 13 and 14 it had been supposed that the boat could be pushed slowly in place without causing an appreciable reduction of the chloride, at least in nitrogen, below the boiling point of zinc. As this was found to be erroneous, the experiment was repeated (No. 15) in an atmosphere of nitrogen, in which the boat was pushed quickly in place after the temperature had

Fig. 23

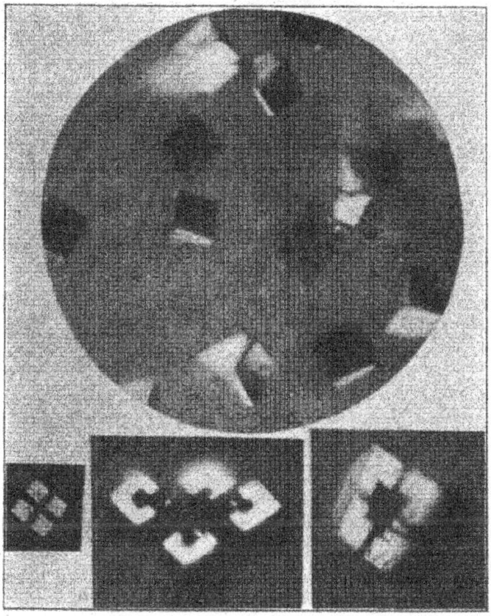

Fig. 24 Fig. 26 Fig. 25

been properly regulated. Experiments 13, 14, and 15, moreover, yielded the same results.

Both methods, Péligot's and Poumarède's, yield crystalline products derived from the same type. The reduction by hydrogen, however, produces isolated crystals rather than dendrites, while the reduction by zinc vapor yields an abundance of dendrites and few crystals.

The results obtained are described below.

Isolated Crystals. — These crystals are again cubes. (Fig. 23, 250 diameters, oblique illumination.)

The cubes are frequently modified at their angles by a^1 truncations. We did not find any well-defined octahedrons, but

Fig. 29

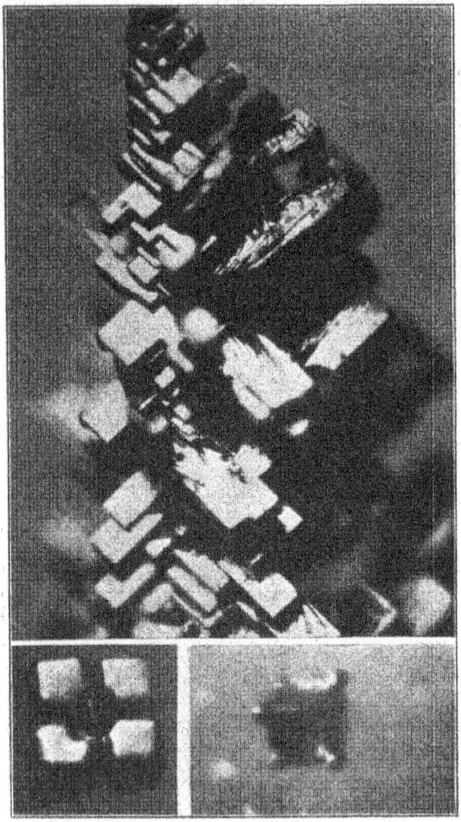

Fig. 27 Fig 28

the cuboctahedron, although rare, was detected. The truncations do not, in general, affect all the angles and are not equally developed, and all sorts of combinations of the octahedron and cube are found. The difficulties of the observations, however, do not permit the affirmation of the existence of a *mériédrie.*

Forms of Association. — 1. Two or more cubes may be united by their faces. Groups of four cubes are found quite frequently. One of these groups is shown in Fig. 24 (250 diameters), under vertical illumination, and revealing, therefore, the true shape of the four horizontal faces. Figs. 25, 26, and 27 show the same associations under oblique illumination; the internal angle is generally grooved (*évidé*), at least in the products obtained by the Poumarède method.

2. Two or more parallel cubes having one portion in common are united, as shown in Figs. 6 and 7, in the case of *Alpha* and *Beta* iron. The axis of the association is probably a binary axis, or, more often, a ternary axis of the structural elements. A part of a cube may be seen in Fig. 28 (250 diameters, oblique illumination), carrying a small parallel cube on each of its free angles. Such arrangement is frequently found in diamonds from the Cape. It is the starting point of more complicated associations, which will be described later on.

3. Two cubes are united by a truncation a^1.

This form of association is, so far, the only one exhibited exclusively by *Gamma* iron, the former two belonging to all the allotropic varieties of iron.

Dendrites. — The types of dendrites are quite numerous. They were studied by the means already mentioned, and the principal ones are described in detail in the following pages.

1. The form of association No. 1 (elongation along a quaternary axis) appears to be rare in our synthetic products.

2. The form of association No. 2, carried on sufficiently far, yields dendrites shown under oblique illumination in Figs. 29 (100 diameters), 30 (250 diameters), and 31 (75 diameters). It is seen that the cubes are frequently grooved, hopper-like (*évidés en trémies*), and that the terminal cube (Fig. 31) may have a reëntering angle corresponding to the reëntering angle formed by the grouping of the subjacent cubes, in accordance with one of the modes of termination previously mentioned. Their succession would give rise to a longitudinal channeling in a dendrite of similar formation and composed of undifferentiated elements.

Some grooves (*trémies*) are also found which belong to the

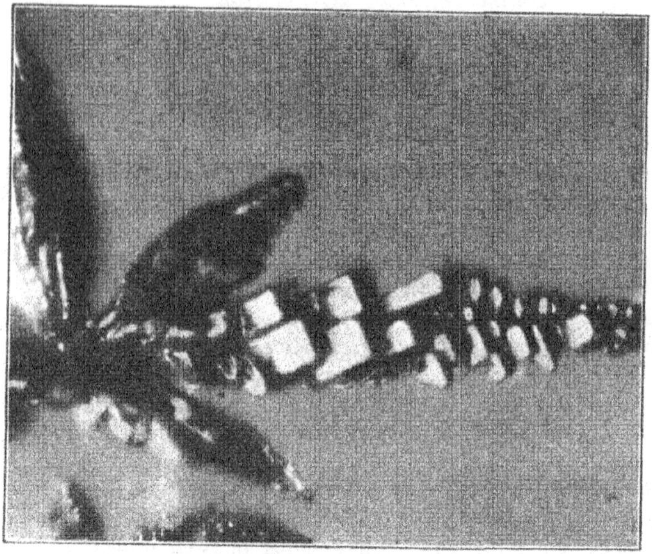

Fig 30

Fig. 31

same form of association and which recall those of sodium chloride (Fig. 32, 250 diameters, oblique illumination).

These groupings, as already stated, are not characteristic of *Gamma* iron, for they are also found in *Alpha* and *Beta* iron, al-

Fig. 32

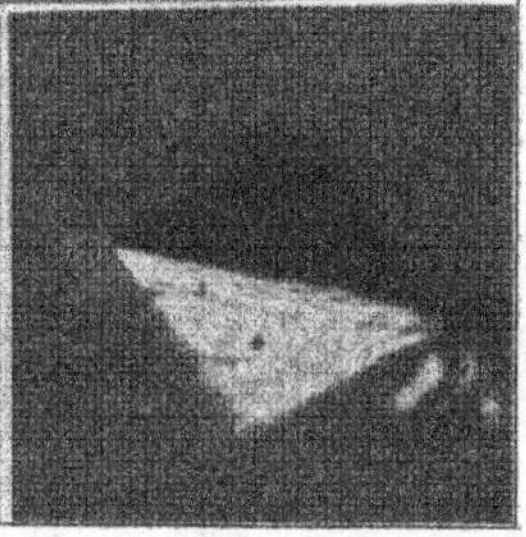

Fig. 34 Fig. 35

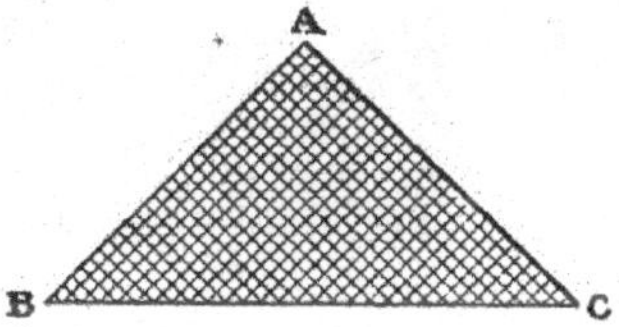

Fig. 33

though generally less sharply developed, probably owing to the lower temperature required for the production of these two varieties.

3. Let us imagine a rectangular, isosceles triangle (Fig. 33), composed of one or more layers of small cubes united by their faces, and limited along AC and BA by the faces p of a row of cubes and along BC by a common truncation a^1. Let us now

Fig. 40

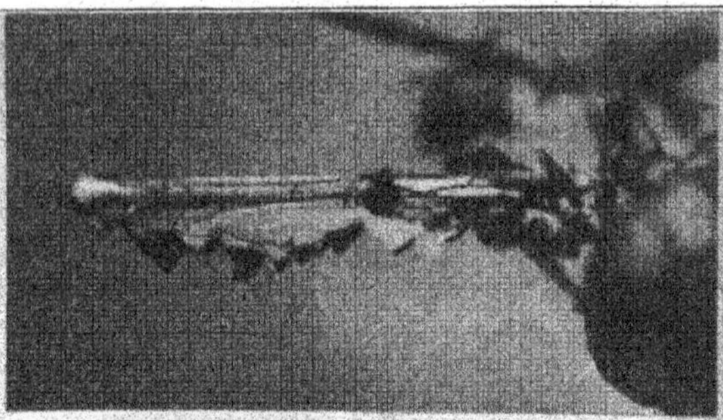

Fig. 39

Fig. 38

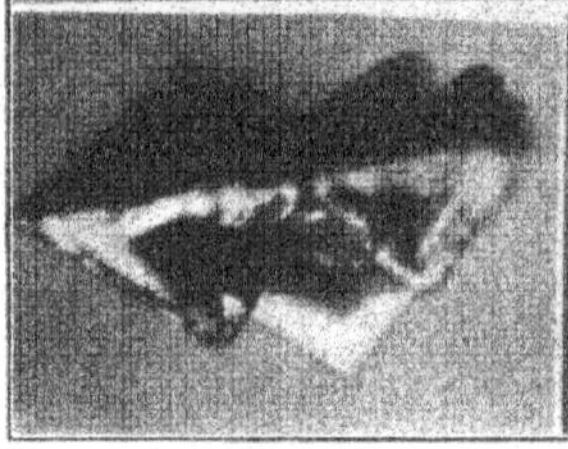

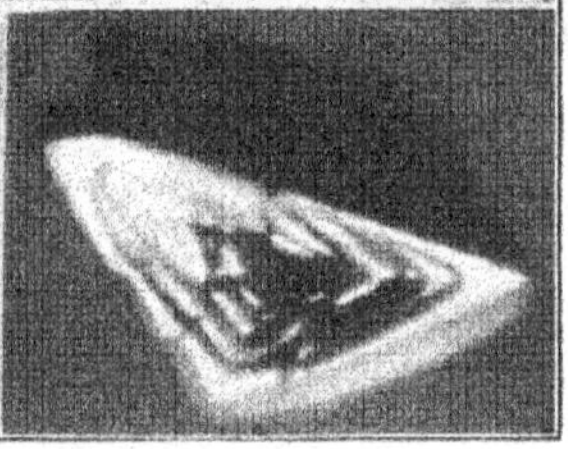

Fig. 37 Fig. 36

suppose that two such triangles are united by the truncations a^1. The result will be an arrangement which Poumarède called hollow tetrahedrons (Figs. 34 and 35, 250 diameters). Fig. 34 was lighted by vertical illumination, and shows only one undistorted face, i.e., a rectangular, isosceles triangle. Fig. 35 shows under oblique illumination the whole of the solid. The faces may be grooved (*évidées*). (Fig. 36, 250 diameters, oblique illumination.)

The faces of the grouping, instead of forming a single rectangular triangle, present frequently the appearance of the teeth of a saw (Fig. 37, oblique illumination).

Two hollow tetrahedrons may have one common face, consisting in such case of two portions of rectangular, isosceles triangles having parallel hypotenuses (Fig. 38, 250 diameters, vertical illumination). This common face corresponds to the bottom of a kind of gutter whose sides are made up by one or more teeth. Fig. 39 (250 diameters), lighted by oblique illumination, shows the bottom of the gutter and one of the sides; Fig. 40 (100 diameters, oblique illumination) shows a more complete gutter, but somewhat injured by the hydrochloric acid used in separating it from a mass of other dendrites which surrounded it; one of the sides is lighted, and the extremities of a few teeth of the other side are seen in black on a lighter background.

In the grouping just described one side of the gutter may in turn be replaced, through the same process, by a portion of a third hollow tetrahedron. The bottom of the gutter becomes then itself a gutter, as shown in Fig. 41, both in cross section and in elevation. The half invisible in the elevation would be similar to the visible half.

In support of this interpretation, we have, on the one hand, the frequent occurrence, on the sides and bottom of the gutter, of striæ, or of parasite crystals, pertaining, through their orientation, to the integrant cubes, and, on the other hand, the measure of the angles A B C, B C D, and C D E (Fig. 41). These angles, if they actually result from the union of two truncations a^1, must measure 109° 28′. A number of these angles were measured by causing the dendrite, placed as horizontally as possible, to turn under the objective and noting the rotations required to bring each face in a vertical position, in both directions successively. The theoretical angle was obtained within 2° or 3°, i.e., within the limits of the experimental errors unavoidable in the handling of such delicate preparations, impossible to isolate mechanically or chemically without some deterioration.

4. Let us suppose a row of parallel cubes piled up and con-

nected by an angle so that the axis of the group be a ternary axis of the elements (Fig. 6). Let us then suppose that a second such row be united to the first one, each element of one group being joined to the corresponding element of the other by a truncation a^1. The resulting arrangement is shown in Fig. 42. Each element, however, has a tendency to give rise to secondary colonies. In the simplest case the two opposite faces of two conjugate ele-

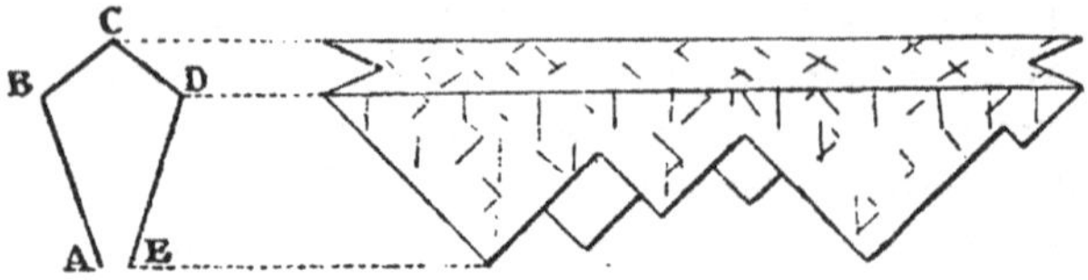

Fig. 41

ments, marked *p* in Fig. 42, serve as supports for two new cubes, each one parallel to the cubes of its own row. These two cubes meet each other as they grow, and this mutual interference causes the formation of a six-faced solid, each face being a rectangular, isosceles triangle, and the right angles being found in groups of three around two opposite apexes. This is a very common disposition. Figs. 43, 44, 45, and 46 (100 diameters,

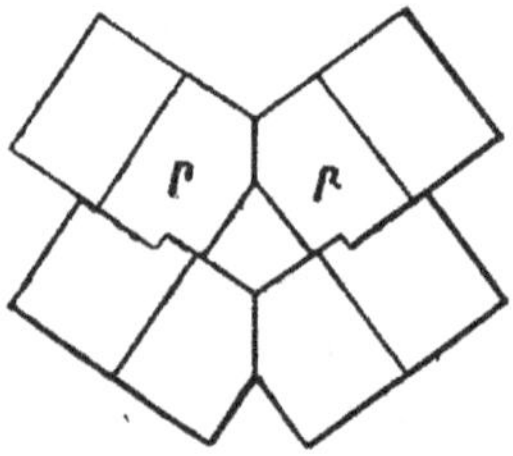

Fig. 42

oblique illumination) show the various appearances of such a dendrite, made to revolve, 45° at a time, around its axis. Since the dendrite has two planes of symmetry, a complete rotation would reproduce Figs. 44 and 46 twice and Figs. 43 and 45 four times.

Figs. 43 and 44 show, moreover, the formation of branches upon some of the elements. The process of generation is always identical.

Figs. 43, 44 and 45

As far as it has been ascertained, this type of dendrite is characteristic of *Gamma* iron.

The terminals of these dendrites correspond to one of the two appearances already described. They are either formed by the gradual decrease in size of the structural elements giving rise to a sharp point (Fig. 47, 134 diameters, oblique illumination), or they consist of a synthetic solid in which the faces of the structural

Fig. 46 Fig. 47 Fig. 48

elements are represented with their true direction. In the case considered, each row of cubes gives three rectangular planes, and the terminal solid is again the same double trirectangular tetrahedron, which forms also the edges of the dendrite.

Forms of Association of the Dendrites. — The relatively single groupings just described form, in turn, more complex associations. Fig. 48 (25 diameters, oblique illumination) shows a dendrite, a little injured by the manipulations, which exhibits some perpendicularly fastened branches. The main dendrite is still composed of parallel cubes piled, step-like, around a ternary axis, but the branches are made by another grouping method of the cubes around a binary axis (Fig. 49, 134 diameters, oblique illumination); these branches themselves carry perpendicularly a series of gutters of decreasing length.

In Fig. 50 (25 diameters, oblique illumination) three main dendrites start from the same centre, forming between them

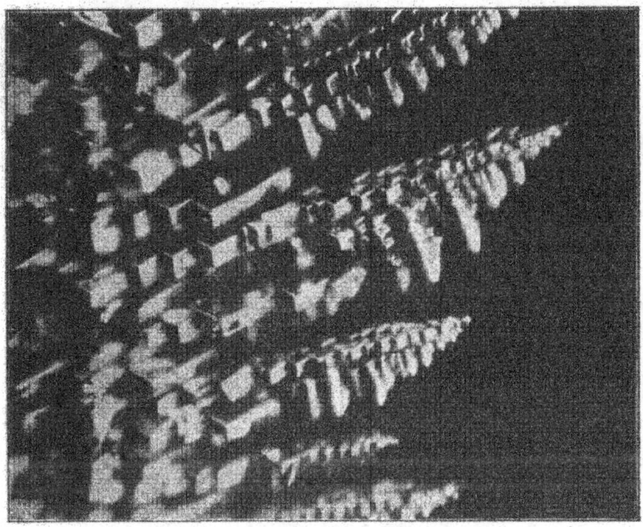

Fig. 49

angles of 120°, while their lateral branches form angles of 60° with the dendrites' axis. A fourth dendrite, projecting laterally, rests upon the plane of the three others, and could not be focussed; it is indicated by a cloudy appearance in Fig. 50.

In Fig. 51 (75 diameters, oblique illumination) will be seen a group of dendrites, of a type common to *Alpha, Beta,* and *Gamma* iron, and also inclined at an angle of 60° on the common axis; the axis is hidden by other branches situated in a plane perpendicular to that of the figure.

Fig. 50

Fig. 51

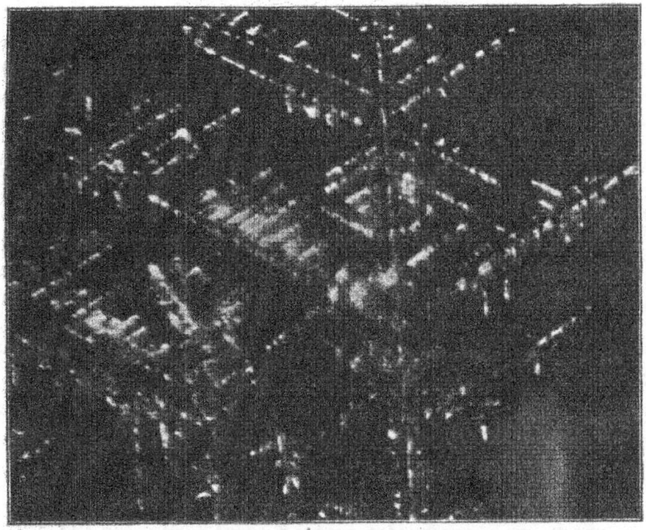

Fig. 52

Fig. 53

The three angles of 120° (or the six angles of 60°, if the extensions are considered) are again seen in Fig. 52 (250 diameters, vertical illumination). Each branch is itself a simplified dendrite passing to the condition of filiform iron, but frequently terminated by a cube, not easily distinguishable in the photograph, whose form and inclination give the key to the internal structure.

Such are the principal associations obtained by synthesis.

It might be suggested that there remains a gap to be filled in order to connect them with the crystalline structures which characterize *Gamma* iron in industrial products, and which are, as stated in the first part of this paper,* skeletons of octahedrons united along the quaternary axis.

As a matter of fact, this gap is only apparent, as may be easily shown by a few instances borrowed from the crystallography of salts.

A solution of chrome alum, not saturated, yields by spontaneous evaporation some violet-colored octahedrons, if the crystallization be slow, and some crystallites of the cubic system, if it be rapid (Fig. 53, 100 diameters, transmitted light). Some terminals will be noted in this photograph, with orthogonal branches and octahedral envelopes, which recall exactly the beautiful steel crystal of Professor Tschernoff, reproduced in Fig. 22 of the first part of this paper.†

Another type of cubic crystallite was also observed on the tube in which the ferrous chloride was reduced by vapor of sodium (Fig. 54, 250 diameters, vertical illumination). It can hardly be anything else but sodium chloride. Certain branches are terminated by a square, one diagonal of which coincides with the axes of the branches, from which it follows that these crystallites have a more regular and advanced organization than would be inferred from their appearance. The same law is always detected: the dendrite's terminal reveals the structure of this dendrite.

If some chrome alum be dissolved hot, forming a concentrated solution, and if the transformation of violet salt into green salt be carried sufficiently far, a thin layer of the *sirupy* liquid,

---

* *Annales des Mines*, Vol. XVII, p. 110. *The Metallographist*, Vol. III, pages 181 and 275, July and October, 1900.

† Also *The Metallographist*, Vol. II, frontispiece facing page 1, January, 1899, and Vol. III, p. 199, July, 1900. — Ed.

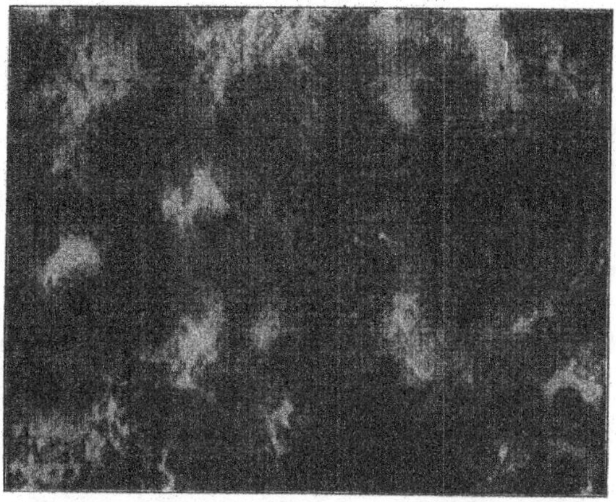

Fig. 54

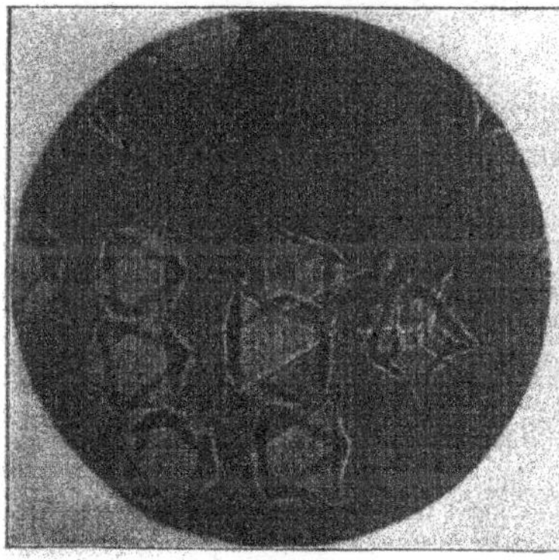

Fig. 55

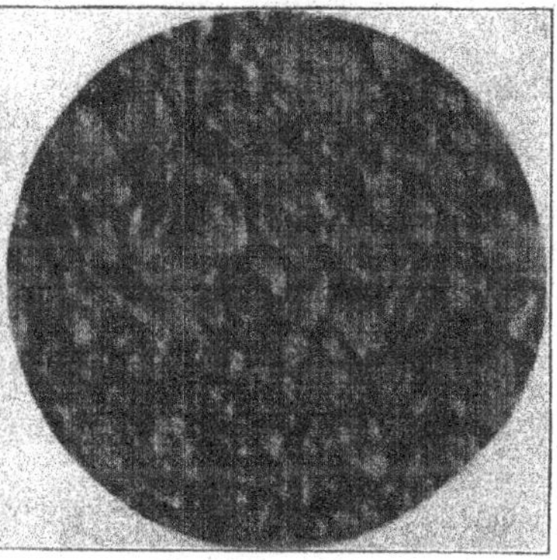

Fig. 58

poured upon a glass plate, will still deposit some perfect octahedrons. After a while, however, say after two or three days, the angles are elongated and the edges appear to assume a hollow rounded shape (Fig. 55, 250 diameters, transmitted light). The elongation of the angles finally causes the formation of a rod, upon which some ramifications are attached. Some of these, however, are perpendicular to the rod, while others formed with it an angle of 60° (Fig. 56, 250 diameters, transmitted light). This difference in the orientation of the ramifications is not accidental; it depends upon the initial position of the generative octahedron. If it be one of the faces of the octahedron which rests upon the glass plate, the angles of the faces parallel to the plate carry some branches with inclined ramifications, while if the octahedron rests upon an edge, the four visible angles carry some branches with ramifications perpendicular to the axis. Both types are seen in Fig. 56.

Fig. 57 (250 diameters, transmitted light) shows a good instance of dendrites of chrome alum with oblique ramifications.

It is seen that with the same salt and in the same preparation dendrites may be produced, side by side, some of which resemble certain of our synthetic dendrites of *Gamma* iron, and the others, those produced spontaneously in metallurgical operations. A small initial difference in the position of the generator suffices to cause the formation of one or the other type.

In the same way, "cœsium" alum, which usually gives classical crystallites of the cubic system, and potassium alum, which yields octahedrons, may, in turn, produce some octahedrons elongated in the direction of the angles, like those of chrome alum, and with dendritic terminals; to bring this about, it suffices to properly thicken their solution by the addition of gum.

Coatings. — The continuous coatings deposited by the reduction of ferrous chloride may present, under favorable conditions, i.e., when superposed dendritical vegetations are not formed, the appearance of Fig. 58 (400 diameters, vertical illumination). Leaving aside the division into grains by a polygonal network, which is not characteristic, some parallel bands may be seen on some grains which recall the preparations obtained by Mr. Saniter in etching iron at about 900° C. in a bath of calcium chloride. These parallel bands are indications of twin crystals, and here again we note the identity of structure between the continuous deposits obtained by the Péligot method at about 1000° and industrial iron at the same temperature.

Conclusions. — *Gamma* iron crystallizes in the cubic system; its crystalline forms are combinations of the cube and the octa-

Fig. 57

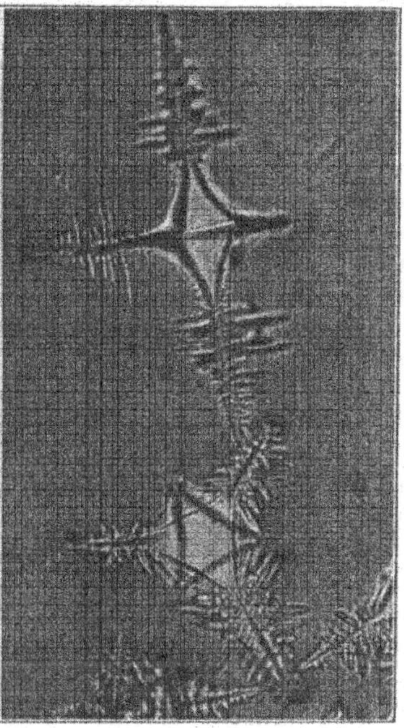

Fig. 56

hedron derived from the cube, and, under favorable conditions, approach the octahedral symmetry. They present frequent associations by contact of truncations a^1. They do not show any modifications on the edges.

(*To be concluded*)

IRON AND STEEL FROM THE POINT OF VIEW OF THE "PHASE-DOCTRINE"*

By F. OSMOND

THE researches started by Professor Tschernoff in what is called in America the physics of steel, and actively pursued during the last fifteen years, led, towards the end of 1896, to the meeting of two currents which had so far run in parallel directions, i.e., scientific metallurgy and physico-chemistry. The contribution of Professor Roozeboom to metallurgical studies was therefore very natural, and even to be expected. Professor Roozeboom was so well prepared for this work, through his exhaustive study of similar phenomena, that he was at once able to move, in a field apparently new to him, with an ease, a steadiness, and an authority that might be envied by professional metallurgists.

If I take the liberty of presenting here a few remarks suggested by his masterly synthesis, it is merely in order to define with greater precision some of the factors which I may have contributed to his genesis.†

To the structural constituents of steel, which until then were classical, ferrite, cementite, pearlite, and hardenite, I added martensite (sub-saturated hardenite), sorbite, troostite, and, later, austenite. Judging from the criticisms which they suggested (I only refer to reasonable criticism), and the additional information I was asked to give, I realized that I had not defined with sufficient clearness two of these constituents, sorbite and troostite. It is because my first conceptions themselves lack clearness. A longer experience has thrown more light in my thoughts, and I may to-day present them with greater precision.

At a high temperature, between the end of the solidification and the beginning of the transformations, iron, leaving aside

---

* *Bulletin de la Société d'Encouragement,* November 30, 1900. These remarks of Mr. Osmond and the following note by Professor Le Chatelier accompany a full French translation of Professor Bakhuis Roozeboom's paper published originally in *Zeitschrift für physikalische Chemie,* XXXIV. Professor Roozeboom presented a comprehensive abstract of his paper to the Iron and Steel Institute, which was reproduced in the October, 1900, issue of *The Metallographist.*

† "Méthode générale pour l'analyse des aciers au carbone." *Bulletin de la Société d'Encouragement,* X, May, 1895.

cementite and graphite, exists in all carburized irons in the condition of mixed crystals of *Gamma* iron. If, through some adequate means, these mixed crystals are preserved unchanged at the ordinary temperature, we have the constituent austenite, which is nonmagnetic. Slow cooling would yield, as a state of equilibrium stable at the ordinary temperature, the system ferrite + cementite.

Between austenite (mixed crystals of *Gamma* iron) and the system ferrite + cementite, iron undergoes two transformations, namely, *Gamma* to *Beta* and *Beta* to *Alpha,* and the iron passes from the dissolved condition to that of a definite compound, Fe_3C.

It is not impossible that during this period of transformations, associations corresponding to phases are formed between *Beta* iron and hardening carbon, for instance, but we do not know it positively; neither do we know whether these supposed phases may be preserved. We are, therefore, forced to consider the transformations collectively, and the three constituents, martensite, troostite, and sorbite, are apparently the successive forms, partially preserved at the ordinary temperature, assumed by carburized iron as it passes from the condition of mixed crystals of *Gamma* iron to the system ferrite + cementite.

The existence of these forms of transitions, both in theory and in fact, is as certain as the existence of an orange shade between the red and yellow in the solar spectrum. The operation of dividing a continuous series, however, is necessarily an arbitrary one, and the number of divisions will vary with different points of view. To continue our comparison with the solar spectrum, for instance, one intermediate shade between red and yellow will suffice for our general needs, while color specialists, such as painters or decorators, designers or milliners, will distinguish in the same range a number of delicate shades.

Is this classification which I propose in the case of the micrographical nomenclature a proper one?

It is the only legitimate question that can be asked in regard to the transition forms of iron. Is the number of divisions too great or too small? Is each division sufficiently characterized by its micrographical and mechanical properties?

In the case of martensite there is hardly any possible objection. Between it and austenite, to say nothing of a difference, apparently secondary, in its crystalline character, there always exists a gap: the transformations are either begun (martensite), or they are not begun (austenite). On the other hand, although I have been able in certain special cases to observe some continuous transformations between martensite and troostite, there exists also, generally, a gap between these two constituents; the topographical limit is well marked. Martensite, therefore, corresponds: (1) from a physico-chemical point of view, to the maximum amount of allotropic iron and of hardening carbon that may be retained by very sudden cooling in carburized irons in which the carbon content of the mixed crystals does not exceed about 1.10 per cent before quenching; (2) from a mechanical point of view, to the maximum of hardness corresponding to such carbon content. In practice, martensite is produced by quenching small samples in cold water.

Troostite represents a transition form which approaches more than martensite the normal state of equilibrium at the ordinary temperature. Micrographically it can generally be sharply distinguished from martensite, as stated above; between troostite and sorbite, however, there is no well-defined demarcation. If a piece of steel containing about 0.85 per cent of carbon (the composition of the eutectic alloy) be quenched in cold water, during the recalescence, the grains which have undergone the transformation consist of pearlite, those not yet transformed of hardenite (saturated martensite), while between these two well-defined constituents all degrees of transformation may be detected. Troostite may be produced either by the sudden cooling of steel during the recalescence period or by a mild quenching (in oil or in boiling water, for instance) of small pieces from a temperature higher than that of the critical range, or, again, by a mild tempering of martensite.

Sorbite approaches still more to the condition of stable equilibrium at the ordinary temperature. It might almost be called *un-segregated pearlite.* Between it and troostite or sorbite there exists no sharp line of demarcation. Sorbite is obtained irregularly by a cooling sufficiently slow to allow the transformations to proceed far and sufficiently rapid to produce an imperfect separation of ferrite and cementite. In practice these conditions are more or less fulfilled in the cooling in air of small samples, in quenching in cold water towards the end of the recalescence, in quenching in molten lead, in certain "*double trempes,*" or in reheating (tempering) martensite to a blue color.

It is seen, therefore, that martensite, troostite, and sorbite, although not always adapting themselves, as might have been expected, to a rigorous micrographical definition, correspond, nevertheless, to micro-constituents of sufficiently unlike appearances as to demand recognition, and possessing mechanical properties which, on the whole, are very different.

The distinctions which I made in the forms of transitions between the mixed crystals of *Gamma* iron and pearlite appear to me, now that I consider them with an experience lengthened by five additional years of work, as very well grounded and justified, both from the point of view of the metallographist and that of the mechanical engineer. Had I not made these distinctions, I would make them now. From the standpoint of the physico-chemist, however, there is no conclusive evidence that they correspond to phases, unless it be to transition phases whose range would be represented by a line of inappreciable width. It is not surprising, therefore, that Professor Roozeboom could not find any place in his system for troostite and sorbite; indeed, it is a proof of the perspicacity of its criticism.

As to martensite, it is in reality given two somewhat different meanings, for the name is applied both to a structural constituent of steel in the cold and to mixed crystals of *Gamma* iron at a red heat. Martensite as a structural constituent of cold steel is always somewhat magnetic, and is not, therefore, identical with the mixed crystals; it only approaches them in constitution. It would, therefore, seem desirable to reserve the name of martensite for the constituent of quenched steel, although I have myself proposed the extension which I now criticise, and which may be somewhat justified, owing to the identity of crystalline forms. If the mixed *Gamma* crystals at a red heat were to be designated by a term already in use, that of austenite would be more accurate, but as it is identified with a special mode of crystallization, it would be better still to simply refer to them as *mixed crystals of Gamma iron.*

If, as I now believe, these mixed crystals and austenite are, from a physico-chemical point of view, the same thing, Figures 5 and 6 of Professor Roozeboom and their description may, for the present, be set aside. I say for the present, because our knowledge is not sufficiently positive to warrant an irrevocable verdict. This portion of Professor Roozeboom's work contains many very

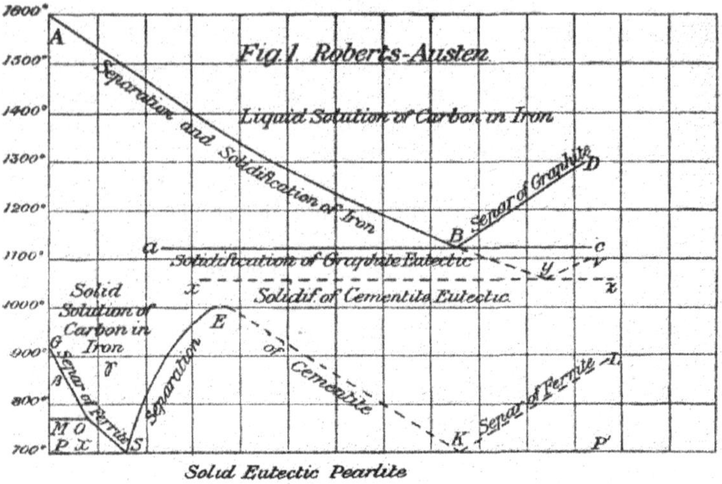
Fig. 1. Roberts-Austen.
1600°
1500°
1400°
1300°
1200°
1100°
1000°
900°
800°
700°
Separation and Solidification of Iron
Liquid Solution of Carbon in Iron
Separ. of Graphite
Solidification of Graphite Eutectic
Solidif. of Cementite Eutectic
Solid Solution of Carbon in Iron
Separation
of Cementite
Separ. of Ferrite
Solid Eutectic Pearlite

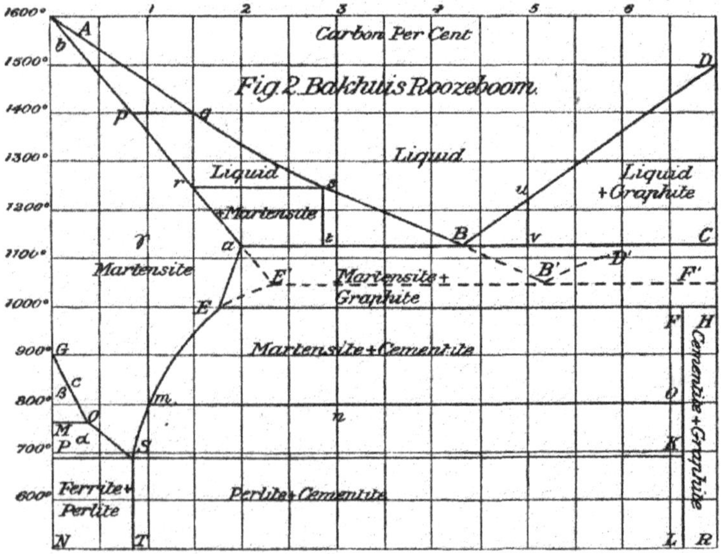
Carbon Per Cent
Fig. 2. Bakhuis Roozeboom.
1600°
1500°
1400°
1300°
1200°
1100°
1000°
900°
800°
700°
600°
1
2
3
4
5
6
Liquid
Liquid
+ Martensite
Liquid
+ Graphite
Martensite
Martensite +
Graphite
Martensite + Cementite
Cementite + Graphite
Ferrite +
Perlite
Perlite + Cementite

ingenious considerations, very well adapted to certain experimental facts, which we shall, perhaps, be glad to take up again in the future. For the present, his Figure 2* appears to me to answer alone all our needs, and it is to this diagram that I shall confine my attention.

I see no occasion for criticism either in the portion of the diagram which is also found in Sir Roberts-Austen's curve, or in reference to the new line Aa added by Professor Roozeboom to represent the end of the solidification of mixed crystals.

The lines a E and E F suggest some difficulties.

Among the experimental facts having a bearing upon these two lines, there are some which appear to favor their existence and others which can only with difficulty be made to support it.

Below the line a B C, representing the solidification of the eutectic made up of mixed crystals and of graphite, the region a B C E F H reserved for this system in the diagram would be frequently missing, if the cooling is not too slow nor the carbon content too high.

As an evidence of this I may cite an experiment of Mr. Royston,† who heated rapidly to 1030° C. and quenched some cast iron of the following composition:

Graphitic carbon	trace
Combined carbon	3.85 per cent
Silicon	0.29 " "
Sulphur	0.03 " "
Phosphorus	0.02 " "
Manganese	0.15 " "

After quenching the metal contained 2.35 per cent of graphite and 1.50 per cent of combined carbon.

Since the quenching treatment suppressed the transformation E F of the system mixed crystals + graphite into mixed crystals + cementite, the composition of this quenched cast iron would also represent its composition at a temperature of 1030° C.; the agreement with our prevision is qualitatively complete and quantitatively sufficient. The same experiment, however, in which the

* Fig. 2 of Professor Roozeboom's paper, as well as Sir Roberts-Austen's curves, are again reproduced here to avoid constant reference on the part of the reader to a previous number of *The Metallographist.*—Ed.

† *Journal of Iron and Steel Institute*, 1897, No. I, page 166.

rapid cooling was replaced by a slow cooling in air, yielded the same proportion of graphite, but as it is accepted by all that the transformations graphite + mixed crystals into cementite must be slow, this fact is not abnormal and could have been anticipated.

It is also certain that the graphite carburizes the ferrite in the region E F S K of Fig. 2.

Among my unpublished experiments I find one which, although incomplete, owing to an accidental interruption in the recording of the pen of the chronograph, may, nevertheless, throw some additional light.

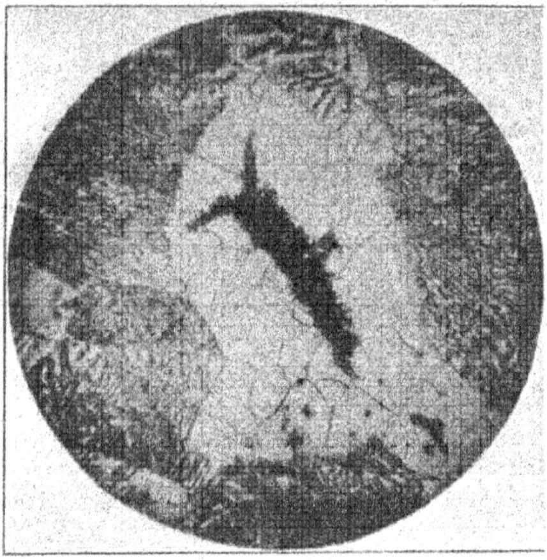

November 28, 1895, Mr. Brustlein sent me a sample of steel twice cemented. After the first cementation, followed by melting and forging, the carbon content was 1.70 per cent, all in the combined condition. The cross ends of the cemented bars were then subjected to a second cementation. At the end of the operation they were found to be malleable and to contain 1.38 per cent of graphite and 0.24 per cent of combined carbon.

The microstructure exhibited nests of graphite surrounded by ferrite containing some minute particles of graphite, and the balance of the structure was made up of pearlite. (See the accompanying photomicrograph, magnified 100 diameters; it was etched for 10 seconds in nitric acid diluted four times with water.)* It was exactly the same structure that Professor Arnold† had obtained in reheating a sample of steel containing 1.47 per cent of total carbon (1.14 graphitic carbon after reheating) for seventy-two hours, packed in lime in a closed cast-iron box.

Mr. Brustlein's steel was subjected to various reheatings, at temperatures increasing from 780° to 950° C. Two small steel bars holding between them the thermo couple of Professor Le Chatelier were placed, between two asbestos plugs, in a porcelain tube surrounded by a layer of asbestos thread. The whole arrangement was then placed in the muffle of a Mermet furnace, previously heated to its maximum temperature (about 1000° C.). The heating of the metal was recorded, and when it had reached the desired temperature, the porcelain tube was withdrawn and the steel allowed to cool to a temperature of about 200°, the rate of cooling being also registered. The metal was then heated to a little higher temperature, and so on, the operation being repeated six times. The results of these heatings will be found in the appended table.‡

In comparing the retardations taking place during these successive heatings and coolings, it will be seen that they gradually increase. The cooling curve, moreover, which should be, with 0.24 per cent of carbon, that of a medium-soft steel (exhibiting two retardations, $Ar_{3\cdot 2}$ and Ar_1) has become identical to the cooling curve of hard steel, with only one critical point. Undoubtedly, therefore, the transformation graphite + ferrite into mixed crystals took place between 780° and 950° C.

As the structure was not examined, however, the transformation graphite + mixed crystals into cementite in the region E F S K is not proved by these experiments. It would seem, however, as if the production of cementite between these limits of temperature was obtained by Margueritte, who heated some finely

---

* The position of the graphite in the centre of the ferrite should be noted. In another piece of steel, however, twice cemented, and containing only 1.50 per cent of total carbon, the graphite, which was less abundant, was in contact with pearlite, no structurally free ferrite being present.

† *Proc. Institution Civil Engineers,* Vol. CXXIII, Part I.

‡ No corrections were made for variations in the temperature of the laboratory during these experiments, which explains, but only in part, the progressive fall of the position of the critical temperature.

Initial Temperature		Time (Intervals in Seconds)											
		Heating 1	Cooling 1	Heating 2	Cooling 2	Heating 3	Cooling 3	Heating 4	Cooling 4	Heating 5	Cooling 5	Heating 6	Cooling 6
		21	778	225	788	198	837	198	877	198	936	198	946
Temperatures	926	—	—	—	—	—	—	—	—	—		23.2	5.7
	917	—	—	—	—	—	—	—	—	20.4		22.5	5.9
	907	—	—	—	—	—	—	—	—	18.3		19.2	4.8
	897	—	—	—	—	—	—	—	—	15.3		16.5	5.4
	887	—	—	—	—	—	—	—	—	14.3		16.8	5.3
	877	—	—	—	—	—	—	—	—	14.0		15.5	5.8
	867	—	—	—	—	—	—	—	—	13.8		13.6	4.8
	857	—	—	—	—	—	—	12.0	8.7	13.5		13.3	4.9
	847	—	—	—	—	—	—	11.2	6.6	12.0		12.0	4.8
	837	—	—	—	—	—	—	10.0	7.4	10.4		12.5	
	827	—	—	—	—	—	—	10.5	6.7	9.8		11.5	
	817	—	—	—	—	—	8.2	9.6	6.2	9.4		10.8	
	807	—	—	—	—	9.0	8.0	8.4	6.1	9.2		9.2	
	797	—	—	—	—	7.7	6.9	8.5	6.4	8.3		9.2	
	788	—	—	—	—	7.2	7.1	7.3	6.7	7.9	Lost	8.0	
	778	—	—	—	—	7.7	6.7	7.4	6.6	7.7		7.8	
	768	—	—	—	—	6.8	7.9	6.7	6.5	7.2		7.1	
	758	8.1	—	6.9	—	6.8	7.1	6.6	6.9	6.3		7.3	
	748	8.8	8.2	7.1	—	6.9	7.5	6.7	7.0	6.8		6.5	
	738	13.8	9.1	7.3	—	6.8	8.0	7.0	7.1	6.7		7.0	
	728	33.3	9.5	8.5	9.0	6.9	7.6	7.6	7.5	6.8		7.8	
	718	11.0	10.3	12.3	8.9	9.0	8.1	9.7	8.0	10.2		7.7	
	708	8.9	9.0	37.5	9.5	43.4	7.9	44.8	7.0	47.2		50.4	
	698	9.0	10.1	9.8	10.4	9.9	9.0	10.5	8.2	10.3		11.3	
	688	8.8	10.4	8.3	13.5	9.9	10.7	8.9	8.6	8.7		8.4	Lost
	678	9.1	11.3	8.2	11.8	8.3	10.7	7.9	9.7	8.5		8.4	
	669	—	25.4	—	21.8	—	—	—	—	—		—	
	669	—	14.2	—	21.7	—	—	—	—	—		—	
	669	7.5	22.1	8.5	15.7	7.7	63.3	7.8	65.5	8.8		8.5	
	659	7.5	8.7	7.8	8.8	8.0	12.3	6.8	13.9	7.4		7.6	
	649	8.2	7.7	7.6	7.4	7.0	7.8	7.5	7.0	6.9		7.9	
	639	7.6	9.2	7.8	8.2	7.4	7.5	7.2	7.8	7.2	7.1	7.4	
	629	6.7	8.0	7.4	8.3	6.6	8.0	6.6	7.9	7.0	7.7	6.9	
	619	6.7	9.3	7.7	9.5	6.4	9.5	6.0	9.1	6.9	8.5	6.9	
	609	6.4	9.8	7.0	9.5	5.9	9.0	6.7	9.0	7.0	8.9	7.0	
	599	6.5	9.3	8.0	9.8	6.0	9.3	6.5	8.7	6.5	9.5	6.9	
	589	6.4	10.0	7.2	9.7	5.5	11.3	6.2	10.0	6.4	10.0	6.2	
	579	6.4	11.1	6.4	10.5	5.8	9.2	5.7	10.1	6.2	10.0	6.8	
	569	6.3	10.8	5.9	11.2	5.8	11.7	5.8	10.6	5.8	10.4	5.7	
	559	5.8	11.1	5.4	10.6	5.4	10.9	5.7	10.9	5.6	10.4	5.8	
	549	5.8	11.4	5.6	11.9	5.7	11.8	5.6	11.9	5.8	11.8	5.7	
	539												
Stop at .		—	—	70	—	—	—	703	—	—	—	—	—
Rise or fall from .		—	664	—	664	704	660	—	660	703	650	701	649
To . .		—	669	—	670	701	668	—	668	701	658	698	657

divided iron (some oxalate) for three hours in a current of carbonic oxide; the amount of carbon absorbed was found by the increase in weight. His results were as follows:

Temperatures	Carbon Per Cent
Incipient cherry red, softening and fusion of glass	6.60
Clear cherry red	6.55
Dark orange, fusion of silver	1.21

A minute division of the iron is necessary; under the same condition bulky masses of iron absorb much less carbon.* It is interesting to note that in the first two experiments the carbon content corresponds almost exactly to the formula Fe_3C. Such a close agreement twice found cannot be accidental, the more so that Margueritte had no preconceived opinion and did not seek the definite compound.

On the other hand, Mr. Royston, in reheating for three hours some malleable cast iron whose carbon was entirely graphitic before reheating, did not succeed in producing more than a trace of combined carbon at 620°; at 720° he obtained only 0.85 per cent, and at 1030°, only 1.50 per cent. Mr. Saniter,† on the contrary, caused some iron wire, 1 millimeter in diameter, to absorb 2.95 per cent of total carbon, 2.42 per cent being combined, by heating it at about 900° C. in charcoal. This amount of combined carbon is certainly greater than the percentage of carbon which the mixed crystals are able to retain in solution at 900°, and indicate the formation of free cementite.

If Mr. Royston's figures were accepted as representing the saturation point of solid iron in presence of graphite at the temperature indicated, and if these figures were plotted on Professor Roozeboom's diagram, the whole of the line a E (extended) would pass to the left of E S, and would end almost exactly at the point S, so that E F would practically coincide with S K.

It is not at all certain, of course, that in Mr. Royston's experiments the saturation point of iron was reached after three hours. There are other facts, however, which point to an inclination towards the left of the line a E (extended). I do not mean to refer to the production of malleable cast iron in a packing of

* *Comptes Rendus,* Vol. LIX, p. 726, October, 1864.

† *Journal Iron and Steel Institute,* 1897, No. II, p. 115. *The Metallographist,* Vol. I, p. 52, January, 1898.

oxide, because the presence of oxygen may greatly complicate the phenomena, and because the products obtained in this way are far from being homogeneous, as was shown by Mr. Stead.* Prolonged reheating of white cast iron, however, also causes the separation of graphite. In a sample of cast iron, for instance, containing 3.27 per cent of combined carbon, no graphite and 0.304 per cent of silicon, Forquignon found 1.21 per cent of graphite after a seventy-two-hour reheating to a light cherry red, and 1.69 per cent after forty-four hours, the temperature being probably below 900° C.† I mentioned on a preceding page the case of a steel with 1.70 per cent of carbon whose proportion of combined carbon was lowered to 0.24 per cent after a second cementation, which caused the carbon to assume the graphitic condition. Indeed, Mr. Royston states that the totality of the carbon may be precipitated as graphite. These facts (and others might be found) lead to the belief that the line a E (extended) should be inclined to the left so as to cut the line G O S in a point whose position is not at present positively known. Mr. Stansfield reaches the same conclusion in his communication to the last meeting of the Iron and Steel Institute.‡

Summing up the preceding remarks, we meet with contradictory results concerning the formation of graphite in carburized iron. We must, of course, take into consideration, not only the time, but also the presence of certain impurities. It is known, for instance, that when silicon or aluminum are present in sufficient proportions, they cause the formation of graphite as soon as solidification begins. A very small quantity of the same elements might, if they are given time, act in the same way. A single atom of silicon represents a high proportion of that substance in a small mass of iron surrounding it; the proportion increases as we consider a smaller surrounding mass, and may be as high as we wish. If this atom be mobile, as might be expected in a solution, and if the redissolution of the precipitated graphite be slow, a state of equilibrium may be reached between the causes inducing the formation of graphite and those inducing the formation of cementite. On the other hand, Forquignon has shown

* *Proc. of the Cleveland Institution of Engineers*, December, 1895.

† "Recherches sur la fonte malleable et sur le recuit des aciers," Gauthiers-Villars, Paris, 1881.

‡ *The Metallographist*, Vol. III, p. 300, October, 1900.

that an atmosphere of hydrogen opposed the formation of graphite, under conditions of time and temperature identical to those which would have produced it abundantly had the same cast iron been heated in contact with oxide or with charcoal.

It would seem, premature, therefore, with the limited knowledge which we now have, to suggest positive conclusions concerning the last difficulties encountered in the study of carburized iron. The best that can be done is to begin some methodical experiments, assisted by the useful advice of Professor Roozeboom.

IRON AND STEEL FROM THE POINT OF VIEW OF THE "PHASE-DOCTRINE" *

By H. LE CHATELIER

THE study of Professor Roozeboom on the application of the law of phases to alloys of iron and carbon is of great interest, although it does not offer any new experimental fact, and consists only in a more systematical grouping of previously known facts. Several efforts had already been made to connect the theory of the constitution of steel with the laws of solution, and had been very fruitful. These theories, however, remained incomplete, and serious errors were sometimes introduced. Professor Roozeboom's study is, scientifically, absolutely correct, and may be considered as conclusive. Some corrections will undoubtedly have to be made in the numerical values of the coördinates of the various points of the curves, none of which are yet accurately determined. Some curves will have to be completed by the addition of branches corresponding to new phases that may be discovered. I shall deal, further on, with some corrections of this character, but there will be no suppression, no essential modifica-

---

* *Bulletin de la Société d'Encouragement,* November, 30, 1900. These remarks of Professor Chatelier and the preceding note by Mr. Osmond accompany a full French translation of Professor Bakhuis Roozeboom's paper published originally in *Zeitschrift für physikalische Chemie,* XXXIV. Professor Roozeboom presented a comprehensive abstract of his paper to the Iron and Steel Institute, which was reproduced in the October, 1900, issue of *The Metallographist.*

tion required in order to include the new facts in Professor Roozeboom's diagram.

The interest of such coördination is as follows: the phases, that is, the various states of the substance in iron, steel, and cast iron are extremely numerous. There are at least six at the ordinary temperature: *ferrite, martensite, austenite, troostite, cementite, graphite,* to which five more must be added for high temperatures: the allotropic varieties, *Beta, Gamma, Delta,* of iron, the non-magnetic variety of martensite, and, finally, the molten metal — eleven different phases in all.

During the successive cooling ranges, from the liquid state to the ordinary temperature, one, two, three, and possibly more of these phases may exist together. The number of possible permutations between these different phases is enormous. For the same combination of phases, moreover, the variation of their relative proportions must be taken into consideration. The experimental study of such complex phenomena is absolutely impossible without a guiding thread; lacking this, the study would be so long that whole generations of scientists would not suffice; of greater moment still, the inevitable experimental errors would not allow an accurate distinction between certain facts, and darkness would go on increasing with the number of experiments.

Roughly stated, the law of phases may be utilized in the present case as follows: At each temperature (the pressure being supposed to remain constant) a system composed of iron and carbon, in any given proportions, tends to assume a well-defined condition, a state of equilibrium. It is the knowledge of these states of stable equilibrium that we must first endeavor to acquire. The law of phases gives us concerning them a certain number of absolutely accurate indications, with which the experimental facts must necessarily agree, if they have been well observed.

1. Four juxtaposited phases will never be found in such a state of equilibrium. When four phases are detected, it is certain that the system has not reached its equilibrium; it is still undergoing a transformation.

2. The coexistence of three phases is possible, but only at a certain temperature and for a constant composition of each of three phases. The relative proportions of the three phases only may vary. Such a system is said to be *invariant;* its graphical representation corresponds to a point.

3. The coexistence of two phases is possible at a number of temperatures; the composition of certain phases (solid or liquid solutions) may vary with the temperature, but is always the same for a given temperature. One of the phases of the system, therefore, but only one, may be made to vary arbitrarily, and the system is called accordingly *monovariant.* The graphical representation of the composition of each phase in function of the temperature will be a curve, and each of these curves will unite two invariant points. This remark is of great practical importance: the various invariant points are generally sufficiently near each other to permit the curves which unite them to be drawn as straight lines without appreciable error. As two points are all that is needed to determine the position of a straight line, the knowledge of the position of the invariant points is, therefore, sufficient to give very accurate information regarding the conditions of equilibrium of monovariant systems. It will be perceived how much the experimental work is thereby simplified.

It may be noted, in passing, that any binary alloy at the ordinary temperature, which is a temperature arbitrarily selected, must correspond to a monovariant system, i.e., in the present case, must contain two different phases. For certain compositions, however, the system may be composed of only one phase. If the metal contains o carbon, for instance, it will be made up of a single phase, ferrite; if it contained a proportion of carbon corresponding to the definite compound Fe_3C, it would, in theory at least, consist only of cementite, but iron cannot be made to absorb in a combined condition the carbon required by this formula. Again, we would have only one phase in the case of relative proportions of iron and carbon corresponding to a solid solution undersaturated, etc.

4. Finally, in the case of a single phase, we shall have a system *divariant,* which means that two of the factors defining the system may be made to vary arbitrarily, the temperature, for instance, and the composition of the phase. These two quantities once determined, all the others will thereby also be determined. Since there is only one phase, however, this determination has no importance from the point of view of questions of chemical composition.

Now that we have recalled these general principles, let us return to Professor Roozeboom's application of these principles.

The service to be asked from the law of phases is a clearer classification of experimental facts, which will allow their consideration as a whole. To accomplish this, graphical representations are very useful. It might be asked, however, whether the mode of representation adopted by Professor Roozeboom fulfils all the desired conditions, seeing the complication of the diagrams, which are still incomplete. Fig. 2* of Professor Roozeboom does not contain any branch of curve pertaining to austenite, and his Figure 5 none of the branches representing states of equilibrium between graphite and martensite.

This complication of the diagrams is due to two causes:

1. The presence of numerous horizontal lines which have nothing to do with the law of phases. Their presence is justified in the representation of experiments by the cooling method; they indicate the stops or retardations, sometimes of long duration, occurring in the fall of the temperature at the end of a transformation, in the solidification of eutectic mixtures, for instance.

2. The curves representing the composition of the various variable phases (molten metal or solid solution) are grouped in the same diagram, so that each point may, according to the point of view, represent the composition of one or the other of the phases of a given system. It is preferable, for the sake of clearness, to construct a separate diagram for each phase, taking care to juxtapose them in such a way that the same temperatures will be on the same horizontal line. The grouping together in the same figure of curves pertaining to different phases is only interesting in illustrating more forcibly the differences of composition of the juxtaposed phases and in facilitating the calculation of their relative proportions. This advantage, however, does not offer sufficient compensation for the confusion likely to result from such grouping. In order to avoid confusion, it would be necessary to give to the various curves some strikingly different appearances, using different colors for different phases, for instance.

Professor Roozeboom's curves will be found divided into several diagrams in Figs. 1 and 2 of the present paper, but before examining them it will be well to examine some of the figures which were used in the construction of these curves.

---

* See page 154. — Ed.

1. *Curves of the Molten Metal.* — The temperatures of the beginning of the solidification of the molten metal, which give the equilibrium curve of the liquid phase, with the various solid phases, were determined on several occasions, notably by Mannesman, Osmond, and Roberts-Austen. Mannesman's determinations, which at first sight differ much from the others, owing to the scale adopted for the temperatures, may easily be corrected, if it be accepted that the temperatures in the neighborhood of 500° C. are accurate, that the melting point of iron is 1550° instead of 2000°, and in introducing a proportional correction in the whole range above 500°.

The results obtained by these three experimentors are summed up below:

Carbon per cent	0	1.3	1.8	2.7	3.5	4.3	5.5
Mannesman	1550°	1400°	1310°	1250°	1200°	1160°	1200°
Osmond	1550	1410	1320	1260	1200	1125	1260
Roberts-Austen	1600	1420	1360	1250	1200	1140	1290

The agreement between these results is sufficiently close to construct accurately the fusibility curve. By plotting the figures just given, it will be seen that the mean curve corresponds closely to the following figures:

Carbon per cent	0	1	2	2.25	3	3.4	4.3	5.5
Temperature	1575°	1450°	1320°	1275°	1230°	1170°	1150°	1250°

It would seem as if, starting from pure iron down to the lowest melting point, the curve might be represented by two straight lines crossing each other when the carbon content reaches 2.25 per cent and the temperature 1275°.

The positive existence of such an angular point would be very important; it would indicate a sudden change in the nature of the solid substance in equilibrium with the liquid. These two solid phases replacing each other would be martensite and austenite. The precision of our experiments, however, is not at present sufficiently great to warrant a positive inference of this character. The results might also be satisfactorily represented by a continuous curve.

2. *Curves of the Solid Solutions.* — The bulk of the facts so far observed leads us to recognize the existence of two solid solutions, martensite and austenite, but their distinction is not yet clearly outlined. The existence of austenite may be ignored

in a first approximation, as was done by Professor Roozeboom, and we may then return to it and discuss the position which it should have occupied in the diagram.

With regard to martensite, the successive states of equilibrium with the molten metal, the graphite, the cementite, and the ferrite should be studied. We have no information concerning the composition of martensite in equilibrium with the molten metal. On the other hand, by the work of Sauveur, Osmond, and Roberts-Austen, we possess some fairly accurate information regarding its equilibrium with cementite and ferrite. These values have been utilized by Professor Roozeboom; there is nothing to be changed in them.

The second branch, pertaining to the state of equilibrium with graphite, was drawn in a manner altogether hypothetical and certainly inaccurate. It was supposed that it crossed the cementite branch, while it always lies above it, for it is known through the production of malleable cast iron, that cementite in contact with martensite is in an unstable condition and has a tendency to break up, liberating some graphite.

The branch of curve representing the equilibrium with cementite corresponds to a supersaturation similar to that of sodium sulphate containing seven molecules of water; it must, therefore, lie in its entirety above the graphite curve, just as the curve of sodium salt with 7 H_2O is above that of the salt with 10 H_2O.

We have, moreover, some experimental results of greater precision, which we owe to the experiments of Mannesman and Royston, to determine the equilibrium curve of martensite with graphite. These results are as follows:

Temperature	700°	400°	900°	1000°	1100°	1130°
Mannesman, Carbon %	0.45	0.55	0.75	1.5	4	5.5
Royston, Carbon %	0.7			1.4		

It is seen that above 1000° the carbon content of the martensite increases very rapidly; it would even finally exceed the carbon content of the molten metal, if the experiments of Mannesman are accurate. This would be a very important fact, which would demand verification, but which will be overlooked for the present.

The bulk of the observations leads to the diagrams shown

in Fig. 2 for the curves pertaining to martensite; the diagram has been placed next to that showing the equilibrium curves of the molten metal, and the same temperatures are on the same horizontal line.

3. *Austenite Branch.* — It remains for us to examine the question of the existence of austenite as a phase distinct from martensite. According to Mr. Osmond, austenite is character-

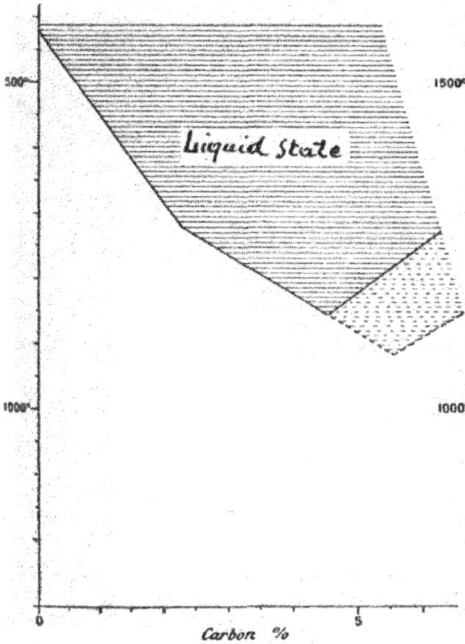

Fig. 1

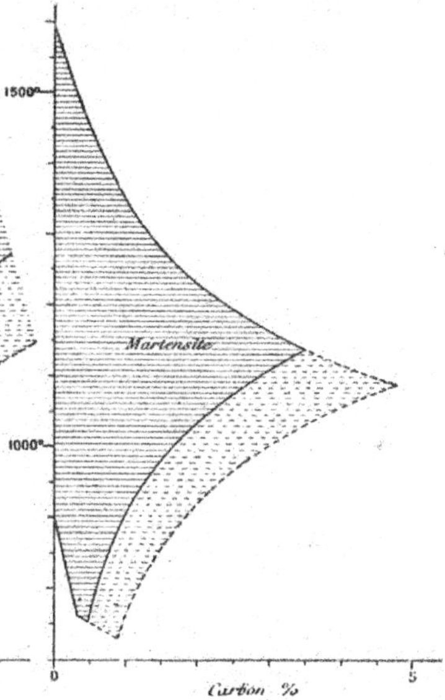

Fig. 2

ized by being less hard than martensite, by a smaller magnetic permeability, and by being less readily acted upon by reagents. These differences justify a sharp line of demarcation between these two substances at the ordinary temperature, but they do not prove that the same differences exist at a high temperature. It is well known that above the point of recalescence martensite is not magnetic; it becomes magnetic during cooling. Austenite might, therefore, be a variety of martensite, retained at the ordinary temperature in a condition identical to that which it possesses at a high temperature, such retention being possibly due to the presence of a larger percentage of carbon. The sharpness of the outlines of austenite crystals, however, renders this hypothesis difficult to accept. It is more natural to infer that there is no continuity between these two kinds of solid solutions and that they belong to different crystalline systems, a phenomenon of

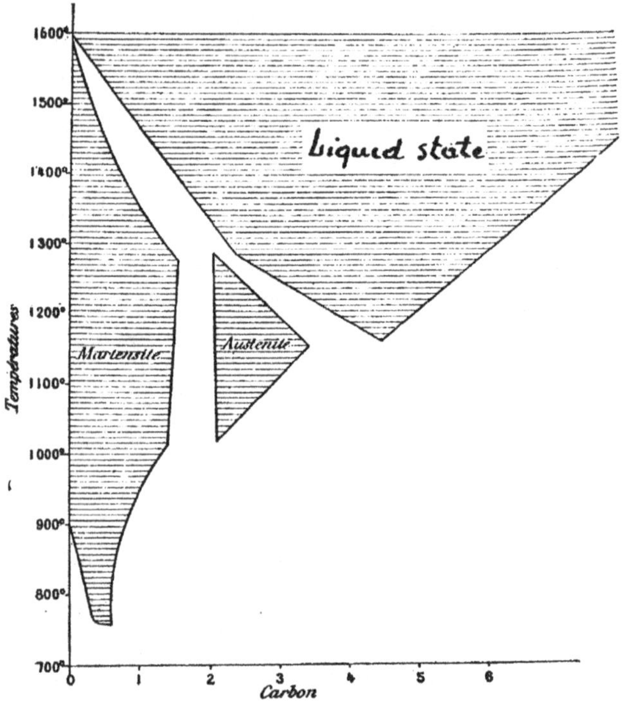

Fig. 3

which there exist a great many instances. The passage of one variety into the other should be accompanied by a sudden change in the carbon content. If such be the case, the diagram of Fig. 2 should be modified, as shown in Fig. 3.

4. *Troostite Branch.* — Professor Roozeboom has ignored the existence of troostite, owing to the lack of positive information regarding it. Some researches, which have not yet been published, have thrown some light upon its conditions of existence.

The recalescence phenomenon is not, as we had believed it, a simple one. It is, in fact, a succession of two distinct phenomena: the transformation of pearlite into troostite, followed by the transformation of troostite into martensite. I ascertained this fact in studying the dilatation of steel. A steel containing 0.9 per cent of carbon, on being heated, when it reaches the recalescence point, exhibits a sudden contraction, immediately followed by an expansion of the same magnitude, and amounting to 0.1 per cent. In order to look more closely into the nature of these two opposite changes, I conducted the following quenching experiments: The metal quenched at the beginning of the sudden contraction was examined under the microscope and found to be made up entirely of pearlite; quenched when the maximum contraction was reached, it was composed almost entirely of troostite surrounding a few particles of pearlite and of martensite; finally, when quenched as soon as it had again assumed its normal dimensions, the metal was found to be made up entirely of martensite. The difference of temperature between these two successive temperatures does not exceed 50°. Their complete study will be very delicate. I noted the same division in two periods in the case of steel containing 0.3 per cent of carbon, but only during cooling. With 1.1 per cent of carbon it could not be clearly detected. I did not obtain any indication concerning the composition of troostite; it is, very likely, a solid solution, differing from martensite and austenite in being magnetic at its temperature of formation.

The diagram shown in Fig. 3 was constructed in taking into consideration all the connections indicated in the preceding pages; the curves pertaining to the various phases have been grouped in the same diagram in order to facilitate comparison with that of Professor Roozeboom. The curves indicating unstable conditions have been omitted to avoid confusion. The troostite branch has not been included, owing to lack of information concerning the composition of that phase.

At first sight, there appears to be an objection to these applications of the law of phases that it will be well to anticipate.

The micrographic examination of alloys shows that the number of phases is very often superior to that indicated by the theory. Cast iron, for instance, frequently contains graphite, cementite, and ferrite, when only two phases should be present instead of three. Sometimes the number of phases is infinitely great, martensite having a composition varying continuously from one point to another, as indicated by the varying width of the plates of the pearlite resulting from the transformation of the martensite.

In all such cases, however, which are generally the result of too rapid cooling, the state of equilibrium has not been reached, and the law of phases should not, therefore, be applied to them.

METALLOGRAPHIC NOTES

The National Standardizing Bureau. — The bill establishing a National Standardizing Bureau was passed by both houses of Congress, March 5, and the Secretary of the Treasury will at once proceed with the organization of the bureau in order that no time may be lost in making its facilities available to the general public. The most important provisions of the bill are as follows:

"That the functions of the bureau shall consist in the custody of the standards; the comparison of the standards used in scientific investigation, engineering, manufacturing, commerce, and educational institutions with the standards adopted or recognized by the Government; the construction, when necessary, of standards, their multiples and subdivisions; the testing and calibration of standard measuring apparatus; the solution of problems which arise in connection with standards; the determination of physical constants and the properties of materials, when such data are of great importance to scientific or manufacturing interests and are not to be obtained of sufficient accuracy elsewhere.

"That the bureau shall exercise its functions for the Government of the United States, for any State or municipal government within the United States, or for any scientific society, educational institution, firm, corporation, or individual within the United States engaged in manufacturing or other pursuits requiring the use of standards or standard measuring instruments. All requests for the services of the bureau shall be made in accordance with the rules and regulations herein established.

"That the officers and employés of the bureau shall consist of a director at a salary of $6000, one physicist at a salary of $3500, one chemist at a salary of $3500, two assistant physicists or chemists each at a salary of $2200, two laboratory assistants each at a salary of $1200, one secretary at a salary of $2000, one clerk at a salary of $1200, one clerk at a salary of $1000, one messenger at a salary of $720, one engineer at a salary of $1500, one fireman at a salary of $720, one mechanician at a salary of $1400, one mechanician at a salary of $1000, one mechanician at a salary of $840, one watchman at a salary of $720, and two laborers each at a salary of $600.

"That the director shall be appointed by the President, by and with the advice and consent of the Senate.* He shall have the general supervision of the bureau, its equipment, and the exercise of its functions. He shall make an annual report to the Secretary of the Treasury, including an abstract of the work done during the year and a financial statement. He may issue, when necessary, bulletins for public distribution, containing such information as may be of value to the public or facilitate the bureau in the exercise of its functions."

The qualifications of the scientific corps of the bureau are set forth in the following memorandum prepared for the Senate Committee on Commerce:

"The plan of organization of the entire scientific corps should be on the basis of a few men of the highest order rather than a large number of poorly paid men, as the questions here involved are questions demanding men of ability and authority to decide.

"The next grade [after the director], that of physicist or chemist, would consist of men of reputation in the scientific world and specialists in some particular branch of physics or chemistry. They should be men such as are now occupying chairs of physics and chemistry in the leading universities of the country. They would be required to perform investigations of the highest order and direct others engaged in investigations, calibrations, and tests of all kinds; to devise and select the best methods of producing and comparing standards; to keep informed as to the work and methods of the standardizing bureaus

* Professor S. W. Stratton of the University of Chicago was appointed director March 9.

of other Governments in the lines along which they are particularly interested.

"The assistant physicists and chemists would be men of considerable reputation, such as are now filling chairs of physics and chemistry in colleges, or assistant professors in the larger universities. They would be required to perform investigations and undertake the tests, calibrations, and comparisons of the higher orders.

"The next or lowest grade in the scientific class would be that of laboratory assistants. These men would necessarily be graduates of some recognized university or technical school, and who have done postgraduate work in the lines along which they intend to specialize. They would be selected with the greatest care and would be in direct line of promotion to all of the grades above."

The plan of the new bureau has been widely indorsed by scientific men and experts representing many important lines of manufacturing, including especially the iron and steel trade. Albert Ladd Colby, metallurgical engineer of the Bethlehem Steel Company, has prepared an interesting memorandum emphasizing the importance of the bureau, and Dr. Charles B. Dudley, chief chemist of the Pennsylvania Railroad Company, addressed a strong appeal to the Senate committee pointing out the advantages that would follow the passage of the bill. Indorsements have also been received from the Brown & Sharpe Manufacturing Company, the Baldwin Locomotive Works, M. C. Bullock Manufacturing Company, the Westinghouse Electric and Manufacturing Company, the General Electric Company, the Crocker-Wheeler Electric Company, F. H. Clarke, mechanical engineer of the Chicago, Burlington and Quincy Railroad Company; W. H. Marshall, superintendent motive power Lake Shore Michigan Southern Railroad Company, and C. H. Querean, assistant superintendent of machinery Denver and Rio Grande Railroad. — *The Iron Age*, March 7, 1901.

It is to be hoped that metallographic investigations will be included in the research work of this important bureau.

The Andrew Carnegie Research Scholarship. — The Council of the Iron and Steel Institute has issued the following circular:

A Research Scholarship or Scholarships, of such value as

may appear expedient to the Council of the Iron and Steel Institute from time to time, founded by Mr. Andrew Carnegie (Vice-President), who has presented to the Iron and Steel Institute thirty-two one-thousand dollar Pittsburg, Bessemer and Lake Erie Railroad Company 5 per cent Debenture Bonds for the purpose, will be awarded annually, irrespective of sex or nationality, on the recommendation of the Council of the Institute. Candidates, who must be under thirty-five years of age, must apply, on a special form before the end of April to the Secretary of the Institute.

The object of this scheme of Scholarship is not to facilitate ordinary collegiate studies, but to enable students, who have passed through a college curriculum or have been trained in industrial establishments, to conduct researches in the metallurgy of iron and steel and allied subjects, with the view of aiding its advance or its application to industry. The National Physical Laboratory, on the governing body of which the Iron and Steel Institute is represented, would for many reasons be a very suitable place in which such a research could be carried out. There is, however, no restriction as to the place of research which may be selected, whether university, technical school, or works, provided it be properly equipped for the prosecution of metallurgical investigations.

The appointment to a Scholarship shall be for one year, but the Council may at their discretion renew the Scholarship for a further period instead of proceeding to a new election. The results of the research shall be communicated to the Iron and Steel Institute in the form of a Paper to be submitted to the Annual General Meeting of members, and if the Council consider the Paper to be of sufficient merit, the Andrew Carnegie Gold Medal shall be awarded to its author. Should the Paper in any year not be of sufficient merit, the Medal will not be awarded in that year.

Metallography at the Technical Schools. — *Harvard University.* — The mining department of the Lawrence Scientific School has now a very complete outfit for metallographic work, including a Schuckert-Zeiss arc lamp, a complete Zeiss optical bench, a special stand, objectives, and accessories, a micro-camera, and a special polishing machine. Some valuable research work is being carried on by several students.

Massachusetts Institute of Technology. — Dr. Henry Fay, formerly instructor of analytical chemistry at the Massachusetts Institute of Technology, has been appointed assistant professor of analytical chemistry and metallography at the same institution. On March 14 Dr. Fay delivered a lecture before the Society of Arts on "The Applications of the Science of Metallography."

A number of students from the chemical and mining departments have selected metallographic subjects for their graduating thesis.

Columbia University. — The department of metallurgy has recently added to its metallographic equipment a Roberts-Austen autographic Le Chatelier pyrometer, a portable Le Chatelier pyrometer, and a Le Chatelier microscope with monochromatic illumination by mercury arcs. There is much activity in the metallographical laboratory.

A new Short Metallurgical Laboratory Course went into effect in January, 1901. Its includes five subjects. The first three which are related to Metallography are described below:

(1) Microscopy. — In this each student prepares for the microscope, first, specimens easy to prepare — the copper silver alloys — and next, specimens of the greatest industrial importance — those of steel. In doing this he learns, first, the methods of grinding and polishing, of developing the structure by heating and by etching separately, and of sketching the microstructure; and second, the laws of the eutexia, the relation of the constitution and structure of alloys to their composition, and the influence of thermal treatment. In short, such skill as he acquired is used in observing phenomena in themselves of great importance. (2) Pyrometry. — Here the student learns the principle of the thermoelectric pyrometer, and acquires a certain degree of crude skill in calibrating the instrument, and in observing the pyrometric phenomena attending the critical temperatures of steel. He thus at once learns the use of the instrument, and in using it becomes acquainted with phenomena of the first technical importance. (3) The Thermal Treatment of Steel. — Here the student performs experiments which illustrate the manner in which the hardening power of steel is acquired and lost, and the relation of these phenomena to those of magnetism; and also a variety of experiments illustrating how injuries to steel arise through improper thermal treatment, and how they can be

cured by either thermal or mechanical treatment, observing also the symptoms of the injury and of its cure.

The fourth subject deals with the Roasting of Copper Ores and the fifth with the Desilverizing of Lead.

The above is the outline for the first year's work; it is expected that the course will be considerably expanded for another year, though following the same general direction.

Other Scientific Schools. — The authorities of Purdue University have under consideration the equipment of a laboratory for metallographic work. Brown University and the School of Mines and Metallurgy of Rolla, Mo., are also considering the advisability of taking up the subject.

Diffusion of Gold in Solid Lead at Ordinary Temperatures. W. Roberts-Austen. (Roy. Soc., Proc. 67. pp. 101–105, October 26, 1900.) — Continuing the experiments on diffusion described in the Bakerian lecture of 1896, certain pure lead cylinders, each with a smooth surface at its base, clamped into close contact with the clean surface of a disc of gold, and so maintained undisturbed at a temperature varying but little from 18° C., were examined after four years of such contact. Successive slices were cut at right angles to the axes of the cylinders, the first being about 0.75 mm. thick, and each of those succeeding it about 2.3 mm. thick. Assays of these slices showed that those in contact with the gold had absorbed a quantity equivalent to about 1 oz. 6 dwt. of gold per ton of lead, whilst the next layers contained less, the next still less, and so on, the amount even in the poorest layer being in the proportion of 1½ dwt. per ton. The results shown by different experiments were not identical (probably because of the difficulty in insuring perfect contact between the gold and the lead surfaces), but were all of the same order. The importance of temperature in relation to diffusion is shown by the fact that the amount of gold that would diffuse in solid lead at the ordinary temperatures in 1000 years is the same as that which would diffuse in molten lead in a single day. Preliminary experiments, in which a current of 1.5 amperes was passed continuously through the clamped joint for 544 hours at 150°, seem to show that a solution of gold in lead does act to a small extent as an electrolyte. The failure of the author, in 1887, to obtain indications of electrolysis in a solution of gold in molten lead by the passage of a current of 300 amperes may have been due to the distribution of the precious metal by diffusion through the mass at the high temperature of the experiment. — *Science Abstracts,* January 25, 1901.

Pearlite and Eutectic Alloy. — In Mr. F. Osmond's biographical sketch, published in the October, 1900, issue of *The Metallographist,* the editor expressed his belief that Mr. Osmond was the first to call attention to the similarity existing between the structure of pearlite and that of eutectic alloys in general. Priority for this important remark is claimed for Professor H. Le Chatelier by one of our correspondents. In support of his claim the following extract is given of a paper by Professor Le Chatelier, published in January, 1897: "Mr. Charpy has ascertained, through some recent experiments, that most of the eutectic alloys are made up of extremely thin lamellæ of the two constituents; their appearance is exactly that of pearlite." *

The Bessemer Medal. — Our contributor, the distinguished metallographist, Mr. J. E. Stead, of Middlesbrough, England, has been awarded the Bessemer gold medal by the council of the Iron and Steel Institute. The readers of this journal are well acquainted with Mr. Stead's important metallographic work, and they will feel that the high honor just bestowed upon him has been well earned. It should be a cause for congratulation to all those interested in metallographic methods.

Journal of the Royal Microscopical Society. — In addition to the subjects already included, it is intended to commence with the new year furnishing the Fellows with brief abstracts of papers on the microscopic structure of metals and minerals, which, it is hoped, may be useful to those engaged in metallurgy and in mining operations. — *Journal of the Royal Microscopical Society,* February, 1901.

---

* "L'état actuel des théories sur la trempe de l'acier." *Revue Générale des Sciences,* January 15, 1897. — *The Metallographist,* Vol. I, p. 59. January 1898.

The Metallographist

A QUARTERLY PUBLICATION DEVOTED TO THE STUDY OF METALS, WITH SPECIAL REFERENCE TO THEIR PHYSICS AND MICROSTRUCTURE, THEIR INDUSTRIAL TREATMENT AND APPLICATIONS.

Vol. IV | JULY 1901 | No. 3

THE CONSTITUTION OF CAST-IRON, WITH REMARKS ON CURRENT OPINIONS CONCERNING IT *

By H. M. HOWE

I. General Statement

IT has seemed to the writer that one important, and indeed sufficient, reason for our slow progress in learning the relation between the chemical composition and the physical properties of cast-iron has been that the problem has not been attacked properly. It would be well to select the most probable working hypothesis which we can find, and proceed to test that hypothesis by comparison with observed facts, and by making new observations specially designed for such a test. Our first step, then, is to select the most probable working hypothesis; and to attempt this selection, to state the hypothesis and its immediate corollaries, and to take some first steps toward testing it, is the object of the present paper. Its chief purpose is to bring the hypothesis to the attention of those who are more familiar with the evidence than the writer can pretend to be, and in a better position both to present new evidence and to discuss it.

The hypothesis here selected is that composition governs properties in the case of cast-iron in the same general way and

* American Institute of Mining Engineers, Richmond Meeting, February 1901.

for the same reasons as in the case of steel, *mutatis mutandis.* This hypothesis, which of course makes no claim to originality, certainly seems reasonable and probable; its exposition and discussion are made easy by our existing knowledge of the constitution of normal or "carbon" steel; it therefore seems worthy of consideration.

Carbon-steel, when cooled slowly, consists essentially of two components:*

(1) *Ferrite,* pure iron, weak, soft, ductile, copper-like.

(2) *Cementite,* a definite carbide of iron, Fe_3C, containing 6.67 per cent of carbon, harder than glass or hardened steel, extremely brittle, and probably strong like glass under gradually-applied axial stress; its carbon is the "combined carbon," the "carbide carbon" of Ledebur.

The way in which ferrite and cementite habitually interstratify to form the composite mass, pearlite, will be considered in Section III.

When the carbon-content reaches a certain high limit, the exact position of which we need not here try to settle (taking it, however, provisionally, as somewhere between 2 and 3 per cent), the metal ceases to be called steel, even if it be nearly or quite graphiteless, and it is henceforth called white cast-iron. But we may here regard both steel and white cast-iron as really one unbroken series, varying progressively from the softest tube-steel with 0.06 per cent of carbon through the die- and file-steels to white cast-iron. The name white cast-iron, applied to the most

---

* For the purposes of a mere bird's-eye view of the subject I have purposely left out of consideration many points which a complete exposition would demand. For instance, I speak of steel as graphite-less, whereas in fact steels rich in carbon often carry an appreciable amount of graphite; yet being relatively graphite-less, it is better for our present purpose to speak of them as containing no graphite. So, too, I speak of the combined carbon of cast-iron as existing in the form of cementite; whereas in point of fact many analyses of cast-irons which appear to have been cooled with at least moderate slowness report, along with the cementite, considerable quantities of combined carbon in the condition of martensite, or "hardening-carbon." But for our present purpose we may disregard this, especially as its effect in a general way is similar to that of cementite, making the metal hard and brittle.

In like manner, I leave out of view in this sketch the effects of silicon, manganese, etc., and of thermal and mechanical treatment.

highly carburetted part of the series, really indicates no break in that series, which is simply like a London street, with different names in different parts, for the convenience of the Londoner and the exasperation of the traveler.

Turning now to common or graphitic cast-iron and regarding it as composed of two distinct and unlike parts, (1) graphite and (2) its metallic part or matrix, which includes everything except the graphite, our natural working-hypothesis is to assume that the properties of any individual cast-iron as a whole depend (1) on the percentage of graphite which it contains, and (2) on the properties of the matrix; and further to assume that this matrix is, both in its constitution and in its properties, substantially like steel, or white cast-iron, of the same carbon-content. Let me here insist on this view that the graphitic cast-irons are by constitution a biniary conglomerate, composed of a foreign body, graphite, and of a metallic part or matrix: and that this matrix is in each case simply some one member of the steel-white-cast-iron series, being soft steel, hard steel, or white cast-iron, according to whether its combined-carbon content is low, intermediate, or high. This conception is the essence of our hypothesis.

For example, in this view, those graphitiferous cast-irons which contain less than 2 per cent of combined carbon are essentially steel of like carbon-content, plus graphite; while such cast-irons with more than say 3 per cent of combined carbon are simply white cast-irons of like combined-carbon-content, plus graphite. Cast-iron with 0.25 per cent of combined carbon and 3.75 per cent of graphite is steel of 0.25 per cent carbon, say structural steel, weakened and embrittled by 3.75 per cent of graphite; cast-iron with 0.50 per cent of combined carbon and 3.5 per cent graphite is steel of 0.50 per cent carbon, say rail-steel, weakened and embrittled by 3.5 per cent of graphite; cast-iron with 1.25 or 1.50 per cent of combined carbon is steel of 1.25 or 1.50 per cent of carbon, say file-steel, weakened and embrittled by graphite. Cast-iron with 3 per cent of combined carbon plus 1 per cent of graphite is essentially a mechanical mixture of (1) 99 parts of white cast-iron containing 3 per cent of combined carbon, and (2) 1 part of graphite.

To recapitulate, we have:

1. The graphite-less or steel-white-cast-iron series, consisting essentially of ferrite and cementite.

2. The graphitic series (the graphitic cast-irons*), the members of which are simply the steel or the white cast-iron (as the case may be) of the graphiteless series, plus graphite.

Let us consider first the graphite-less series, the simpler because the influence of the ferrite and cementite is not here obscured by that of graphite.

If we plot a great number of cases, with carbon as abscissæ and tensile strength, ductility and hardness as ordinates, as in Fig. 1,† we find that, whereas the ductility diminishes and the hardness increases continuously as the carbon increases, the tensile strength increases only until the carbon reaches about 1 per cent, when it reaches a maximum, and then in turn decreases. This law we may regard as firmly established for steel under normal conditions. The reason for the general features of these curves will be discussed in Section III.

Let us now turn to the graphite series, which includes both gray and mottled cast-iron, and let us consider first (A) cases in which the graphite is constant and the combined carbon varies; then (B) those in which combined carbon is constant and graphite varies; and then in II. those in which both vary.

A. *Graphite Constant, Combined Carbon Varies.* — Here, since variations in the combined carbon, i.e., in the carbon of the matrix, should cause the same effects in that matrix as similar carbon-changes cause in steel, therefore, as the combined carbon increases the hardness of the mass as a whole should increase, and its ductility should decrease — in both cases continuously. But the tenacity of the matrix, and through it the tenacity of the conglomerate whole, should increase until the combined carbon reaches about 1 per cent, and should in turn decrease as the combined carbon further increases.

B. *Combined Carbon Constant, Graphite Varies.* — The graphite by itself should weaken, soften and embrittle cast-iron.

* Strictly speaking, any high-carbon steels and any cast-irons which, in spite of having a white fracture, yet contain a really important quantity of graphite, belong here; but for the purposes of our bird's-eye view we may neglect them.

† We have no sufficient quantitative data to enable us to plot the hardness-curve. That given in Fig. 1 is simply sketched in by eye, to show the results of general observation; and the dotted parts of the tenacity- and ductility-curves make no pretence whatever to quantitative accuracy, but are simply conjectured as the result of general observation.

Fig. 1

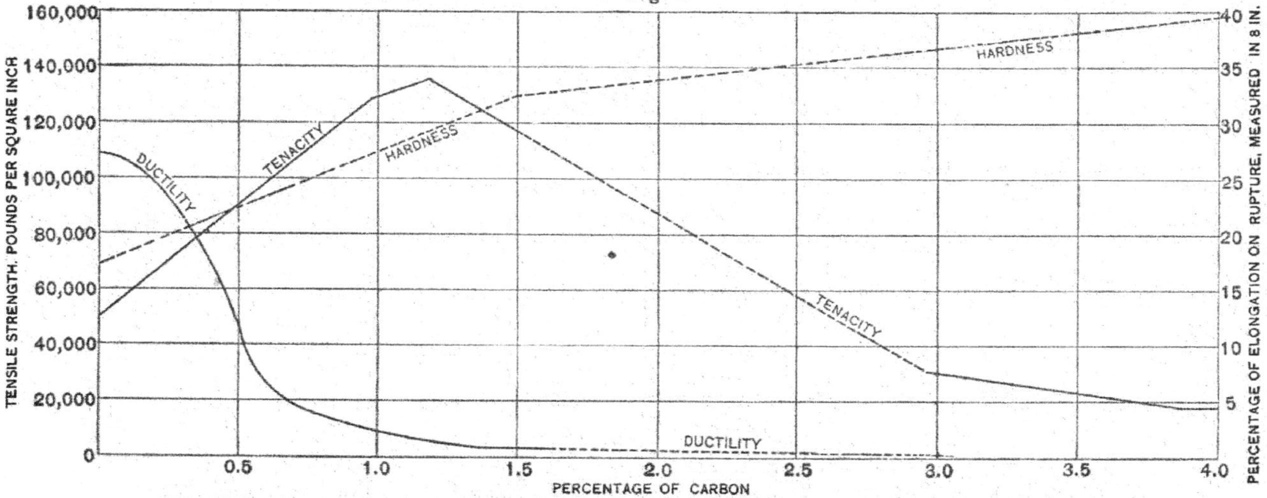

TENACITY, DUCTILITY AND HARDNESS OF IRON, AS AFFECTED BY THE PERCENTAGE OF CARBON.

(The position of the parts of these curves which appear in broken lines is conjectural.)

Scattering through the metal sheets of a weak substance like graphite should in effect break up the continuity of the mass, and this should both weaken and embrittle the mass as a whole; and it should soften the mass somewhat, because graphite itself is so soft. Hence as the graphite increases the strength, ductility and hardness should all decrease continuously.

II. Commercial Foundry Pig-Irons

The total carbon of these pig-irons as they run from the blast-furnace usually varies relatively little, so that as the combined carbon increases, graphite usually decreases. That is, one is in general approximately the complement of the other. For the sake of simplicity and definiteness, we will through the present section consider the influence of variations in the combined carbon in a series of cast-irons all containing exactly 4 per cent of total carbon, and to fix our ideas we will consider the influence of a progressive increase in the combined-carbon content. Such an increase has a twofold aspect, the progressive increase in the cementite-content of the matrix, and the complementary decrease in the quantity of the foreign body, graphite.

Tensile Strength.— Starting with combined carbon at zero and graphite 4 per cent, we must have a very weak cast-iron; for its matrix is essentially ferrite, the weakest member of the ferrite-cementite series in the steels, mixed with the maximum quantity of graphite, in itself a most weakening substance.

As the combined carbon progressively increases from zero up to one per cent, the tenacity should increase continuously for a double reason: (1) the matrix is growing stronger and stronger because it is growing richer and richer in carbon, passing from the softest and weakest steels through the structural steels up to the tool-steels and those which yield the strongest plough-steel wire (see Fig. 1); while (2) at the same time the proportion of graphite, in itself a weakening substance, is growing less and less. So far all is simple.

But as the combined carbon increases beyond one per cent the case is less simple, for our two influences now oppose each other. The matrix is growing weaker as its carbon rises above one per cent (see Fig. 1); but the simultaneous decrease in the proportion of graphite present is strengthening the mass as a whole, simply by lessening the amount of this weakening substance. So in the range lying between 1 and 4 per cent of combined carbon, i.e., between 3 per cent and zero of graphite, we cannot tell from these synthetic data where the strongest cast-iron will be found.

Hardness.—As regards hardness the case is most simple. We find in the steel-to-white-cast-iron series that the hardness increases progressively and without limit as the carbon increases; indeed it could hardly be otherwise, for we can hardly understand how progressively replacing a copper-like substance, ferrite, with a glass-hard one, cementite, should fail to increase the hardness progressively. Hence (returning to our present series of cast-irons with 4 per cent total carbon, and progressively increasing combined carbon) the hardness of the matrix should increase continuously as the combined-carbon content increases; and the corresponding decrease in the quantity of so soft a substance as graphite should in itself progressively harden the conglomerate as a whole. In short, in both its aspects the progressive transfer of carbon from the graphitic to the combined state should increase the hardness of the whole.

Ductility.—As a little reflection shows, the decrease in the ductility, like the increase in the hardness, should be continuous as the combined carbon increases from zero to four per cent. In the steel-to-white-cast-iron series, as the carbon increases, the very brittle cementite progressively replaces the ductile ferrite, a change which certainly must lessen ductility continuously; and this effect should take place likewise in the matrix of our present series, our graphitic cast-iron; and, while the simultaneous decrease of graphite should in itself tend to increase the ductility by lessening the quantity of this continuity-destroying substance, yet this effect should be less than that of the increase of cementite in the matrix. For the change of one per cent of carbon from the graphitic to the combined state simply lessens the quantity of graphite present by one per cent; but it increases the quantity of the very brittle cementite by 15 per cent, and the loss of ductility due to this feature of the change certainly should outweigh the gain of ductility due to removing one per cent of graphite.

Indeed, we may note in passing that, in general, a given change in the combined-carbon content should have a far greater effect on the hardness and ductility than a like or opposite change in the quantity of graphite. Suppose that, by introducing silicon or otherwise, without changing the total carbon, we increase the graphite by 2 per cent, thereby lessening the combined carbon by 2 per cent and the cementite by 30 per cent. Now while we cannot understand how the introduction of 2 per cent of a foreign substance like graphite, in thin flakes somewhat widely scattered, can make the grains of iron between those flakes much more malleable, much easier to machine, much softer under the file or under the diamond; yet we readily understand how removing from the matrix of the iron 30 per cent of a glass-hard substance like cementite, which has been scattered through the whole mass in microscopic particles,— most intimately mixed, flesh of its flesh,— and replacing it chiefly by the soft ductile ferrite, should greatly lessen the hardness and brittleness of the mass.

But this of course does not imply that the same holds true of tensile strength, because the influence of combined carbon on this property is not constant, but reverses when the combined-carbon content rises past one per cent.

III. Explanation of the Influence of Carbon on the Physical Properties of Cast-Iron

Hardness and Ductility.—As any increase in combined carbon in effect substitutes the glass-hard extremely brittle cementite for the soft, ductile, copper-like ferrite, the continuous increase of the hardness and the continuous decrease of ductility as the carbon increases need no further explanation.

Tenacity.— In order to understand this much less simple case, we may well consider the usual mode of arrangement of the ferrite and cementite. In any individual steel these two substances habitually interstratify in thin sheets, in the ratio of about 7 parts of ferrite by weight to one of cementite; and the composite mass or conglomerate thus formed is called pearlite. As cementite contains 6.57 per cent of carbon, and as the ratio in which it interstratifies with ferrite is about 1 : 7, it follows that pearlite must contain about $\frac{6.57}{7+1} = 0.80$ per cent of carbon. It is a true eutectic; the two substances, ferrite and cementite, which compose it, though present in nearly constant proportions, are wholly separate and distinct from each other, quite as the grains of quartz, feldspar and mica are in a granite. Nevertheless, since though composite it is of constant composition, it is convenient to treat it as a distinct entity.

If the steel as a whole contains just 0.80 per cent of carbon, i.e., $\frac{0.80 \times 100}{6.57} = 12.5$ per cent of cementite, so that the ferrite (which must form the remainder of the mass, and therefore amount to $100 - 12.5 = 87.5$ per cent) stands to the cementite in the ratio of 7 : 1, then all the ferrite and all the cementite usually interstratify to form pearlite, during slow cooling.

If there be more than 0.80 per cent of combined carbon there will be more than 1 of cementite to 7 of ferrite, i.e., more cementite than can interstratify with the ferrite present in its fixed 7 : 1 ratio to form pearlite; and, as happens in all eutectiferous masses, the excess of cementite simply separates out within the mass as distinct "free," "massive" or "structurally free" cementite; so that steel containing more than 0.80 per cent of carbon consists of pearlite plus free cementite. It is "Hyper-eutectic" steel.

If, however, the steel contains less than 0.80 per cent of carbon, and hence less than 12.5 per cent of cementite, i.e., less than is needed to form pearlite with the whole of the ferrite present, there will now be an excess of ferrite over and above that which can unite with the cementite present in the fixed ratio 7 : 1; and this excess will separate out within the mass, in the form of "massive" or "structurally free" ferrite. The steel is "Hypo-eutectic."

To recapitulate, while steel consists essentially of ferrite mixed with cementite, the latter increasing with the carbon in the ratio of 15 per cent of cementite to every 1 per cent of carbon, these two components usually interstratify in the ratio of 7 : 1 to form pearlite. Hence steel consists either of pearlite alone, or of pearlite mixed granitically with the excess of either ferrite or cementite, according to whether it contains just 0.80 per cent, or less, or more, than that quantity. The same is probably true (*mutatis mutandis*) of the "matrix" in cast-iron.

While the existence of this special mode of arrangement of the ferrite and cementite does not disturb our inference that every increase of combined carbon, and hence of cementite, must increase the hardness and lessen the ductility of the mass taken as a whole, yet it may well have a very marked influence on the relation between carbon-content and tensile strength.

When two substances differing as greatly as ferrite and cementite in hardness, strength and ductility, are thus intermixed, it is indeed not easy at first to predict how they will behave under stress. But we can understand that the particles of each must bear a share of the stress and transmit it to their neighbors. One of them, ferrite, flows under stress, and is strengthened by this flow, as the strengthening effect of wire-drawing, cold-rolling, and all other forms of permanent distortion of low-carbon steel and wrought-iron testifies; the other, cementite, appears to be absolutely rigid and non-ductile. We can understand that they should behave under tensile stress like distinct entities, because of the wide difference in their properties; and that, in a way and within limits, the cementite should mechanically support the ferrite and prevent its excessive flow and ready yielding. For we must recollect that the mixture of the two is very intimate. Not only are the ferrite and cementite which interstratify to form pearlite most intimately mixed, but the excess of ferrite is penetrated by the pearlite, which dowels its particles together.

But even in view of these conditions, the increase of tensile strength caused by carbon seems in some cases too great to be readily explained by this purely mechanical action of cementite. For instance, in steel containing 0.21 per cent of total carbon,* the cementite is contained partly in a network of pearlite and partly in some plums of pearlite swimming in the matrix of ferrite. That the network, usually very thin and probably often discontinuous, should have a material strengthening effect is certainly not self-evident; that the plums, widely separated from each other as they are, should have such an effect is even less evident. In passing these islands rupture should follow the path of least resistance, around them and through the matrix of ferrite; and this path should not be very greatly lengthened by having to pass around these islands. Indeed, the pearlite rather suggests strong links set ineffectively among the weaker ones

* See Mr. Sauveur's sketch of such steel when slowly cooled, *Trans.*, xxvi, 863, Plate I, Fig. 7. — *The Metallographist,* Vol. I, page 39, Fig. 7, January 1898.

of a chain of ferrite, and adding nothing to the general strength. Yet such steel is nearly 50 per cent stronger than pure ferrite, as we find it in the most nearly carbon-less iron. So, too, in steel of 0.14 per cent of carbon,* the cementite present is contained in a network of pearlite, apparently so very thin and so much broken up that we hesitate to ascribe to it any great strengthening-effect. Further, the increase of tensile strength which accompanies progressive increase in carbon-content seems at first too regular to be accounted for fully by the corresponding but varying changes which occur in the mode of distribution of the pearlite. Hence, while it is wholly possible that this mechanical explanation really covers the chief or sole cause of the strengthening-effect of carbon, yet its lack of cogency leads us to seek some contributory cause. And such a cause I now suggest, drawn from an obvious extension of the allotropic theory. It is clear that, above the critical temperature Ac_2, the iron itself becomes allotropic; and one of the prominent theories of the hardening of steel is that the sudden cooling preserves this high-temperature allotropic iron, by denying the time required for its change back to the normal state — that during slow cooling this allotropic iron changes back to normal iron; but that if the steel be cooled very suddenly there is not time enough for this reverse change to occur, and we thus retain in the cold steel the allotropic iron, of which one variety (β) is intensely hard and brittle, and the other (γ) is harder and stronger than normal or *Alpha*-iron. In order to explain the facts, it has been found necessary to assume that the carbon present in the steel acts as a brake to retard this reverse change from allotropic to normal iron; hence the hardening of high-carbon steel, in which, thanks to this brake-action, the allotropic iron is preserved in sudden cooling; hence also the non-hardening of low-carbon steel, in which even sudden cooling, lacking the brake-action of the carbon, fails to retain the allotropic iron.

Now, I would invoke this brake-action to help explain the increase of tenacity of even slowly cooled steel which accompanies the increase of carbon from zero towards one per cent. A late study of the magnetic properties of steel led me to suppose that, even with slow cooling, an appreciable quantity of

* See Osmond, *Méthode Générale pour l'Analyse Micrographique*, Figs. 101 and 162; also Sauveur, *op. cit.*, Plate III, Fig. 4.

allotropic iron remained in the cold metal, without changing back to normal iron. It may well be that this brake-action of carbon applies to a limited extent even to slow cooling; and hence that in the ferrite even of slowly cooled steel there should be present some of the hard, strong, allotropic iron, and that the proportion of this, and hence the tensile strength of the conglomerate whole, should increase with the carbon-content.

The same assumption would give an additional explanation of the increase of hardness and decrease of ductility caused by carbon.

We thus have two explanations of the increase of tenacity as the carbon increases from zero towards one per cent, either or more likely both of which may be true.

Of the reversal of this curve, and the decrease of the tenacity as the carbon rises beyond one per cent, the microstructure offers a reasonable explanation. Until the carbon reaches 0.80 per cent the cementite usually exists in minute sheets as part of the pearlite; and though when the carbon first rises beyond 0.80 per cent there is a little free cementite not included in the pearlite, this free cementite is in small and discontinuous particles. As the carbon increases still farther, the excess of free cementite rapidly increases, forming thicker and thicker masses, tending to form a continuous skeleton or network.

Thus, among Roberts-Austen's Brymbo steels* those containing 1.461 per cent and 1.38 per cent of carbon respectively already have the cementite in a continuous skeleton, encircling the polygons of pearlite; and one containing 1.80 per cent of carbon has the cementite in a like skeleton, but very much thicker. Indeed, a continuous skeleton of cementite is shown by Mr. Sauveur† in a steel containing only 1.20 per cent of carbon.

Now we can readily understand that, so long as the free or "excess" cementite is in discontinuous minute particles, each of which is suspended in the pearlite and free to move in it as in a menstruum, the cementite should strengthen the mass, for instance, by opposing its tendency to flow. At the same time

---

* Fifth Report to the Alloys Research Committee, Figs. 43 and 45, Plate II, and Fig. 99, Plate XXI. — *The Metallographist*, Vol. II, page 186, July 1899.

† *Trans.*, xxvi, 878, and Plate I, Fig. 14. — *The Metallographist*, Vol. I, page 27, January 1898.

we can understand that, when the cementite had reached such proportions that it formed a considerable continuous skeleton, or relatively thick sheets, the fact that these can no longer accommodate themselves to the yielding and flow of the pearlite would concentrate upon them an undue part of the stress. The consequence of this might well be that, being unyielding, non-ductile, they break under this undue stress, and that the fine cracks, the solutions of continuity, thus set up would lead to the break-down of the mass as a whole. The more of this free cementite there is, within limits, the greater this tendency towards the break-down of the mass as a whole, through the starting, in the unyielding cementite, of cracks sufficiently important to spread rupture.

In some such way we can readily understand both the increase of tenacity as the carbon rises to 0.80 per cent and its decrease as the carbon in turn rises beyond, say, 1 per cent.

IV. What will the Strongest Cast-Iron Be?

This fact, that the strongest iron-carbon compound is steel with about one per cent of carbon, and that the tenacity decreases as the carbon either decreases or increases from one per cent, leads us to suppose that the less total carbon cast-iron contains, the stronger can it be made, and that one important line of progress will be the production of cast-iron castings lower in carbon than our present ones.* I purposely say " can it be made," because the strength depends not only on the total carbon but on its distribution between the states of graphite and cementite, as determined by the quantity of silicon, manganese and sulphur present, by the casting-temperature, the thickness of the casting, and other things which influence the rate of cooling, etc., etc. For every percentage of total carbon, however, there is a " best proportion " for this distribution, a proportion which yields greater tenacity than any other proportion for that particular total percentage. Hence my thesis is that, " the lower the total

---

* I am well aware that some who ought to know hold the opposite opinion. One prominent writer asserts "that the strongest and most ductile cast-iron is that which contains a maximum of carbon..." (*The Iron Trade Review*, Oct.4, 1900.) It is not in spite of their opinion, but because of it, that this paper is written.

carbon, the stronger should the cast-iron 'of the best proportion' be." This is a proposition to which I ask careful attention. On turning it over and over I find no escape from it as a reasonable probability. Certainly if to steel of one per cent of carbon we add combined carbon, the tenacity curve of Fig. 1 shows that we weaken it; the addition of graphite, which is a weak, foreign, continuity-destroying body, ought to weaken it; if we add both combined carbon and graphite, the latter, as before, should weaken it, and the former, by adding carbon to the metallic matrix, should, according to Fig. 1, weaken that matrix, and thus weaken the conglomerate mass. And if every addition of carbon in excess of one per cent weakens, every reduction towards one per cent should strengthen, always assuming that in each case the carbon is distributed in the "best proportion." The "best proportion" for one per cent of total carbon is: combined carbon, 1 per cent; graphite, zero. It is hard to escape the conclusion that for every increase of the total carbon beyond one per cent the best proportion must yield a weaker conglomerate; for if this new best proportion changes the combined carbon, it must carry the matrix away from the maximum-strength carbon-content of one per cent; and if it increase the graphite, that in and by itself should be a weakening change. And, as before, if every increment weaken, every decrement should strengthen.

I fail to conceive any definite case in which the opposite effect is to be expected. Perhaps the case in which this hypothesis seems most likely to fail is one in which the change from the "best proportion" for a given total carbon-content to the "best proportion" for a lower one takes place through a relatively large decrement of combined carbon which is initially below one per cent, and a simultaneous smaller increment of graphite. Such a decrement of total carbon should be weakening, since both the steps which compose it are in themselves weakening. But it is hard to believe that such a condition of affairs can exist. That the assumed initial distribution can have been "of best proportion" is hard to believe, because, while denying to the matrix the one per cent of carbon which would give it the greatest strength, it liberates, in the directly weakening form of graphite, the carbon thus denied to the matrix. And, if we assume that this initial distribution was "of best proportion," it is for like reasons difficult to suppose that these steps could lead from an initial to a final " best proportion."

The verification of such a hypothesis must be extremely difficult, owing to the number and importance of the other variables. Even if these were eliminated, verification of disproof would probably need, as a first condition, that we learn what is the " best proportion " for different percentages of total carbon, i.e., the " law of the best proportion."

It will probably be extremely easy to present an enormous quantity of evidence which at first sight opposes this hypothesis, but which, like that of Mr. Keep, discussed in Section V., is wholly incompetent, either because of our ignorance as to the law of the " best proportion," or for other reason.

I have no cogent evidence for or against this hypothesis; I present it for others better placed to test. There is, however, a consideration which, though it certainly is not cogent evidence, may well be presented here as an aid to any who may attempt a more thorough verification of the present hypothesis. It is that two of the most excellent classes of cast-iron, air-furnace castings* and charcoal cast-iron, are just those likely to be particularly poor in total carbon.

* I take this occasion to record an important bit of evidence as to the superiority of air-furnace over cupola-castings made under closely like conditions. Mr. William Metcalf, writing of his priceless experience in the manufacture of cast-iron guns and projectiles at the Fort Pitt Foundry in Pittsburg, during the War of the Rebellion, says: " During the rush of the war we melted all of our projectile iron in large cupolas, every pound of our reverberatory capacity being required for guns. We used for shot and shell good Hanging Rock, Iron Mountain and some Pennsylvania charcoal irons, and no scrap except our own projectile scrap. General Rodman was always...demanding 25,000 lbs. tenacity in projectiles, but we could not get it from the cupola without using gun-iron, of which we had not a pound to spare. Finally, when the rush was over, he laid down the law, all projectile iron must be over 25,000 lbs. We then calculated that it was cheaper to melt our projectile iron in reverberatories than to melt gun-iron in the cupolas, so we shut down the cupolas and melted our projectiles in the reverberatories. There was no change whatever in the irons; the only change was in the melting. The cupola projectiles averaged 22,000 to 23,000 lbs. tenacity; the reverberatory projectiles averaged 26,000 to 28,000 lbs., a clear gain of about 4000 lbs., or, say, from 3 to 5000 lbs. Why? I do not think sulphur and dirt from the coal had any more influence than the air; I believe oxygen and nitrogen blown through the iron in the cupola do as much harm as the sulphur, probably more." (Private communication, November 26, 1900.) The nitrogen and hydrogen

The cupola-furnace is a carburizing apparatus, the air-furnace a decarburizing one. If the iron when first melted in the cupola is low in carbon, whether from the use of low-carbon cast-iron or even of steel scrap,* or from decarburization in front of the tuyeres, it greedily absorbs carbon as it runs across and through the column of incandescent coke, which stretches down through the zone of fusion and rests on the very bottom of the furnace. Those who have seen the almost instantaneous carburization of molten steel by Darby's process will recognize how powerfully carburizing the conditions in the bottom of the cupola must be. Indeed, our wonder rather is why cupola castings, and indeed all cast-iron as it issues from the blast-furnace, are not nearly saturated with carbon.

In the air-furnace, however, not only can the metal gain no carbon, but it must lose some, and it may lose much. I do not find many analyses which are explicitly given as representing air-furnace castings, but the following statements are of interest. Mr. T. Ulke reports† the average total carbon-content of 15 air-furnace gun-irons as 3.03 per cent. Mr. C. H. Vannier reports:‡ "With equivalent mixtures I find air-furnace metal to carry in round numbers 3 per cent carbon, against 3.5 per cent in cupola metal." Dr. Richard Moldenke gives the following as a typical analysis of fine gray air-furnace castings, together with the usual range of composition of such castings: §

blown across the iron in the cupola might well be harmful; but it is not so easy to suppose that the atmospheric oxygen should affect so highly carburetted a body as cast-iron, except by oxidizing and removing its carbon, silicon, etc.

* Ledebur points out that steel and malleable iron melted in a cupola are thereby necessarily converted into cast-iron. (*Eisen- und Stahlgiesserei*, 1892, page 144.) On remelting pig-iron in a cupola six times, Jungst found that the total carbon rose on the first fusion from 3.10 to 3.33 per cent, and remained practically constant thereafter. Manganese was removed rapidly at first, then more slowly; silicon was progressively removed after the first melting. So, too, the carbon-content of three pig-irons, each containing initially more than 4 per cent of carbon, and each remelted by Ledebur separately in a cupola four times, fell in the first (or first two) fusions to about 3.50, and thereafter varied little, sometimes increasing, sometimes decreasing. (*Das Roheisen*, 1891, pages 72-3.)

† *The Iron Trade Review*, Dec. 1, 1898, page 8. ‡ *Id.*, Oct. 25, 1900.

§ Private communication, November 24, 1900. It is proper to refer here to the many analyses of cast-iron guns, which I confidently believe were cast from the air-furnace, recorded in the "Reports of Experiments

	From Per Cent	To Per Cent	Typical composition Per Cent
Total carbon	2.75	3.25	2.81
Silicon	1.90	2.40	2.09
Manganese	.40	.80	.47
Phosphorus	.25	.60	.523
Sulphur	.02	.05	.045

The upper limit for carbon in air-furnace castings here given is approximately the lower limit for common cupola castings, according to my observation. Looking through a large number of analyses of such castings I find only three with less than 3.11 per cent of total carbon, and some with over 4 per cent. In a group of 75 analyses of car-wheels, presumably made of cupola cast-iron, all but four have at least 3.30 per cent of total carbon. Ledebur, while admitting the decarburization in the air-furnace, thinks that not more than one-sixth of the total carbon is thus removed.*

on Metals for Cannon," 1856. In the "Final Report" on composition are given thirty-two analyses of such iron, with the total carbon varying between 2.60 and 4.60 per cent, and averaging 3.72 per cent. The results of this monumental work thus seem at first to strike at the root of my evidence, by showing that air-furnace castings are habitually rich, and may be very rich, in carbon. But after careful examination I must reject these results as certainly unworthy of consideration. In this final report we find that one gun, which endured 111 service rounds, contained 5.21 per cent of manganese; and, in spite of having only 2 per cent of silicon, it yet incongruously contained 2.80 per cent of graphite. One gun contained 0.23 per cent of calcium, and another 0.69 per cent of aluminum. A previous report in the same volume, by the same chemists, gives most of the guns as containing over 6 per cent of manganese; two of them had nearly 16 per cent, and one of them 18 per cent of that element. One had 3.46 per cent of aluminum; two had more than 6 per cent of phosphorus. Average results of a whole class of iron gave 14.84 per cent of carbon. The graphite in one specimen was 16.25 per cent. In the light of our present knowledge these results, which referred to cast-iron guns of high quality, are absurd: the data as to carbon-content rest on the same basis of authority, and hence deserve no weight. That such conscientious work should have led to such gross errors is hard to understand, even in view of the time when it was done. One at first wonders whether one reads aright, especially in view of the confusing way of reporting the composition, not in percentages but in fractions of unity; but I am confident that these errors cannot thus be explained away.

* *Eisen- und Stahlgiesserei,* 1892, page 101. He indeed asserts that if the temperature be purposely kept low, and sufficient manganese and

It is true that the cupola can be managed so as to decarburize materially, by cutting down the coke, and thus doing a considerable amount of Bessemerizing. Thus Mr. Ulke* even reports that the cupola gun-iron of the Niles Tool-Works contains from 2.4 to 2.6 per cent of total carbon; and, on the authority of S. S. Knight, that certain cast-iron of the Rarig Engineering Works (improperly called "semi-steel"), made by melting pig-iron with a very large addition of boiler-plate scrap in a cupola, contains 1.71 per cent of total carbon. But this fact, instead of weakening the force of our present consideration, supports our hypothesis. For our present consideration is based on the superiority of air-furnace castings over common, i.e., high-carbon, cupola castings; and the fact that when, by special steps, cupola-castings are made low in carbon they are exceptionally strong, is just what our hypothesis calls for. Thus the Niles Tool-Works gun-iron, just referred to, had a tenacity of 28,000 to 37,000 lbs. per sq. in., with an elongation of 0.50 to 0.60 per cent; and the Rarig cast-iron had a tenacity of 33,680 to 35,730 lbs. per sq. in.

So, too, another excellent class of cast-iron, namely charcoal-iron, has as a whole less carbon than the classes of coke-iron comparable with it, i.e., low-sulphur foundry-iron; though I must admit that the force of this fact is greatly lessened by the large proportion of exceptions to the rule. The argument is as follows:

If we assume that the carbon-distribution of charcoal-irons as a class is as near the "best proportion" as that of coke-irons as a class, the fact that charcoal-irons as a class are stronger, and also contain less total carbon than coke-irons as a class, is consistent with the hypothesis that low carbon in cast-iron makes for tenacity. That fact is, of course, explicable on other grounds also; this can, therefore, at best be but corroborative evidence. If it should be found that the high-carbon charcoal-

silicon be present, the removal of carbon can be wholly prevented; but his warrant for this, viz., the occasional lack of decarburization in the melting down of the puddling process (*id.*, page 100) — is hardly sufficient, first, because of the difficulty of accurate sampling under these conditions, and, secondly, because the basic puddling slag leads to the preferential removal of silicon instead of carbon, to a degree not to be expected with the acid slag of the air-furnace, as used for cast-iron.

* *Loc. cit.*

irons are as strong as the low-carbon ones, the fact would oppose the present hypothesis.

Let me present evidence as to the usual lower total-carbon content of charcoal- than of comparable coke-irons.

On plotting the composition of many hundred pig-irons, the majority of which were coke- or anthracite-irons, I found the total-carbon generally above 3.50 per cent, and in a very large proportion of cases between 3.75 and 4.25 per cent.*

In a series of 225 coke pig-irons and 66 charcoal ones, 37 per cent of the charcoal-irons and only 14 per cent of the coke-irons have less than 3.25 per cent of carbon, while 16 per cent of the charcoal-irons, and only 4 per cent of the coke-irons have less than 3 per cent of carbon.†

Among 30 coke and anthracite foundry-irons and 16 charcoal-irons, 44 per cent of the charcoal-irons have 3.19 per cent carbon or less, against only 17 per cent of the coke-irons.‡

Among 18 coke- or anthracite-irons (ferro silicons and irons with 3.5 per cent silicon excluded) and 18 charcoal-irons, 61 per cent of the charcoal-irons and only 28 per cent of the coke-irons have 3.50 per cent carbon or less; and 39 per cent of the charcoal-irons, against only 17 per cent of the coke-irons, have 3.19 per cent carbon or less.§

In a group of 27 charcoal pig-irons 13 have less than 3.50 per cent of carbon.||

In a group of seven analyses of charcoal pig-iron, 5 have less than 3.50, and 3 have less than 2.90 per cent of carbon.¶

That charcoal pig-iron should contain less carbon than coke-foundry, i.e., low-sulphur, pig-iron, is natural. In coke-furnaces the removal of the sulphur introduced by the coke requires that the slag be very calcareous; calcareous slags have a very high melting-point; hence the temperature must be extremely high so as to melt them. In short, the coke-furnace needs an espe-

---

* "The Influence of Silicon and Sulphur on the Condition of Carbon in Cast-Iron," *Trans.*, xxx, Fig. 1.

† *Analyses of Pig-Iron*, by S. R. Church.

‡ Vosmaer, *Iron and Steel*, pages 172-175.

§ Ledebur, *Eisenhüttenkunde*, 1889, pages 377-379.

|| Henderson and Davis, *American Engineering and Railroad Journal*, Jan. 1, 1899.

¶ *Report of the Tests of Metals*, Watertown Arsenal, 1899, page 369.

cially high temperature in order that it may desulphurize. Charcoal-furnaces, at least when treating low-sulphur ores, have not this need, because, thanks to the freedom of the charcoal itself from sulphur, they need not desulphurize. Now, the especially high temperature may well lead to especially high carburization, since the solvent power for carbon may be expected to rise with the temperature.

An observation which is very suggestive in this connection is that, while charcoal pig-iron yields excellent castings when melted in the air-furnace, its superiority diminishes, or even disappears, if it be remelted in the cupola. Now, in so far as the initial superiority of charcoal-iron is due to its freedom from sulphur, that superiority should persist after cupola-fusion; for both coke and charcoal irons in such remelting should absorb sulphur about equally, so that the relative merit of charcoal-iron should remain. So of the removal of silicon. The loss of superiority on cupola-fusion, then, being insufficiently explained by absorption of sulphur, or removal of silicon, must have a further explanation; and one is supplied by the present hypothesis. For if the usual initial low carbon of charcoal-iron is really one cause of its superiority, this cause ought to disappear on cupola-fusion, since the carburizing action of the cupola ought to make the final carbon of castings made from charcoal-iron substantially the same as that of coke-iron castings. This phenomenon of the loss of superiority of charcoal-iron or cupola-fusion, then, is inexplicable on the hypothesis that that superiority is due solely to low sulphur, or silicon, or both, but is consistent with the hypothesis that it is due in part to low carbon-content.

In this view the capital defect of the iron blast-furnace, as an instrument for making foundry cast-iron, is that it nearly saturates the iron with carbon, or at least gives it far more carbon than is consistent with the greatest strength, or the greatest combination of strength with ductility. It is hard to see how this fault is to be remedied within the furnace itself when the fuel is coke, since the desulphurizing action of the furnace, which cannot be dispensed with, is apparently part of the carburizing action. Both apparently take place in the crucible of the furnace, and both depend upon the same conditions, viz., that the bath of molten iron in the crucible is penetrated by a solid column of incandescent coke, with the carbon of which the molten iron rapidly saturates itself fully, or almost fully. This same carbon removes the sulphur from the iron, apparently, by the reaction

$$FeS + x\,CaO,\ SiO_2 + C = Fe + CaS + CO + (x - 1)\,CaO,\ SiO_2.$$

So, too, in this view, the capital defect of the cupola-furnace, as an instrument for making strong castings, is that the molten iron in like manner greedily absorbs carbon from the fuel over which it runs at the bottom of the furnace; so that, no matter how little carbon the charge contains initially, as it issues from the cupola it is over-rich in carbon.

To sum this discussion up, the hypothesis, that, assuming the carbon to be distributed in "the best proportion" between the states of cementite and graphite, then the lower the total-carbon the greater will be the strength, is in itself reasonable, and derives color from the fact that two of the best classes of cast-iron, charcoal cast-iron and air-furnace castings, are just those with habitually low-carbon content; and while this excellence is wholly explicable by other causes, which indeed doubtless do contribute to it, yet the loss of excellence of charcoal cast-iron on cupola-remelting is not readily explicable by these other causes, but is wholly in harmony with our hypothesis. This hypothesis, then, is offered tentatively, as worthy of further test by comparison with observed facts.

V. Remarks on Current Opinions Regarding Cast-Iron

The notion that the addition of ferro-silicon increases the quantity of scrap which a cupola-charge can carry has been ridiculed. This, I think, is unfair. The proposition may or may not be found actually true in practice. My understanding is that it has been found true. But whether actually true or false, it is not ridiculous, but on the contrary eminently reasonable. When scrap-iron is melted in the cupola-furnace it tends to lose silicon without correspondingly losing carbon, for the simple reason that, while the silicon which is burnt away as the iron drops past the tuyeres cannot be readily restored in the cupola, the carbon simultaneously burnt away is so restored, as has been explained in Section IV.

Now the removal of the silicon tends to diminish the graphite in the resultant casting, and thereby to increase the quantity of combined carbon; thereby to increase the cementite; and thereby to make the resultant castings harder and more brittle. The addition of ferro-silicon raises the silicon in the castings; thereby increases the graphite; and thereby diminishes the combined carbon. Or, it is probably more accurate to say that the addition of ferro-silicon diminishes the amount of combined carbon by lowering the solvent power of the solidifying iron for carbon, thus forcing part of the carbon out of combination into the form of graphite, and this reduction of the amount of cementite, or of combined carbon, softens the castings. One would naturally suppose that even a charge of scrap cast-iron alone could readily be converted in good gray castings by the addition of a sufficient amount of ferro-silicon. Such a charge of scrap, when melted, would in any event have sufficient carbon to make it brittle and intensely hard, provided no silicon at all were present. The addition of ferro-silicon should diminish the combined carbon in such castings, and this itself would make the castings softer.

The confusion, I think, arises from the notion that it is the graphite itself which chiefly affects the hardness of the castings; whereas it is the combined carbon which affects the hardness of the castings chiefly, and the presence of much graphite is simply a sign that the combined carbon is correspondingly low. You cannot eat your cake and keep it; and carbon which has separated as graphite can no longer remain in combination as cementite to harden and embrittle the metal.

Mr. W. J. Keep* finds "*the most conclusive proof that the existence of combined carbon has no influence, unless to weaken a casting by making it brittle,*" and as regards graphite, that "*strength or weakness seem to be absolutely independent of this element.*" These inferences he draws from a great number of cast-irons varying but little in total carbon. In such a series, as the combined carbon increases, the graphite must decrease; so if both the combined carbon (in excess of 1 per cent) and the graphite weaken and embrittle the cast-iron, the effect of the increase of one in such a series as his might be masked by the effect of the simultaneous decrease of the other. Thus one might very readily be led to Mr. Keep's conclusions, and fail

* *Trans. Am. Soc. Mech. Engrs.*, xvi, pages 1103-4, 1895.

to note that they are probably true only for approximatively constant total-carbon. While it may be true that, for any given percentage of total-carbon, as in his data, variations in the condition of that carbon have relatively little influence (the state of cementite injuring the strength about as much as that of graphite), this is no evidence that each of these substances by itself is harmless. It would be about as logical to infer that neither graphite nor combined carbon weighed anything, because changing carbon from one state to the other did not affect the weight of the cast-iron as a whole, as to infer that neither separately affected the strength or ductility, from the fact that shifting the carbon from one state to the other does not greatly affect those properties. The resultant of two equal and opposite forces, no matter how powerful, is zero.

IRON AND PHOSPHORUS

By J. E. STEAD

(Continued from page 115)

Part II

INFLUENCE OF CARBON ON IRON CONTAINING PHOSPHORUS WHEN MELTED TOGETHER

THE study of simple binary compounds is comparatively easy compared with the research relating to compounds containing three or more elements, and in giving the results of a large number of experiments with compounds of iron with carbon and phosphorus, I do not wish it to be supposed that the research has been more than just commenced.

In studying the compounds containing both separated massive carbide and phosphide of iron, existing as they often do side by side, it has been found that the etching methods will not develop indications sufficient to enable the microscopist to tell one from the other. Mr. Osmond and I have independently discovered that the very beautiful method, employed first by Professors Behrens and Martens, which consist of simply heating the polished metals till they are more or less colored by oxidation, enabled them to do so. Carbide of iron on heating takes the oxidation tints of straw-yellow, yellow, brown, red, purple, blue, and silver-gray in sequential order. Phosphide of iron passes through the several ranges of color, but not so rapidly as the carbide.

When a piece of metal which contains carbide and phosphide of iron is heated (by placing it on an iron plate which is heated by a flame under it) until the section assumes an orange color to the naked eye, and is then rapidly cooled by placing it on a bath of mercury to check further oxidation, on examination under

Fig. 7a. Illustrating the penetration of phosphide eutectic liquid into solid steel.

the microscope the carbide will be seen to be red, and the phosphide of a pale yellow color; or if heated till the carbide is blue, the phosphide will be brown or red salmon color.

It requires a little practice to heat the metal to the exact degree, but as the surface is not corroded in any way, if failure follows the first attempt, it is easy to repeatedly re-polish and heat again the same specimen until success is attained.

Objects prepared in this way are not only most valuable, but, owing to the gorgeous coloring, may be classed as very beautiful.

The phosphide is a little softer than the carbide, and when the two bodies are juxtaposed in massive form, polishing removes the former relatively a little more rapidly than the latter, leaving it in relief, but the difference is not sufficiently marked to be of any certain value; and although softer than the carbide, it is too hard to cut or saw with the hardest steel blade.

The Effect of Carbon on Saturated Solutions of Fe_3P in Iron

In prosecuting this research, the first aim was to ascertain what effect increasing additions of carbon would have upon saturated, or nearly saturated, solutions of phosphide of iron in iron.

In making the trials, a compound was used containing 1.75 per cent of phosphorus, equal to 18 per cent of the phosphide eutectic, or about 12 per cent of phosphide of iron, Fe_3P.

This was effected by melting it with increasing proportions of well-calcined charcoal in magnesia-lined plumbago crucibles. The mixtures were covered with lids of magnesia, consolidated by an admixture of magnesium chloride, and over this a layer of graphite was laid, and finally the crucible was closed with a fire-clay cover, and the mixtures melted in a coke fire.

The results proved that the carbon at the point of solidification throws out of solution a certain proportion of a phospho-pearlite eutectic, the softer constituent juxtaposed with the phosphide consisting of pearlite instead of iron phosphide solid solution.

The proportion so thrown out steadily increases with the carbon. When the carbon is small, the phosphorus eutectic is found at the junctions of the grains more or less completely enveloping them, and surrounding it the pearlite forms most perfect fringes, well illustrated in the two photographs (Nos. 8 and 9), where 1.75 per cent of phosphorus is associated with 0.125 per cent of carbon. With 0.7 per cent of carbon the mass of the grains consist of pearlite, but there still exist areas of the phosphide saturated solid solution. (Photo No. 10.)

As the carbon increases, the amount of the phosphorus eutectic and the pearlite also increase in quantity, until, when about 0.8 per cent of carbon is present, the grains consist entirely of pearlite.

When the carbon is increased to 1.4 per cent the eutectic of phosphide becomes surrounded by a solid layer of cementite or carbide of iron, which increases in thickness as the carbon is in-

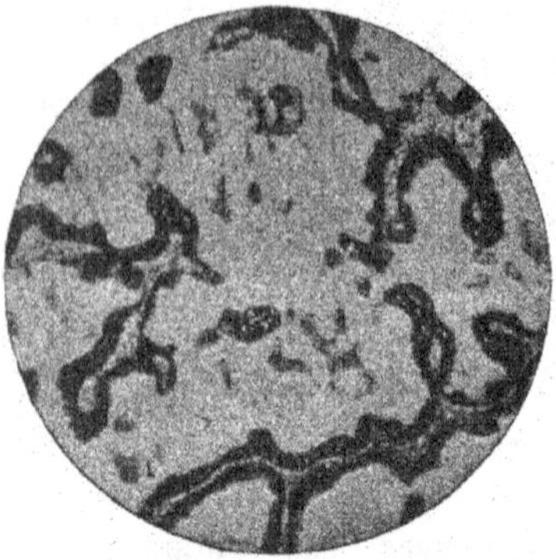

Fig. 8. Ingot containing 1.74 per cent phosphorus and 0.18 per cent carbon. The mottled portions surounded by dark borders are the eutectic of phosphide of iron. The borders are pearlite. The white ground mass is the solid solution of phosphide of iron in iron.
Magnified 60 diameters.

Fig. 9. A portion of the last more highly magnified, showing the fringed character of the pearlite and broken-up structure of the eutectic.
Magnified 350 diameters.

Fig. 10. Ingot containing 1.7 per cent phosphorus and 0.71 per cent carbon, showing a large grain surrounded by the eutectic. The body of the grain consists partly of solid solution of phosphide of iron in iron (white), and the greater part of pearlite (banded).
Magnified 250 diameters.

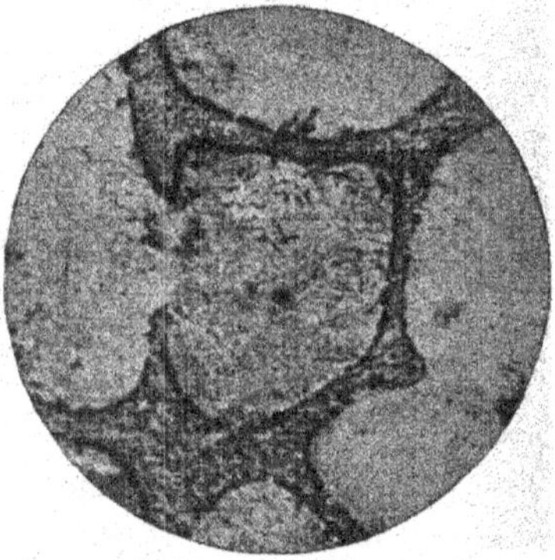

Fig. 11. Ingot containing 1.7 per cent phosphorus and 1.4 per cent carbon, polished and heated till the eutectic was tinted pale yellow. The dark band is carbide of iron. The pearlite grains were blue with brown laminæ of cementine.
Magnified 250 diameters.

creased. The envelopes of carbide are solid and do not consist of two constituents, whereas the enclosed matter has the usual broken-up duplex character of a eutectic. The white carbide and phosphide appear to run into one another.

When a section of the metal was heated till the eutectic assumed a yellow tint, the cementite borders asumed a fine red color; it was easy then to see that the carbide was separate from the phosphide at a most clearly marked junction. (Photo No. 11.)

On re-heating it to about 900° C., and allowing it to cool moderately rapidly, the greater mass of cementite was found no longer as an envelope to the phosphide eutectic areas, but now existed as independent cells, cutting up the original large grains of pearlite into a number of smaller grains. The phosphide eutectic retained its original position.

This observation appears to demonstrate that whereas the phosphide eutectic does not diffuse at 900° C., the carbide passes into solid solution at that temperature, and is, on cooling, redistributed between the newly formed pearlite grains. In other words, it behaves in a similar manner to what it does in high carbon steels free from phosphorus. The eutectic, although retaining its original position at 900° C., undoubtedly must have become liquid, for it had altered its character; the two constituents were in a much finer state, and required much higher magnification to see that they were separate.

The following table gives analyses of a series of the metal containing about 1.75 per cent of phosphorus, and various proportions of carbon:

Nos.	Carbon	Phosphorus		
		In Free Fe_3P	In Solution	Total
	Per Cent	Per Cent	Per Cent	Per Cent
1	Nil	Nil	1.75	1.75
2	0.125	0.18	1.37	1.55
3	0.180	0.59	1.18	1.77
4	0.70	1.00	0.75	1.75
5	0.80	1.06	0.70	1.76
6	1.40	1.16	0.60	1.76
7	2.00	1.18	0.55	1.73
8	3.50	1.40	0.31	1.71

These results were obtained on small specimens, rather rapidly solidified. It is probable that if the mass in each case had been greater, and the cooling through the solidifying points slower, the relative proportion of phosphorus thrown out of solid solution would have been different.

The method of determining the free phosphide, although the best at command, gives approximate results only; but the errors should not exceed 5 per cent on the quantities given.

It was expected that a larger proportion of the phosphide would have been thrown out of solution by the higher percentage of carbon.

The results are interesting as proving that carbon will not throw the whole of the phosphide into the separate state, that a persistent residuum still remains in solid solution, even although the carbon is as high as 3.5 per cent.

When the carbon is still further increased, a large quantity of massive cementite is present, and an entirely new eutectic is found. The original phospho-pearlite eutectic dissolves a portion of the carbide of iron, which does not separate, but solidifies with it, forming a complex compound mixture, in which thin parallel plates of carbide of iron, cutting up the phospho-pearlite eutectic, are easily detected under the microscope.

This eutectic does not appear to be formed till the carbon approaches 2 per cent.

It is always present in white phosphoretic pig irons. Its exact composition has not yet been determined. A sufficient quantity of the compound free from other elements, such as silicon, manganese, etc., was difficult to obtain, on which to squeeze out the eutectic by hydraulic pressure.

A small quantity of metal yielded a fusible liquate of the following composition, viz.:

	Per Cent
Phosphorus	7.05
Carbon	1.30
Iron	91.60
	99.90

This was not pure, and is only given here as representing the approximate composition of the eutectic.

The Effect of Carbon on Iron Containing Less than the Saturation Proportion of Phosphide Phosphorus when Melted Together

When the phosphorus is in much smaller quantity than 1.7 per cent, if the carbon is in sufficient quantity it still causes a separation of phosphide.

An excellent example of this has been afforded by Mr. E. H. Saniter, who discovered in the hearth of an open-hearth basic furnace a piece of partially converted metal which had remained in the bed of the furnace when it was cooled down for relining.

An analysis of this metal yielded in my laboratory:

	Per Cent
Carbon (by combustion)	1.23
Manganese	0.45
Silicon	0.06
Sulphur	0.18
Phosphorus	1.38

A portion broken off from the same material yielded the following result by Mr. Saniter's analysis:

	Per Cent
Carbon	0.98
Manganese	0.42
Silicon	0.055
Sulphur	0.018
Phosphorus	0.84

These analyses are far from concordant, and are evidence of the great difficulty of obtaining two pieces alike from a mass of metal when large quantities of phosphorus are present and the cooling has been done very slowly. The following remarks, therefore, apply to the specimen supplied to me by Mr. Saniter, about 3 cm. square.

A portion of this was crushed to coarse powder in a steel-crushing mortar, and determination made of the free and insoluble phosphide, with the following results:

	Per Cent
Phosphorus as free phosphide	0.76
Soluble phosphide	0.62

The phosphide of iron was exceedingly brittle, and was located in greatest measure on the outside of the grains, so that on crushing the sample the greater part of it was found in the finer portion of the crushings. These were sieved off, and analysis made of the coarser material left on the sieve.

	Per Cent
Phosphorus as free phosphide	0.222
Phosphorus as insoluble phosphide	0.606
Total phosphorus	0.828

a result approximating closely to the previous one, and showing that the pearlite grains themselves contained about 0.61 per cent of phosphorus.

The microscope revealed the presence of large crystalline grains of pearlite surrounded by very thick envelopes of what appeared to be massive carbide of iron (cementite). There were also globular-shaped masses of phosphide eutectic surrounded with cementite in the center of the grains, and idiomorphic crystals of Fe_3P.

The envelopes were not of the character of phosphide eutectics, but were generally massive and homogeneous, excepting at the points where three grains joined together, when they were more or less cut up by inclusions of pearlite.

On coloring the specimen by heat-tinting, the exterior portions of the sections of the envelopes were colored red, but in the central portion there was seen a thick band colored pale yellow, a clear proof that the envelope was compound in character, and that it consisted of phosphide of iron in the center and carbide of iron on the outside. The thickened portions of the envelopes at the junction of three grains, where inclusions of pearlite were present, consisted mainly of phosphide of iron. (Photo No. 13.)

If the mass had been cooled less slowly, there can be no doubt the central portion of the envelopes would have shown the compound structure of the eutectic of phosphorus and iron; but, as is very common in alloys, when the actual mass of the eutectic is small and the cooling very slow, the harder constituent had coalesced or segregated and become massive, whilst the softer portion had been thrown out and united with the greater mass of the surrounding metal.

When this specimen was heated to 900° C., and was allowed to cool in a few minutes from that temperature to 500° C., the envelopes in the polished and tinted section were seen under the microscope to have the true eutectic structure. They were considerably thicker, and occupied more space than the original phosphide areas. The eutectic so formed resembled that in white phosphoretic pig metals.

This complex eutectic is what is always found in white cast iron containing phosphorus.

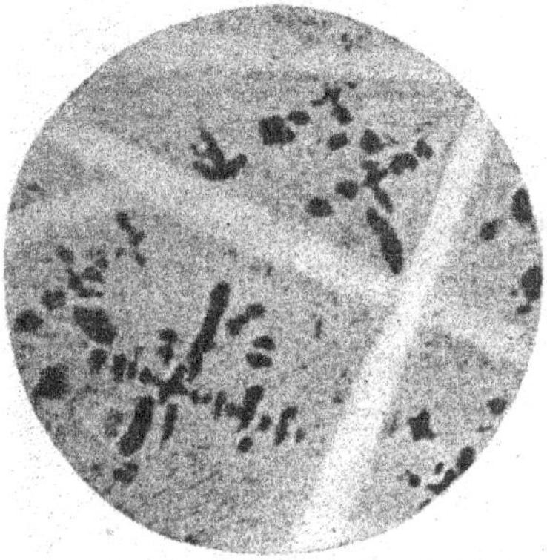

Fig. 12. Metal containing 6 per cent phosphorus and 1.6 per cent carbon, showing white sections of cementine plates, the ground mass of the eutectic, and the black pearlite areas. Magnified 100 diameters.

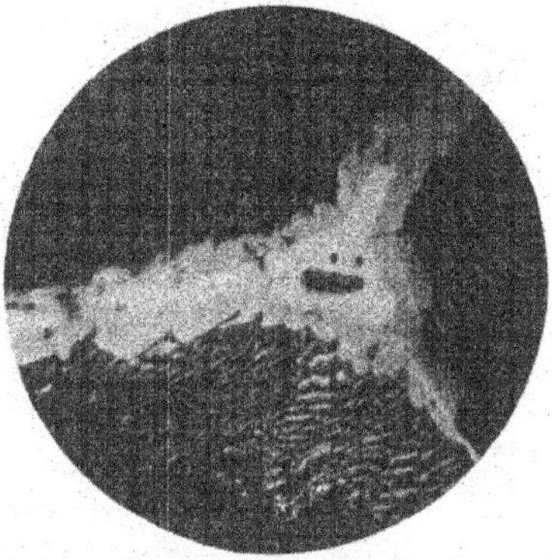

Fig. 13. Mass of slowly cooled steel from the bottom of an open-hearth furnace containing 1.2 per cent carbon and 1.3 per cent phosphorus. The section was first etched with iodine to darken the pearlite. It was then heated till the borders of the cell walls assumed a pale yellow tint. The interior portion of the cell walls are phosphide of iron (white), the exterior parts cementite (half-tone), the dark parts pearlite. Magnified 250 diameters.

On breaking a section of the original metal, which had been previously polished and tinted, the line of fracture passed along the phosphide, in the center of the cell walls.

Mr. Saniter, previous to the systematic micro-study referred to, when he was chemist to the Wigan Coal and Iron Company, having microscopically examined the specimen, and having found that the cell walls were considerably thicker than was justifiable with so little carbon as 0.98 per cent, and knowing that the envelope was much more brittle than the body of the grains of pearlite, he endeavored, by crushing and sieving, to separate the more easily powdered envelope from the greater bulk of pearlite. Of course, such a crude mechanical process could not be expected to give perfect separation, but, with the assistance of chemical methods in addition, he obtained some most valuable results.

By Mr. Saniter's permission his original results are here given in his own words:

" As was anticipated on examination under the microscope, it was evident that brittleness was caused by the segregation of a hard substance with the appearance of cementite along the granular junctions. With a view to ascertain its composition the metal was subjected to crushing, and the more brittle portion separated by sieving from the more malleable, that is to say, the intergranular from the granular matter, with the following results, viz.:

Metal partly crushed and the crushing fractionated by sieving

Would not pass 40-mesh sieve (liner) = 97.3 per cent = This was crushed as fine as possible		Passed 40-mesh but not 80-mesh 1.2 per cent	Passed 80-mesh sieve 1.5 per cent
Would not pass 80-mesh 76.5 per cent This contained:	Passed 80-mesh 20.8 per cent		
Carbon . 0.64 per cent	. .	. .	1.10 per cent
Silicon . 0.005 "	. .	. .	0.093 "
Sulphur . 0.014 "	. .	. .	0.472 "
Phosphorus 0.620 "	0.88 per cent	2.33 per cent	8.200 "
Manganese 0.430 "	. .	. .	1.510 "
Phosphorus soluble in 1.1 sp. gr. H_2SO_4 } = Nil	. .	. .	0.20 per cent

" Disregarding the intermediate products, the analysis on the left is that of the pearlite grains, and that on the right of the table the easily crushed intergranular matter. Considering this latter first, the point of most interest is that phosphide as well as carbide of iron is expelled when in excess of a certain point in carbon-phosphorus-iron mixtures, which no doubt accounts for the more injurious effect of phosphorus on steels with high carbon.

"A portion of the intergranular matter was treated with a large excess of 1.1 sp. gr. sulphuric acid for 18 hours in the cold, and the insoluble portion was analysed with the following result:

	Per Cent
Carbon	2.07
Phosphorus	10.77
Iron	85.00
	97.84

"This works out approximately to a formula of two molecules of Fe_3P to one of Fe_3C.

"If we examine the eutectic containing 0.64 per cent of carbon and 0.62 per cent of phosphorus, we find that, after allowing for iron to carbon in the ratio $Fe_{24}C$, the composition of ordinary pearlite, the remaining iron is in the ratio $Fe_{25}P$, a figure very interesting, as it is so close to that for the carbon eutectic. This may be only a coincidence, and it would require a considerable amount of investigation before it could be determined whether or not phosphorus forms a similar eutectic to that of carbon with iron. Calculating on these lines, the pearlite grains consist of

	Per Cent
$Fe_{24}C$	71.11
$Fe_{25}P$	28.77 with a formula 2.6 $Fe_{24}C$, $Fe_{25}P$.

These results of Mr. Saniter, obtained some years before my researches were completed, coincide closely with those obtained at Middlesbrough. The very low carbon, however, which he found in the finely powdered residue may be due to the fact that some of the comparatively coarse carbide plates of the pearlite escaped the sieve.

It is possible that, when there is a considerable quantity of phosphorus present, the pearlite may contain about 0.6 per cent of phosphorus, and the smaller proportion found dissolved in the compound containing 3.5 per cent of carbon given on p. 203 may be owing to the pearlite present representing only about 41 per cent of the mass.

If the carbon in the pearlite is 0.75 per cent, and that in the massive cementite 6.666 per cent, by calculation the sample with 3.5 per cent of carbon should have the following constitutional analysis:

	Per Cent		Carbon per Cent	Phosphorus per Cent
Carbide of iron (massive) .	48 ×	.0666 =	3.19	—
Pearlite	41 ×	.0075 =	0.30	—
Phosphide of iron (free) . .	9 ×	.1556 =	—	1.40
" " (dissolved)	2 ×	.1556 =	—	0.31
	100.00		3.49	1.71

The pearlite would in such a case contain 0.75 per cent instead of 0.60 per cent of phosphorus. If, however, we remember that in one case the metal was cooled very much slower than the other, it may account for the difference.

It is satisfactory in so far as the calculation gives about the same amount of dissolved phosphide as was actually found in the metal containing 1.77 per cent of phosphorus and 0.80 per cent of carbon, and which consisted of pearlite and no free massive carbide. The amount in solid solution varied according to the condition of solidification, and discredited the hypothesis that there exists a definite chemical compound of the chemical formula $Fe_{25}P$.

In order to ascertain what would result if Mr. Saniter's sample was decarburized by heating with ore in a furnace for the manufacture of malleable castings, a portion was inserted inside of a wrought iron tube, and was packed with fine scale from a wire-rolling mill. The ends of the tube were closed with fireclay, and the whole was then passed through a furnace with the ordinary charges. When cold, it was found that a considerable quantity of the scale had been reduced to the metallic state and adhered very strongly to the metal. On analysis the following interesting results were obtained:

	Before Per Cent	After Per Cent
Phosphorus	1.38	0.91
Carbon	1.23	traces
P in Fe_3P (free)	0.76	0.18
" (solution	0.62	0.73

The loss of phosphorus in the conversion, amounting to 33 per cent of the whole, was sufficiently startling to merit a searching investigation.

In Part I analyses are given showing the result of decarburizing a piece of metal containing 1.8 per cent of phosphorus and 0.18 per cent of carbon. In that case the micro section resembled No. 9 Photo. The phosphide areas were not continuous, but terminated in obtuse points. Where there was direct connection between these eutectic processes and the outside of the metal, the eutectic simply ran out leaving cavities. Had the eutectic been continuous and connected throughout the whole piece, probably much more would have escaped in that way. In the sample in which so much greater quantity of phosphorus had disappeared, the large crystalline grains were entirely enveloped with phosphide and carbide of iron, and were continuous like the cells of a honeycomb. As soon as a temperature of about 980° C. was reached, these envelopes or cells would combine with a portion of the metal in the adjoining grains to form a fusible liquid eutectic. This must have run out of the sample, and the grains falling together as the supporting cells flowed away, united to form a homogeneous whole. The micro sections in this case showed no empty spaces; no cementite or phosphide enveloped the grains; but on polishing on parchment, broad reticulated bands, corresponding to the original positions of the cells, remained bright long after the intercell structure had become tarnished, showing the existence of parts more highly charged with phosphorus in the regions where the free phosphide had originally existed. Etching with acid and heat oxidation all confirmed that conclusion. It is evident, then, that some of the phosphide had diffused into grains, but had not penetrated to any great distance. The amount actually so diffused is shown by the analysis to be 0.11 per cent. In one or two parts of the section comparatively large solidified "pools" of the eutectic were located, thus accounting for the free phosphide found by analysis. The same pools or areas of massive segregations were present in the original metal, and were generally located at the junctions of several grains, and it appears reasonable to conclude that the surrounding metal under such conditions could not fall in and join together, and so force out the liquid eutectic; it would, therefore, remain imprisoned at the very time during which the portions of the eutectic between the granular faces were flowing out of the metal.

If, then, the eutectic was actually expelled, it should be found in the scale in which the metal was packed. Fortunately a portion of this had been reserved. It was tested together with a portion of the scale before it was packed in the tube. The following results were obtained:

	Before Per Cent	After Per Cent
Iron	74.6	96.3
Phosphorus	0.06	0.24

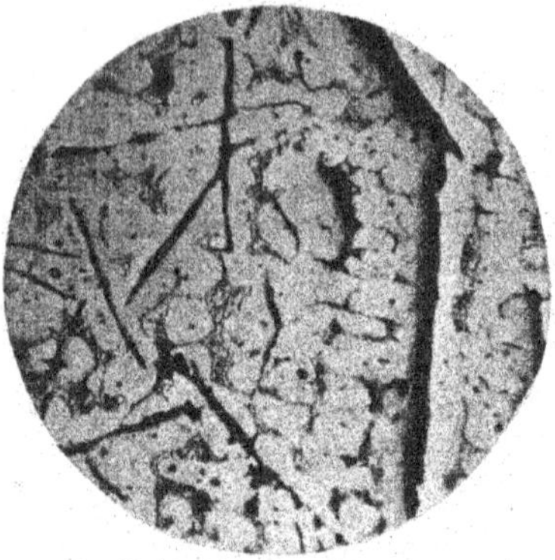

Fig. 14. Cleveland grey Pig Iron showing graphite plates (black), a ground mass of ferrite and the phosphide eutectic in irregular patches of mottled appearance. Magnified 50 diameters.

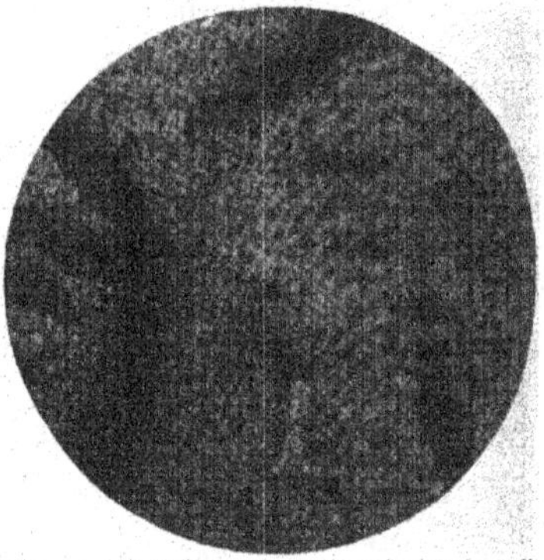

Fig. 15. East Coast hæmatite, showing small phosphide area in centre, the section of the graphite plates surrounding it. Magnified 250 diameters.

The quantity actually used was considerable. No record was kept of the relative weights of the metal and scale, so that it is impossible to make a balanced account of the phosphorus.

The fact, however, that the phosphorus in the scale was greatly increased is strong proof that it had obtained it from the metal.

One other interesting observation was made on the sample. It was noted that the metal immediately adjoining the scale, 1 mm.

in thickness, behaved under the action of acids almost like pure iron. It did not darken on digesting with dilute acid, whereas the whole mass below became soot-like in appearance. The only explanation possible is, that after the carbon had reduced the scale to the metallic state, a portion of the phosphorus actually diffused into it from the outer crust.

The quesion still not satisfactorily settled at this stage is, Does the carbon in commercial steels, amounting to about 0.9 per cent and above that quantity, when solidifying from the liquid state cause the phosphide present to be thrown out of the solution?

To answer that question, a steel containing 1 per cent of carbon and only 0.02 per cent of phosphorus was melted in several portions, with increasing doses of phosphide of iron sufficient to give from 0.06 per cent to 0.24 per cent of phosphorus.

To effect this, and at the same time to ensure strictly comparative results, a large plumbago crucible was filled with plastic magnesia. With a cork-borer cylindrical prisms of the magnesia filling were removed at equal distances near to the sides of the crucible, leaving in this way a series of vertical holes. The crucible so prepared was heated to dry the magnesia, and into the four holes weighed quantities of steel and phosphide were introduced. The holes were then plugged with magnesia stoppers, a covering of plumbago applied, and over all a crucible lid. It was then placed in a hot-wind furnace, and after an hour the fire was allowed to cool down slowly so as to retard the solidification. When cold, the metals were removed and weighed to ascertain whether there had been any gain or loss. For comparison, one of the pieces of steel which was melted had not any phosphide added to it. The following results were obtained:

Nos.	Steel Grammes	Phosphide	Phosphorus	Weight Before	Weight After	Loss or Gain
1 . .	11.810	None =	0.02	11.810	11.814	+ 0.004
2 . .	13.070	0.050 =	0.06	13.120	13.120	Nil
3 . .	12.950	0.100 =	0.11	13.053	13.048	0.002
4 . .	11.630	0.200 =	0.21	11.830	11.830	Nil

ANALYSES OF METALS

	Carbon	Phosphorus
No. 1	1.05	0.020
No. 2	1.06	0.058
No. 3	1.05	0.118
No. 4	1.06	0.245

Micro-Structure

No. 1 was slightly meshed with cementite, but no phosphide could be detected.

No. 2 was more meshed than the last, and phosphide of iron was visible after heat-tinting.

No. 3 was more meshed than the last, and a greater proportion of phosphide was visible.

No. 4 had phosphide in strong evidence.

Although the phosphide was in strong evidence judging from the appearance, only a very small proportion of the whole seemed to be in the free state. In the case of No. 4, if all had separated there would have been about 3 per cent of the eutectic visible, whereas probably not more than one-third of that amount was really present. A further trial was made on a larger scale, but in this case a charge of about 50 lbs. of molten steel was poured successively into five different small fireclay moulds. After the first mould was filled a stick of phosphorus was dropped into the fluid steel in the crucible, where it rapidly dissolved. After mixing a second mould was filled. A second stick of phosphorus was then added to the crucible and a third mould filled. After each mould was filled a further addition of phosphorus was made to the remaining steel till all the five moulds were filled.

The analyses of the steels when cold were as follows:

Nos.	Carbon	Phosphorus	Phosphorus in Free Phosphide
	Per Cent	Per Cent	Per Cent
1	0.95	0.037	Nil
2	0.96	0.099	0.002
3	0.95	0.122	0.035
4	0.96	0.347	0.065
5	1.02	0.548	0.163

No free phosphide could be detected by the microscope in No. 1. In No. 2 it was doubtful if any was present. No. 3 contained traces. Nos. 4 and 5 contained distinct quantities. No. 5 ingot was very red-short, and broke to pieces with the first blow of the hammer.

It seemed reasonable to believe that just as the pearlite in the highly phosphoretic metals dissolves and retains in solution

0.60 per cent of phosphorus, so a pearlite steel would not throw off free phosphide till that amount was exceeded. The above results prove, however, that this is not the case. Steel with about 1 per cent of carbon, in solidifying, does evidently throw off a portion of the dissolved phosphide, even although the proportion of phosphorus is only 0.06 per cent.

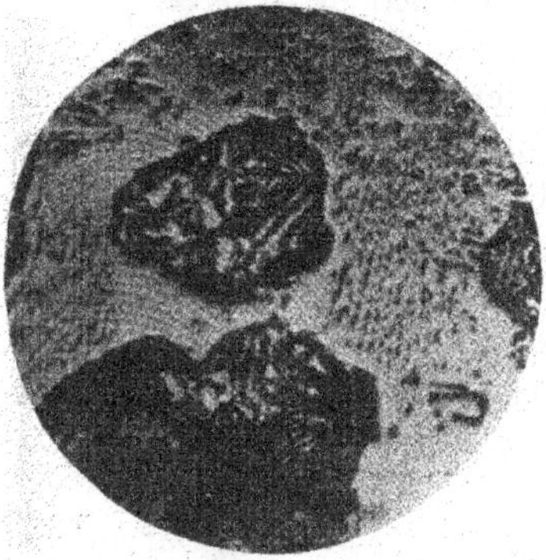

Fig. 16. White Clarence pig iron etched with iodine. The white parts represent the carbide and phosphide areas. The dark areas represent the pearlite. Magnified 250 diameters.

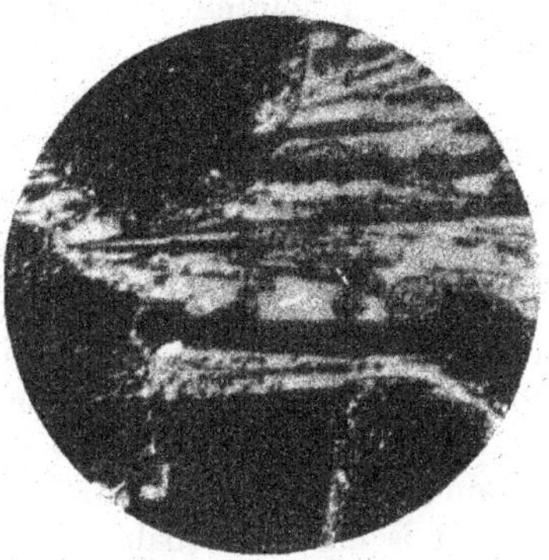

Fig. 17. White Clarence pig iron, etched to darken the pearlite, then heated till the cementite became yellow, the phosphide of iron remaining white. The dark parts with irregular borders on the top and on the right are pearlite. The large fat dark area on the left is massive cementite. The black and white areas in the middle are the compound phosphide eutectic. Magnified 250 diameters.

The rate at which it solidifies appears to influence the result, for in the laboratory experiments, when solidification was retarded, more free phosphide separated than in the trials in which it was unavoidably accelerated.

It was noted that just as high phosphorus carbonless irons in the state of drillings dissolved in dilute sulphuric and hydrochloric acids more rapidly than those containing little, so the carbon steels containing phosphorus dissolved much more rapidly than the same material containing practically none.

Five grammes each of the drillings from the five steels above mentioned, on placing in mixtures of 20 cc. strong sulphuric acid and 250 cc. of water, dissolved at very different rates.

No 1, containing practically no phosphorus, took ten hours to completely dissolve, whereas No. 5, with 0.548 per cent of phosphorus, was dissolved in two hours. The other steels passed into solution in periods intermediate between the two extremes.

A dark soot-like residue was left behind, the proportion appearing to coincide with the quantity of phosphorus present.

Many commercial structural steels containing between 0.10 per cent and 0.6 per cent of carbon, have been examined, but in no case have I detected free phosphide of iron, even when the phosphorus exceeded 0.1 per cent. This is what would be expected, for in such partially carbon saturated irons there is more or less free ferrite, and this strong solvent for phosphide would retain it in solid solution.

It is highly probable, judging from the behaviour of dilute acids upon steels, more or less highly charged with carbon, that the carbon causes the phosphide to be concentrated either in the massive ferrite or in the ferrite of the pearlite, or in both.

Before discussing this question, reference must be made to the valuable results given by ([14]) Messrs. Osmond and Werth.* These gentlemen, on treating various metals, irons, and steels with hydrochloric acid under the same conditions, found that the proportion of the phosphorus given off as phosphine gas (PH_3) varied with the amount of carbon present. I cannot do better than give a short abstract from their most valuable paper. The phosphine evolved was absorbed in an ammoniacal solution of silver nitrate.

" We have determined the quantity of Ag reduced, i.e., the proportion of PH_3 evolved by a series of very different products, passing the gas first through an ammoniacal solution of $CdSO_4$, where H_2S is absorbed.

" Appended are the results obtained on dissolving 4 grammes of each metal in 35 cc. concentrated hydrochloric acid and gradually heated until completely attacked:

---

* *Théorie Cellulaire*, Mém. de l'Artillerie de la Marine, 1887, Vol. ii, page 273.

Material	Analysis					P. Evolved as PH_3 per 100 Samples	Parts per 100 P. as PH_3
	C.	Mn.	Si.	T.	P.		
	Per Cent	Per Cent	Per Cent	Per Cent	Per Cent	Per Cent	Per Cent
Acid Bessemer Steel—							
Before the addition of spiegeleisen	—	—	—	—	0.065	0.044	68
After such addition	—	—	—	—	0.065	0.028	43
Basic Bessemer Steel—							
Before the addition of spiegeleisen	—	—	—	0.038	0.046	0.030	65
After such addition	—	—	—	0.022	0.046	0.018	39
Open-hearth Steel—							
Before the addition of spiegeleisen	—	—	—	—	0.033	0.022	67
After such addition	0.49	0.37	0.075	0.024	0.041	0.014	34
Hardened	0.49	0.37	0.075	0.024	0.041	0.013	32
Very soft steel	0.18	0.10	—	0.060	0.070	0.049	70
"Diamond" steel No. I	1.17	0.18	0.44	0.018	0.033	0.005	15
Ordinary Bessemer steel	0.50	0.57	0.11	0.042	0.065	0.030	46
Ordinary Bessemer steel	0.49	0.74	0.23	0.022	0.065	0.026	40
Pig iron	—	—	—	—	0.055	0.041	38
Burnt iron from the Moselle, in pseudo-crystals	0.11	trace	0.058	0.032	0.810	0.147	18
Spiegeleisen	4.00	19.84	—	—	0.145	0.004	3
Basic Bessemer pig iron	3.00	2.16	1.71	0.13	2.500	0.037	1.5
Forge pig iron	3.00	0.07	1.37	0.48	1.750	0.038	2.2

The author's remark:

"This table clearly shows that there is no relation between the total P and that freed as PH_3. The latter depends entirely on the percentage of carbon; indeed, more phosphoretic metals only allow an insignificant amount of PH_3 to be evolved, whilst soft (mild) irons and steels evolve under the same circumstances comparatively important quantities, the same steel giving more PH_3 after decarburization than after recarburization. Manganese is without effect, seeing that spiegeleisen and Moselle iron behave after the same manner.

"The quantity of PH_3 evolved also varies, all other conditions being equal, with the strength of the acid.

"A fraction of the P then seems to be combined with the iron, with which it forms a phosphide partially decomposed by HCl, whilst the surplus is engaged in a complex combination with the carbon of annealing cement carbon."

Weyl's method gave identical results.

"If we treat the P in the residues which this method separates, we find in the sample containing:

	Per Cent
Carbon	0.490
Silicon	0.075
Sulphur	0.024
Phosphorus	0.041
Manganese	0.037

Steel	Forged and Carefully Cooled (Chilled).	Forged and Annealed	Tempered in Cold Water.	Tempered and Annealed	Average.
	Per Cent	Per Cent	Per Cent	Per Cent	Per Cent
Dry residue per 100 parts steel	3.310	4.110	1.620	4.140	—
Phosphorus treated per 100 parts of dry residue	0.840	0.640	1.940	0.760	—
Phosphorus calculated on 100 parts of steel	0.028	0.026	0.031	0.031	0.029

	Per Cent	Per 100 P. Per Cent
Average of same steel (P)	0.041	67
Liberated as PH_3	0.0135	33"

These results are most instructive, and have been fully confirmed by my own researches.

I have failed to meet with any steel containing much carbon which yielded more than a small fraction of the phosphorus in the state of PH_3, and on the other hand from steels low in carbon the preponderating proportion of the phosphorus passed off as that gas on dissolving them in strong hydrochloric acid.

Baron Jüptner also, in his researches, shows that in high carbon metals only a very small proportion of the total phosphorus passes off as PH_3. He does not, however, appear to have examined steels with more than 0.30 per cent of carbon.

The following table gives Baron Jüptner's results, and in addition the percentage proportion of phosphorus given off as PH_3, which I have calculated from them, viz.:

No.	Carbon per Cent by Weight	Phosphorus per Cent by Weight	Phosphorus per Cent: Parts per 100 Sample, as PH_3	Phosphorus per Cent: Parts per 100 P, as PH_3	Remarks
					A. *Pig ioon.*
1	2.385	0.1315	0.0263	2.00	White pig iron.
2	3.418	0.0744	0.0026	3.5	Dark gray pig.
3	3.418	0.0744	0.0058	7.8	Dark gray pig, dissolved in sulphuric acid diluted with equal volume of water.
4	2.296	1.048	0.0383	3.6	So-called "Panzerguss" for fire-bars.
					B. *Wrought iron and steel.*
5	0.2254	0.0970	0.0054	5.50	Short.
6	0.1800	0.0596	0.0020	5.30	Short.
7	0.1510	0.0690	0.0012	1.71	Short.
8	0.239	0.1010	—	—	Good material.
9	0.259	0.0860	0.0015	1.70	Good material.
10	—	0.0964	0.0162	16.00	Very short.
11	0.273	0.0469	0.0054	11.00	Slightly cold-short.
12	0.308	0.1242	0.0243	20.00	Cold-short.
13	0.343	0.5649	0.0248	4.0	Cold-short.
14	0.122	0.6212	0.0190	3.0	Cold-short.
15	0.375	0.0847	0.0171	20.0	
16	0.217	0.0677	0.0184	27.0	
17	0.168	0.0704	0.0177	25.0	Increasing in cold-shortness.
18	0.137	0.0648	0.0218	32.0	
19	0.220	0.0829	0.0218	26.0	
A	0.110	0.442	0.355	80.33	Welds easily. Very great cold-shortness.
B	0.123	0.456	0.289	63.35	Welds easily. Great cold-shortness.

The actual tests made in a testing-machine of A and B samples were as follows:

	A	B
Tensile strength, tons per square inch	27.00	22.40
Elongation on 50 millimetres, per cent	8.40	12.00
Induction in area, per cent . . .	29.32	25.60

The proportion of phosphorus as PH_3 given off after the samples were annealed and hardened in water was

83.13 per cent in A, and 68.75 per cent in B,

an increase of about

3 per cent in A, and 5 per cent in B.

The results show generally that when the phosphorus is high, little PH_3 is formed. Nos. 12 and 13 are notable instances of this. The observations of Osmond and Werth, Baron Jüptner, and myself agree in this respect. The proportion evolved from the low carbon steels in the Baron's experiment, however, is very different to the result of the other observers. Possibly the steels may have been of a unique character, and we may not have met

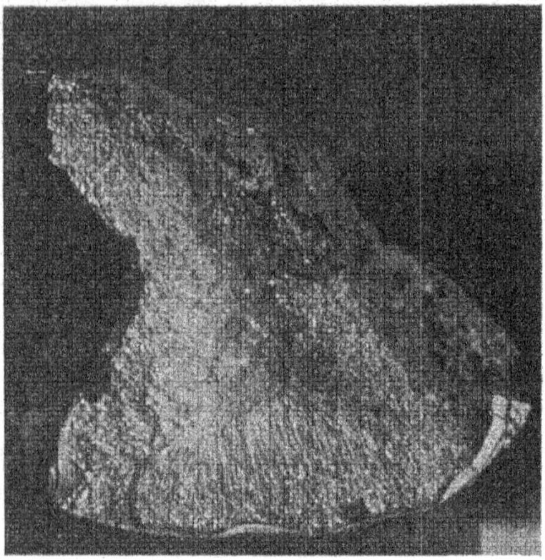

Fig. 18. Natural size of the fractured surface of a piece of metal containing about 1.8 per cent phophorus after carburising by cementation, showing crystalline columns radiating from the central kernel of unaltered metal formed when the iron changed from the *Beta* to the *Gamma* state.

with any of the same kind, or possibly the methods employed may have been responsible for the great discrepancies. It may be accepted, however, that when working in the manner described by Messrs. Osmond and Werth and myself, the carbon has an important influence generally in preventing the formation of PH_3 when steels containing phosphorus and carbon are dissolved in acid.

It is important to ascertain why that should be.

The answer to that question appears at present to be beyond the region of direct experiment, and one is obliged to endeavor to reason by induction, basing hypothesis on known facts.

It has been shown in Part I, that as the phosphorus becomes more and more concentrated, the quantity of phosphine evolved correspondingly decreases. There is no evidence that phospho-carbon compounds are produced or have existence. All evidence negatives such a supposition.

It is most probable that in steels containing carbon and phosphorus, when in the fluid state, there exists a compound solution intimately mixed, one of carbide, the other of phosphide of iron in iron, and that the carbide solution is the more persistent, taking what iron it requires, leaving what is in excess for the phosphide, which forms a more or less concentrated solution with it, the extent of the concentration depending on the quantity of carbon present. When the steel solidifies, the two solutions become solid solutions. On cooling through Ar_1, the carbide separates, leaving ferrite, and ferrite more or less saturated with Fe_3P.

This conclusion is in conformity with the admitted fact that phosphorus is more embrittling in high carbon steels than it is in those with less carbon. If free phosphide is liberated, as it has been proved it is with 0.06 per cent of phosphorus and about 1 per cent of carbon, as it forms meshes round the grains, it must materially reduce the strength of the mass. If the phosphide is concentrated in a part of the iron when the carbon is lower, say between 0.5 per cent and 0.8 per cent, that part of the structure is likely to be more liable to give way easily under sudden stress than if it was not so highly charged with phosphorus.

It must not be at this stage concluded that because generally the higher the carbon the less PH_3 is given off, and *vice versa*, that it always is the case. Possibly variations in the rate at which the liquid steels are cooled, and in the manner it is heated previous to forging, and also whether or not it is annealed, may have influence in determining in what condition of concentration the phosphide remains in the finished steel.

This section of the subject is worthy of most serious research, and it is to be hoped that the chemists in the many steelworks throughout the world will endeavor to clear away the darkness which surrounds this very complicated branch of the subject.

The hypotheses advanced at the present stage of the research can only be tentatively held, and it is not desired that they should be accepted as revealed facts.

Does Quenching Steel after Heating to High Temperatures Affect the Phosphorus Condition?

([13]) E. D. Campbell and S. C. Babcock have made certain investigations to determine whether or not the phosphorus condition is different in the same carbon steel when it is annealed and hardened. The method employed consisted in dissolving the steels in slightly acid solution of mercuric chloride, and determining the amount of phosphorus which remained in solution. The results obtained were as follows:

The analyses of the three steels experimented upon are as follows:

	1 Per Cent	2 Per Cent	3 Per Cent
Carbon	0.100	0.37	1.22
Phosphorus	0.119	0.16	0.098
Manganese	0.484	0.82	0.780
Sulphur	—	—	0.068
Silicon	—	—	0.058

The following proportions of phosphorus were made soluble after varying thermal treatment:

No. 1. *Steel Carbon* 0.10 *per cent*

	Phosphorus on 100 of Sample Per Cent	On 100 of Phosphorus Per Cent
Annealed	0.099	83.2
Heated to 719° C. and quenched .	0.081	68.08
" 825° " .	c.079	66.4
" 928° " .	0.080	67.2
" 1028° " .	0.086	72.2

No. 2. *Steel Carbon* 0.36 *per cent*

	Phosphorus on 100 of Sample Per Cent	On 100 of Phosphorus Per Cent
Annealed . . . . , .	0.137	85.6
Heated to 728° C. and quenched .	0.110	68.8
" 827° " .	0.066	41.2
" 923° " .	0.048	30.0
" 1027° " .	0.049	30.6

No. 3. *Steel Carbon 1.22 per cent*

	Phosphorus on 100 of Sample Per Cent	On 100 of Phosphorus Per Cent
Annealed	0.098	100.00
Heated to 719° C. and quenched .	0.087	89.8
" 750° " .	0.051	52.0
" 825° " .	0.018	18.3
" 923° " .	0.015	15.3
" 1023° " .	0.016	16.2

These results are very instructive, but vary somewhat from what would be anticipated from what has preceded, for it would appear that the phosphorus is more soluble, in the special reagent used, in high carbon steels than in those containing little. The authors, however, are of the opinion that much remains to be done before we shall be able to draw reliable conclusions as to the forms in which phosphorus may exist in iron aud steel, the conditions under which the different forms are produced, and the influence of the different forms on the physical properties of the metals.

I intend, as soon as time will permit, to work with the method described by those gentlemen, and shall communicate the results at a later date. The conclusion they apparently have arrived at is, that varying carbon condition in a steel is coincident with varying phosphorus condition.

On the Effect of Carbon Introduced into Phosphoretic Iron by the Cementation Process

([4]) Behrens and Van Linge, in studying the constitution of Dannemora cemented bar containing only 0.02 per cent of phosphorus, found that the phosphide separated with the carbide, and according to ([11]) Baron Jüptner, they found that the phosphide was inside the carbide. On consulting the original paper, it would appear that on dissolving the cement bar, the authors found some phosphorus associated with the insoluble carbide, and calculated that if the whole of that element existed as Fe_3P there should have been about 0.2 per cent insoluble.

It appeared to me that bars with so little phosphorus as 0.02 per cent were not the best upon which to determine the reactions which take place in the cementation furnace. With the most valuable assistance of Mr. T. W. Sorby of Sheffield, without whose aid no practical results could have been obtained, I have had many pieces of phosphoretic bars, etc., carburized by cementation, and these were carefully examined.

The first sample, for which I am indebted to Mr. A. Cooper and Mr. C. H. Ridsdale, consisted of a piece of iron taken from metal during the afterblow in a basic converter, when about 0.6 per cent of phosphorus remained.

A second sample consisted of a piece of puddled bar containing about 0.5 per cent of phosphorus.

The following are the analyses after carburization:

Basic Iron (Cemented)

	Per Cent
Carbon	1.16
Phosphorus	0.63
Phosphorus as free phosphide	0.02
Phosphorus in solid solution	0.61

The microscope proved that, except at the external parts, no free phosphide was present. The carbon being highest in that region, it was natural to expect free phosphide there if anywhere.

Puddled Bar (Cemented)

This sample was far from homogeneous, and had blistered slightly. It had the following analysis:

Description	Carbon	Total	Phosphorus	
			As Free Fe_3P	In Solid Solution
	Per Cent	Per Cent	Per Cent	Per Cent
Outside layer, 3 mm.	1.35	0.495	0.054	0.441
Second layer, 6 mm.	0.96	0.490	0.028	0.462
Third layer, 9 mm.	0.75	0.500	Nil.	0.500

It will be seen that, excepting near the outside where the carbon is in greatest quantity, practically no phosphide had separated.

The microscope exactly confirmed the chemical analysis.

These two metals were then melted in magnesia crucibles and cooled slowly, in order to find how much phosphide would separate during solidification. The following results were obtained:

Description	Carbon	Total	Phosphorus	
			As Free Fe_3P.	In Solid Solution
	Per Cent	Per Cent	Per Cent	Per Cent
Basic iron	1.16	0.63	0.27	0.36
Puddled bar	1.10	0.50	0.21	0.29

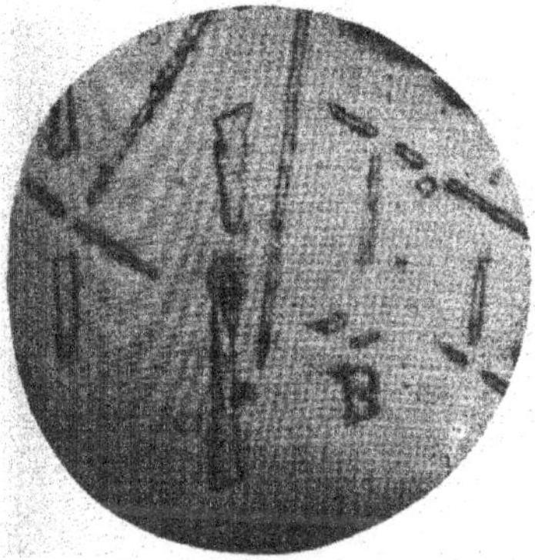

Fig. 19. Metal containing 1.95 per cent phosphorus and 0.18 per cent carbon after repeated annealing in a Malleable Castings furnace, showing prismatic crystals of free phosphide of iron. Phosphorus in solid solution = 1.06 per cent P, in free phosphide 0.88 per cent P.
Magnified 60 diameters.

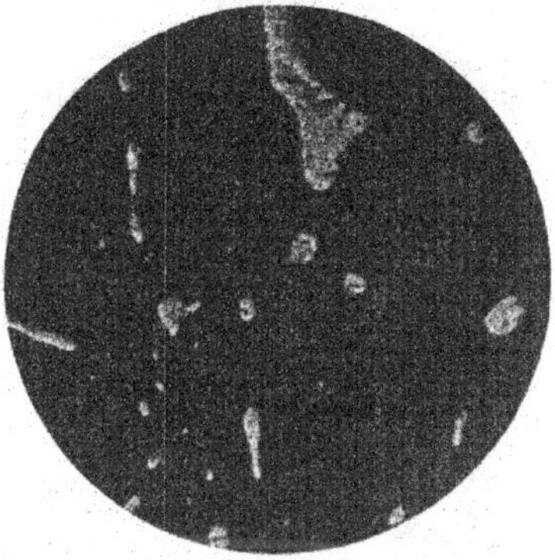

Fig. 20. The same as the last after heating to about 1100° C. Phosphorus in solid solution = 1.01 per cent P, in free phosphide = 0.95 per cent P.
Magnified 60 diameters.

In the next trial a piece of metal containing 1.96 per cent of phosphorus was experimented upon. It was of very irregular shape and about 6 cm. in diameter at the thickest part. When it came out of the cementation furnace, it was much scored on the outside, as if fluid metal had been run over it in thin streams. The lower part had adhering to it a solidified drop of metal which had evidently liquated from the specimen. It was detached for analysis. On fracturing the metal, a most instructive structure was revealed. (Photo No. 18.)

The metal before treatment was coarsely crystalline, with cleavage faces 3 cm. across. After carburization, excepting in a small kernel where the metal had escaped carburization and where the original structure remained intact, the metal was entirely altered.

The external parts of the structure were coarsely granular, but between these and the kernel the crystalline masses were arranged in columns clearly illustrated in the photograph.

Portions were taken from fixed positions, and these were separately analysed with the following results:

Description	Carbon	Total	Phosphorus		
			As Free Fe_3P.	In Solid Solution	Lost
	Per Cent	Per Cent	Per Cent	Per Cent	Per Cent
Kernel, centre	traces	1.96	0.94	1.02	None
Do. average	0.23	1.94	0.94	1.00	None
1	0.89	1.52	0.86	0.66	0.42
2	0.93	1.30	0.61	0.69	0.64
3	1.20	1.05	0.38	0.67	0.89
4 external layer	1.31	0.91	0.29	0.62	1.03

The drop of metal from the bottom of the piece contained

Phosphorus . . . . 4.86 per cent.

The evidence is conclusive that, as the carbon passed into the metal, its dominating influence threw out the phosphide, equivalent to all excepting about 0.65 per cent, and in doing so produced sufficient *Gamma*-iron to effect a complete rearrangement of the structure. The liquid phospho-carbide eutectic thrown off to the boundaries of the grains, being continuous and extending to the exterior, simply liquated away.

As the carbon passed into the *Beta*-iron containing the solid solution of phosphide, an allotropic change under the influence of the carbide must have been gradually effected, and coincident with that change a reorganisation of the crystalline structure. The crystalline development must also have been gradual, growing steadily from the exterior to the center, coincidentally with the penetration of the carbon. White iron, zinc, and other metals crystallize in similar radial columns on casting in cold moulds, when the change of state proceeds from the cold surfaces towards the center.

The change from the allotropic condition of *Beta* to that of *Gamma*-iron, although both are in the solid state, is almost as violent as when the metals above referred to changed from the liquid to the solid condition.

([17]) I have elsewhere shown that when steel containing carbon is decarburized at a temperature under 800° C., a similar columnar structure is developed in the solid metal, but in that case the gradual crystalline growth, although from the exterior to the interior, is caused by an allotropic change in the iron from the state of *Gamma* to *Beta*-iron, exactly the reverse of what took place in the cementation furnace.

Microstructure of Cemented Metal

A section of the metal revealed a most instructive structure when examined under the microscope. The central part, where there was no carbon, was traversed and cut up by plates or prisms of free phosphide of iron, and resembled the same metal after it left the malleable casting-furnace, illustrated by Photo No. 19, a coincidenece which removes the supposition that oxidizing influences were responsible for the reorganization of the phosphide from the condition in which it existed in the metal before treatment. Near to the junction between the kernel and the columnar grains, the eutectic was bordered with a beautiful fringe of pearlite. The cementite laminæ radiated from the junction towards the center, the base of which rested upon a layer of the phospho-pearlite eutectic. The kernel was more or less spherical, and was surrounded by a shell of the same eutectic. This brittle envelope made it easy to break away the surrounding metal, leaving the kernel intact. The radiating columnar masses were encased by a layer of the phosphide eutectic, and the latter was surrounded by a toothlike formation of cementite. The mass of the columns consisted of pearlite in which were embedded oval, elliptical, and globular masses of the eutectic, some of which were encased by cementite. The amount of the phosphide eutectic decreased with the distance from the center. The columnar structure terminated at a distance of 1½ cm. from the center, and was replaced by ordinary polygonal grains, the joints of which contained very little phosphide, but sufficient carbide to completely envelop them. Imprisoned in the center of these grains were globular masses of eutectic, and at wide distances apart irregular-shaped masses of the same substance were segregated at the junction of three grains.

The consideration of these observations leads to the following conclusions:

That as the carbon diffused into the mass in a forward direction, the phosphide eutectic fell out of solution in the rear. and at first the liquid eutectic so formed would be thrown entirely out of the metal and escape. As the carburization progressed, the metal in the rear would change its allotropic state from *Beta* to *Gamma*-iron, and would recrystallize into the well-known polygonal grains. In front of these a layer of fluid eutectic would continually be forming, but it would, as rapidly as it was formed, flow backwards along the crystalline junctions to the outside of the metal, leaving a residuum behind. The grains at first formed would in all probability be smaller than if they were left in the metal when cold, but under the long-continued action of heat would re-arrange their crystalline orientation and fall together so as to form a smaller number of larger grains or columns; but in doing so the intervening eutectic could not be completely expelled, and would remain imprisoned more or less completely, and not having any free passage to escape, would at once begin to coalesce, forming at first oblong drops, which in course of time would contract to the forms more or less globular, actually present in the cold specimen.

If the specimen carburized had been perfectly globular and the carburization regular at all times during the process, there would have been a spherical fluid area between the center and the outside, dividing a globe of solid metal in the middle from a solid shell at the exterior, more or less cut up by fluid cells through which the liquid eutectic would continually flow, and which would eventually leave the mass of metal and fall into the charcoal surrounding it. Although the specimen was not globular, the fluid envelopment must have been developed and continual liquidation of the eutectic been effected.

These interesting inductions are fully justified by the observed facts and data obtained during many years' study of metallography, and is another instance of the valuable aid the microscope can give in the study of metallurgy.

Does Phosphorus in Iron Limit the Amount of Carbon Capable of being Taken up in the Iron?

That question has been answered in the affirmative by the less carbon invariably found in corresponding grades of Bessemer, Cleveland, and other phosphoretic irons.

The following comparative average analyses of East Coast hæmatite and Cleveland pig irons show a difference of about 0.35 per cent of carbon less in the phosphoretic iron:

	Number 1 Brand	
	Hæmatite	Cleveland
	Per Cent	Per Cent
Combined carbon	0.15	0.12
Graphite	3.95	3.63
Manganese	0.85	0.75
Silicon	2.80	2.80
Sulphur	0.03	0.03
Phosphorus	0.04	1.56

The 1.56 phosphorus in the Cleveland iron is equivalent to close upon 10 per cent of phosphide of iron. If this was deducted, the remainder would contain

4.16 per cent total carbon, against
4.10 per cent in the hæmatite.

Basing our calculations on the assumption that it is the Fe_3P which monopolizes the iron, and prevents it combining with carbon under the conditions which reign in the blast-furnace, we find by calculation that with pure iron on the one hand, and pure phosphide on the other, the carbon capable of being taken up would be about 4.6 per cent in the pure iron and none in the phosphide, and that the carbon in the metals intermediate between the two extremes would depend on the phosphide present in them.

The following table will make this clear:

Pure Iron	Phosphide	Phosphorus	Carbon, calculated as Capable of being Absorbed
Per Cent	Per Cent	Per Cent	Per Cent
100	None	None	4.60
75	25	3.89	3.45
50	50	7.78	2.30
25	75	11.67	1.15
None	100	15.58	None

In order to ascertain if the result of actual experiment would coincide with the calculated results, mixtures of phosphide and pure iron were made so as to approximate to the above proportions of phosphorus and iron, and these were melted side by side in a charcoal brasqued crucible containing five separate compartments. The heat was regulated so as not to exceed the temperature of the blast-furnace, and was certainly not high enough to melt pure iron.

The analyses of the products were as follows, viz.:

Analysis of Mixture		Carbon in Melted Metals	Fracture
Iron	Phosphorus		
Per Cent	Per Cent	Per Cent	
100	None	4.15	White
96	4.10	3.25	"
93	7.90	2.00	"
87	13.00	0.70	"
83	16.00	Nil.	"

The results, although deviating from what was calculated, nevertheless are sufficiently close to show that the basis on which the calculations were made is fairly correct. It is well known that even from the blast-furnace the No. 1 pig irons will vary very considerably in carbon, from causes not at present fully understood. The temperature, however, *ceteris paribus,* undoubtedly has the greatest influence in determining the differences.

A second series of comparative trials were made with phosphoretic metals of known phosphorus content by carburizing in the manner above described, but at a lower temperature than in the previous trial, with the following result:

Phosphorus in Metals Per Cent	Carbon in Carburized Metals Per Cent
1.85	3.50
4.17	2.50
6.03	1.50
10.15	0.65

Here the total carbon is much less than in the previous records, but the influence of the phosphorus in preventing the absorption of carbon is manifest.

All of these metals were free from graphite, even the first sample containing only 1.85 per cent of phosphorus.

A most interesting metal actually produced in an East Coast blast-furnace many years ago, before the value of basic slag was recognized, was made on blowing out the furnace with basic slag instead of limestone. The analysis of this metal was as follows:

	Per Cent
Iron by difference	64.927
Combined carbon	traces
Graphite	Nil.
Manganese	4.550
Silicon	0.396
Sulphur	0.050
Phosphorus	17.910
Vanadium	1.711
Chromium	0.446
	100.000

That the heat was great is evidenced by the large quantity of manganese, vanadium, and chromium which were reduced, yet the carbon was practically absent — a convincing proof that phosphorus in sufficient quantity prevents carbon from retaining its combination or from making any union with the iron, in which respect it closely resembles the effect of 20 per cent of silicon.

Does Phosphorus in Iron Tend to Prevent the Separation of Graphite and Produce White Iron?

This question has often been asked, but practical experience with furnaces making hæmatite and phosphoretic Cleveland irons has shown that 1.5 per cent of phosphorus, at any rate, has no influence in preventing the separation of graphite.

A careful observer can, however, note a difference in the grayness of equally open highly phosphoretic pig iron and one practically free from phosphorus; but this is not caused by the presence of combined carbon, but is due to the fact that the free phosphide present gives a white fracture, and that the two irons contain different quantities of graphite. The combined carbon is about equal in each.

In order to ascertain the effect of large quantities of phosphorus in presence of silicon, mixtures of silico-phosphide of iron and iron in various proportions were melted in charcoal-lined partitioned crucibles. The temperature was very intense, certainly higher than that of the blast-furnace. The results were as follows:

	1	2	3	4	5
	Per Cent	Per Cent	Per Cent	Per Cent	Per Cent
Combined carbon	1.10	0.56	0.11	Nil.	Nil.
Graphite . .	2.62	1.73	1.88	1.69	0.83
Silicon . .	0.92	1.96	1.96	2.84	3.36
Manganese . .	traces	traces	traces	traces	traces
Phosphorus . .	0.21	4.95	6.85	8.35	12.86

It was observed that the size of the graphite plates mechanically suspended in the metals were largest in the most phosphoretic metal, and less so in those with lower percentages.

No. 1 metal had the grayest fracture; the others were less gray, the grayness decreasing with the increase of phosphorus.

The analyses prove that this decrease was not due to the presence of combined carbon, for the least proportion of that substance is found in the most phosphoretic sample; it therefore must be due to the decreasing proportion of graphite.

It is unfortunate that the silicon was not equal in each case for the sake of strict comparison; but for all that, the results are instructive, considering that the variable element silicon was not in greater proportion in the more phosphoretic metals Nos. 4 and 5 than is commonly present in gray furnace irons; and that even although not excessive, combined carbon was absent, and all the carbon capable of being absorbed was in the state of graphite.

Summary of Part II

1. That on melting saturated solid solutions of phosphide of iron in iron with carbon, the latter causes a separation of the phosphide near to the point of solidification, which appears in the solid metal as a eutectic in irregular-shaped areas, if the carbon present is small and in envelopes, increasing in thickness with the amount of carbon present, but is incapable of throwing the whole of the phosphide out of solution even when 3.5 per cent of carbon is present. A residuum always remains in solid solution. This residuum is smallest, however, when the carbon is at a maximum.

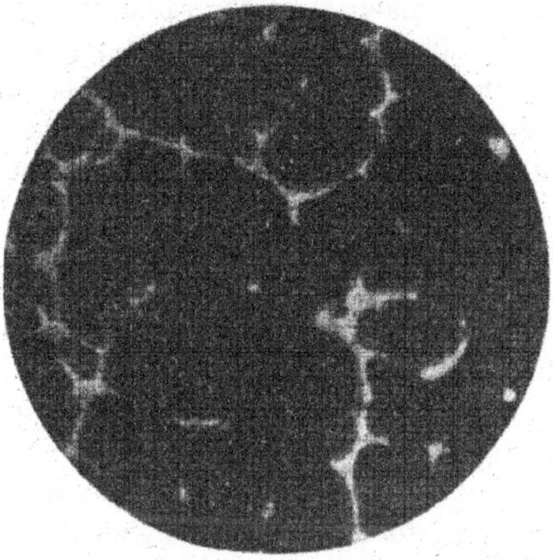

Fig. 21. The same as the last after melting in a magnesia crucible. This contained 1.74 per cent P in solid solution and 0.20 per cent in the free phosphide. Magnified 60 diameters.

2. The residuum appears to be retained in the pearlite, which, when much phosphorus is present, amounts to about 0.60 per cent when the mass has been very slowly cooled, and about 0.75 per cent when cooled more rapidly.

3. That a portion of the phosphide of iron in steels containing under 0.10 per cent of phosphorus is thrown out of solution by carbon when it exceeds 0.9 per cent, and the phosphide so separated is liable to form a brittle cell structure enveloping the grains, yielding a more or less fragile mass. In such cases,

however, the larger part of the phosphide remains in solid solution.

4. That when phosphoretic iron is carburized by the cementation process the carbon at once causes a separation of phosphide from the solid iron if it is saturated with phosphide and continues to cause it to separate until the residue contains 0.6 per cent of phosphorus and 1.3 per cent of carbon. If the iron contains about 0.6 per cent of phosphorus to commence with, no phosphide is thrown out of solution until the carbon is increased above 1.2 per cent, a very different result than is obtained by melting and solidifying material of the same composition.

5. That if highly phosphoretic steel containing much carbon is heated to above the melting-point of the phosphide eutectic in contact with some absorbent material, the eutectic simply liquates out of the mass and is retained in the surrounding substance; practically only that part of it in solid solution remaining behind.

6. That when the phosphide eutectic is expelled from the iron by the absorption of carbon in the cementation furnace, a large part of it liquates out of the mass and falls in the liquid state into the surrounding charcoal.

7. That when saturated solid solutions of phosphide in iron are heated or cooled, they show no thermal critical point at Ar_3, and the structure is not broken up even when the temperature exceeds 1000° C.; therefore there can be no allotropic change from the *Beta* to the *Gamma*-modification by heating to 1000° C. When, however, carbon passes into the mass, a large proportion of the phosphide is expelled, and the carbon absorbed converts, at the temperature of the cementation furnace, a dominating proportion of the *Beta*-iron into *Gamma*-iron, and coincidentally the structure is entirely reorganized and recrystallization is effected. The recrystallization can only proceed coincidentally with the penetration of the carbon from the exterior towards the center.

Under favorable conditions the slowly developing crystalline grains develop longitudinally into columns, which radiate to the middle of the bar.

A similar condition of crystalline development follows the change of allotropic state, but in the reverse direction, when carburetted steel is decarburized in ore or limestone at about 750°. At that temperature in presence of diffused carbide the carbon is in the *Gamma*-state, but as the carbon is removed, *Beta*-iron is formed and crystallization of the iron proceeds from the exterior to the center of the mass, leaving a most perfect columnar structure.

8. The proportion of phosphine, PH_3, gas given off on dissolving phosphoretic irons free from carbon is approximately inversely in proportion to the amount of phosphorus.

When 1.7 per cent is present, only about 4 per cent of it escapes as gas, whereas when there is only from 0.03 to 0.10 per cent, nearly 70 per cent of it passes off as PH_3.

9. When carbon is introduced into irons containing from 0.03 to 0.10 per cent of phosphorus, the proportion of phosphine liberated on solution of the metal steadily decreases with each increment of carbon, until, when the latter reaches about 1.2 per cent, only 15 per cent of the whole escapes as PH_3. This behavior leads to the conclusion that there are two solutions in steel above its solidifying point, a solution of carbide and a solution of phosphide. That the carbide solution is the dominant one, and that the solution of phosphide is concentrated more and more into that portion of the iron not dominated by the carbon. In this way is explained the behavior of acids. The more the solution of phosphide becomes concentrated by the carbon, the less proportion of it is capable of yielding PH_3 when it is treated with acid.

10. The amount of carbon capable of being absorbed in the blast-furnace and crucible by metal containing phosphorus depends upon the proportion of phosphide of iron present. Pure iron free from phosphorus will combine with nearly 5 per cent.

When the compound consists entirely of Fe_3P, no carbon will enter, and the amount capable of being absorbed in mixtures of iron and phosphides approximates inversely to the percentage of Fe_3P which may be present.

11. Phosphorus does not appear to cause the separation of graphite from metals high in carbon or conduce to the retention of that element in the combined state. Its action appears to be inert in that respect.

12. When iron containing 1.7 per cent of phosphorus and no microscopically visible free phosphide is dissolved in either sulphuric or hydrochloric acid, a dense black residue is left behind, insoluble, non-magnetic, and of very complex constitution. It is not a simple definite compound of iron and phosphorus,

but most probably a decomposition product of phosphide of iron. A similar substance is slowly formed when pure Fe_3P is acted upon by the same acids.

(*To be concluded.*)

ON THE CRYSTALLOGRAPHY OF IRON *

By F. OSMOND and G. CARTAUD

(*Concluded from page 149*)

Crystallization on Reheating

IF we have studied so closely the crystallography of iron, insisting upon details which will probably be of but little interest to professional mineralogists, it is in the hope that our researches may be of some value in the study of metallurgical products and of their treatments. After all, we are only following, at a distance, the traditions created by Prof. Tschernoff.

Among the questions dealing directly with the crystallography of iron, there are hardly any which are more important and more attractive than the one pertaining to the reheating and the accompanying transformations of the grains. It is one of Prof. Howe's questions of predilection.

We are taking up this subject here by two methods partly new or rather renewed. One of these methods, borrowed from Mr. Saniter, consists in etching the steel, at a certain desired temperature, with gaseous hydrochloric acid. The metal is heated and cooled in an atmosphere of hydrogen, the treatment being accompanied by a greater or less superficial decarburization. The only advantage of this method is to be found in the fact that the reagent may be introduced or withheld, at will, by an arrangement too simple to require description, and acting only at a temperature accurately regulated, neither above nor below.

The second method consists simply in reheating in hydrogen, at a certain known temperature and for a certain length of time, a polished section of the metal under examination. This method, applied by one of us as early as 1888,* had yielded only un-

* *Annales des Mines,* Vol. xiv, page 62.

satisfactory results, but this was due to the insufficiency of a micrographic technology which to-day appears very crude, and especially to imperfect polishing. The principle was not wrong.

ETCHING BY MEANS OF GASEOUS HYDROCHLORIC ACID. — The etching method was applied both incidentally and directly.

In some of our experiments, to cause the crystallization of iron by the Péligot method, a plate of nearly pure iron, polished on one side, was introduced in the tube, in front of the boat. A deposit of reduced metal was formed on the iron plate near the boat, but as the end of the tube is approached, the atmosphere becomes loaded during the reduction period, with hydrochloric acid, and as the reaction is reversible, a region was found beyond which the iron plate remained bare and had been etched.

In the experiments resulting in the production of *Alpha* and *Beta*-iron, the etching of an iron plate between 650° and 820° C. yields appearances identical to those obtained by etching the same sample in the cold, for a sufficient length of time, with dilute nitric acid or with some double chloride of copper and ammonium. These structures are also identical, therefore (at least qualitatively), to those resulting from the deposition of reduced iron in the same experiments, and which are shown in Figs. 21 and 22. It follows from these considerations that the known structure of ferrite at the ordinary temperature is also the structure, leaving aside the variations in the size of the grain, possessed by *Alpha* and *Beta*-iron during their whole period of stability.*

In order to ascertain the structure of a mass of *Gamma*-iron, a sample of very soft steel, containing 0.125 per cent of carbon, 0.03 per cent silicon, 0.025 per cent sulphur, 0.019 per cent of phosphorus and 0.31 per cent of manganese, was heated, after being polished, in a porcelain tube, through which a slow current of hydrogen was circulating. After three hours the maximum temperature of the Mermet furnace, about 1000°, was reached, and gaseous hydrochloric acid was then made to pass

* This would complete the proof, were it necessary, that the joints between the polyhedra, revealed by the action of acids on polished sections, indicate the limits of crystalline developments and not necessarily sections through tension surfaces. The existence of a tension network remains, however, quite possible, but there is no conclusive reason to suppose that its meshes would coincide with those of the joints network. Both networks might be of very different dimensions.

through the tube for five hours, at a speed of two or three bubbles per second. The supply of hydrochloric acid was then discontinued, and the metal allowed to cool with the furnace in hydrogen. Owing to the decarburizing action of hydrogen and the small initial amount of carbon, the etched surface of the metal must be nearly pure iron. It was found divided in polygonal grains with deep joints. Some grains exhibited parallel striations, that is indications of twinnings (Fig. 59, magnified 250 diam., vertical illumination). The appearance is in every way similar to the coatings formed on the tubes, at the same

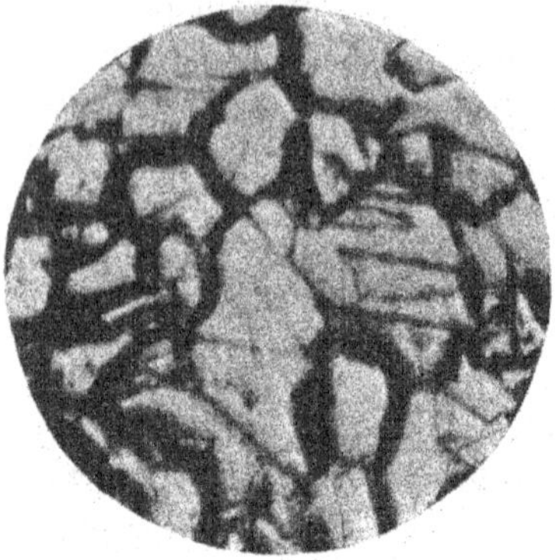

Fig. 59.

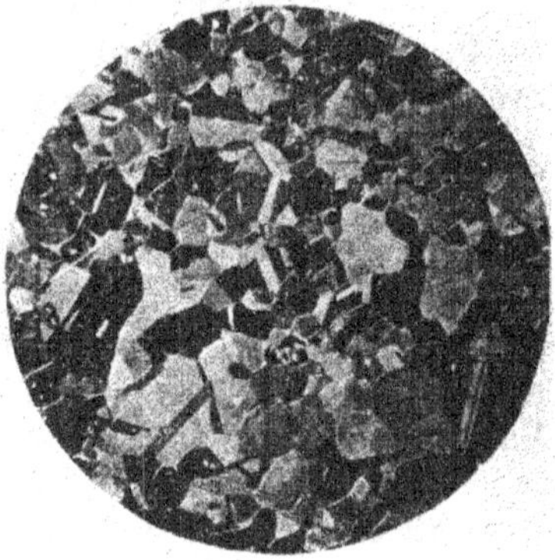

Fig. 60.

temperature, during the reduction of ferrous chloride (Fig. 58), and to that of soft iron etched with calcium chloride, according to Mr. Saniter's method.

To ascertain the influence of carbon, a sample of iron, carburized by the Demenge method, forged, and whose carbon content varied progressively from 1.60 to about 0.35 per cent, was polished and introduced in the tube of a Mermet furnace previously filled with hydrogen, and heated to 1000°. A current of hydrochloric acid was passed through the tube for five minutes, two or three bubbles per second. The hydrochloric acid was then replaced by hydrogen, and the heating continued ten minutes in order to expel the ferrous chloride, after which the furnace was allowed to cool.

Some portions of the preparation were found covered with a black coating, due probably to some carbon which was isolated during the operation, and which failed to be volatilized by the hydrogen. Many parts of the polished surface remained unsoiled, however, and could be examined.

THE CARBON CONTENT HAS PRACTICALLY NO INFLUENCE. — It is true that the surface was partly decarburized by the hydrogen, but in examining a cross-section etched by the ordinary methods, it was found that the decrease of carbon near the surface did not exceed about 0.20 per cent.

Fig. 60 (oblique illumination) shows the structure under the low magnification of 25 diameters. Abundant twin crystals will be noted, limited or not by the joints of the grains in which they occur. These twins which are here larger, for the same temperature, than those of very soft steel, are always characteristic of *Gamma*-iron crystallized in bulk and corresponding to what Mr. Saniter calls allotropic structure. This structure has been found repeatedly:

1. In the continuous coatings deposited on the tube by the reduction, at 1000°, of ferrous chloride in hydrogen.

2. In very soft steel etched at 1000° by gaseous hydrochloric acid.

3. In nickel-steel containing 25 per cent of nickel or more, at the ordinary temperature.

4. In non-magnetic manganese steel, also at the ordinary temperature.

For a thorough examination, however, higher magnifications are required.

At 1000 diameters, the matrix shows some small crystalline particles united to some striæ or cleavages which are uniformly oriented in the same region. As this crystallization becomes more pronounced, the apparance of the surface, amorphous at first, becomes gradually finely granular, and the coloration, under vertical light, varies accordingly from white to brown. Long needles are frequently found presenting the usual characteristics of martensite (Figs. 61 and 62, magnified 250 diam., and Fig. 63, magnified 1000 diam.).

The surface is divided into four networks as follows:

Network A. — This is a polygonal, regular network with small meshes, generally ill defined as if partly erased, and in most cases altogether absent. It appears to correspond, judging from the dimension of the meshes, to the initial structure of the metal before heating, indicating the traces of a past condition.

Network B. — This network is fine and sharp, very irregular, and twisted and frequently interrupted. It may be seen in Figs. 61 and 62. A careful examination of this network in all parts of the preparation shows that it has a tendency to dis-

Fig. 61.

Fig. 62.

appear in the region which contains no martensite, and that it frequently marks the limit of the needles groups, separating them from each other and from the surrounding mass. It is not uncommon, however, to find this network running at random through the martensite.

Network C. — This network is made up of parts of straight lines, and is therefore crystalline. These lines are not furrows, but merely the limits of different crystalline orientations. (See Figs. 64 and 63.) Very frequently two parallel lines indicate the limits of a crystalline band having a constant orientation, that is of twins.

The network *C* does not generally coincide with the martensite. Martensite is not found in those regions where the crystals are well developed, and inversely, the crystals are absent in the parts where the martensite is well developed. Some regions of transition exist, however, in which martensite and the crystals are to a certain extent found together (Fig. 64). Upon close examination of these parts, it is found that the needles of the martensite have frequently the same direction as the crystalline bands close by, which leads to the conclusion that martensite (acicular structure of Saniter) is a preparatory step towards

Fig. 63.

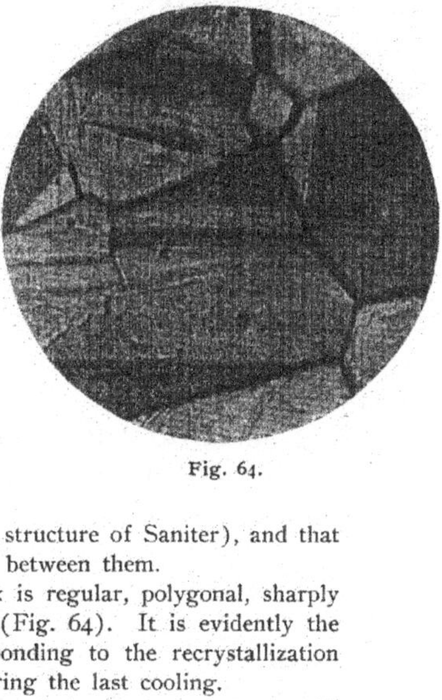

Fig. 64.

perfect crystallisation (allotropic structure of Saniter), and that there is no absolute demarcation between them.

Network D.—This network is regular, polygonal, sharply outlined, and has large meshes (Fig. 64). It is evidently the actual network, the one corresponding to the recrystallization under the *Beta-Gamma* state during the last cooling.

The relation existing between these four networks is still to be ascertained, but it is preferable to describe first the results of annealing experiments of previously polished surfaces, not subjected to any etching treatment. Both methods are, moreover, practically equivalent.

Reheating of Polished Surfaces. — A piece of boiler-plate, of unknown composition, was reheated in hydrogen, for two hours, at 900°. The previously polished surface exhibits a double network, one of which, less sharp, is the network which existed before heating, and the other the actual network (Fig. 65, 250 diam., vertical light). Small parallel striæ indicate the beginning of twin crystals whose limits are frequently independent of the networks. This result confirms that obtained by Mr. Saniter with pure iron etched at the same temperature (900°) with molten calcium chloride. It was not, therefore, necessary to attribute, as we did in the first part of this paper (*The Metallographist,* Vol. III, p. 181), the production of twins to the preliminary reheating at a high temperature to which Mr. Saniter had subjected his samples; such interpretation remains possible, but it is seen that twin crystals begin to appear at 900°.

Another sample of very soft steel, containing 0.125 per cent of carbon, 0.03 per cent of silicon, 0.025 per cent of sulphur, 0.019 per cent of phosphorus, and 0.31 per cent of manganese, was reheated for two hours in a Leclercq and Forquignon furnace, in a tube through which a slow current of hydrogen was made to circulate, and at a temperature estimated to be 1150° or 1200° C. A previously polished surface assumed an appearance very similar to that of the steel-plate with a varying carbon content, described above, and which was etched at 1000° with gaseous hydrochloric acid. The forms of martensite, however, are vague or absent, and the crystalline structure is on the whole more developed.

Here again we find the four networks already described:

1. *Network A.* — It is the network corresponding to the structure before heating. It has been fairly well preserved in certain portions, and may be easily detected in the light parts of Fig. 66 (150 diam., vertical illumination).

2. *Network B.* — This network is fine, distorted and discontinued. It may be detected in some portions of Fig. 67 (127 diam., vertical illumination), and we naturally identify it with a transitory passage through the martensite condition, according to the results obtained with a more carburetted sample in which martensite had been preserved.

3. *Network C.* — This network is crystalline, and is made apparent through the changes in coloration and in crystalline

orientation in Figs. 68, 67, 66, and 69. Figs. 68 and 67 correspond to the same regions, seen respectively under oblique and vertical illumination; although Fig. 67 was by mistake enlarged

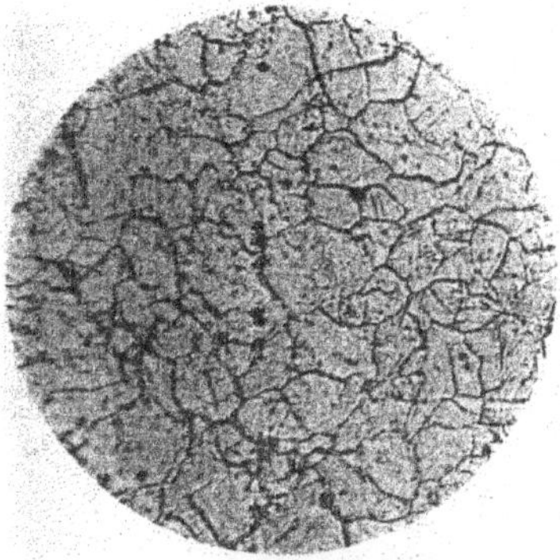

Fig. 65.

Fig. 66.

Fig. 67.

Fig. 68.

127 diam., while Fig. 68 was magnified 100 diam., and although the centers do not coincide exactly, the parts common to both will be easily detected. The light regions of the photographs

appear brown by vertical light, and dark blue by oblique light, illumination, and bluish by oblique light, while the dark regions appear grown by vertical light, and dark blue by oblique light, owing to the different levels resulting from the crystallization. This crystallization results in the formation of small crystalline particles, and in striæ variously oriented from one crystal to the next (see Fig. 69, 1000 diam., vertical illumination). When they have not been interfered with by surrounding developments, the external forms of these crystals are straight. A group of bands will be noted in Figs. 68 and 67, respectively parallel to the three

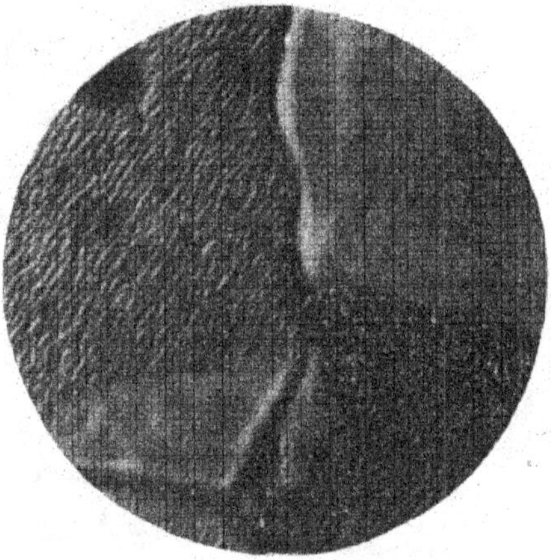

Fig. 69.

sides of an equilateral triangle, and indicating the existence of two twinned individuals. The form of the equilateral triangle frequently found in these preparations, suggests that the section is, in the region considered, parallel to a face of the octahedron. These figures recall, on a small scale, those of Widmannstaetten.

4. *Network D.*— This is the actual network. It is polygonal and made up of wide furrows, detected at once in Figs. 67 and 69.

The preparation was sawed, as shown in Fig. 70, to a short distance from the polished side *AB*. The remaining metal was then broken by bending to ascertain what network the fracture would follow. Owing to the great softness of the metal, however, the distortions, in spite of the small thickness of the broken section, were still considerable, and the experiment did not yield any conclusive result.

One of the broken halves was then etched for forty minutes with warm sulphuric acid diluted with four volumes of water. The etching was very slow, owing to the purification by hydrogen of a steel already quite pure before reheating. As this etching proceeds, the network *A* disappears, the networks *B* and *C* become fainter, while the network *D* becomes more pronounced, thus confirming our interpretation. The network *D* is undoubtedly the network of the actual joints.

A small bar of the steel containing varying amounts of carbon, previously polished, was heated under the same conditions as the bar of very soft steel. The results were qualitatively the same as those yielded by the etching of the same sample of steel with gaseous hydrochloric acid at 100°. They have been described elsewhere.

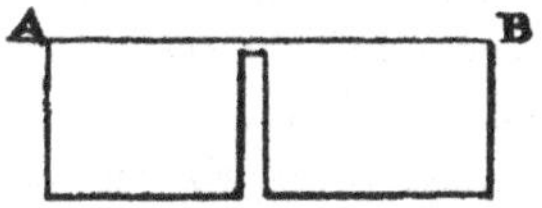

Fig. 70.

INTERPRETATION OF THE EXPERIMENTAL FACTS. — The polished surface of a steel-plate heated in hydrogen is like a sensitive plate upon which are found recorded all the variations of structures that took place. Upon developing this plate, that is, when the final preparation is examined, the successive images, once very distinct, are superposed in space, resulting in a complex image which it is not always easy to decipher.

We may, it seems, suggest the following explanation:

Heating between the ordinary temperature and the point Ac_3 causes the network *A* to appear. This network is then the actual network of *Alpha* or *Beta*-iron, and it remains visible, although blurred, after heating to a higher temperature.

When the point Ac_3 has been exceeded, the iron passes to the *Gamma* state and assumes the condition of martensite. The dimensions of the structural elements of martensite increase with the temperature and with the time, and its various transitory arrangements leave the network *B* whose distorted and interrupted outlines indicate the series of successive joints between the growing grains of martensite.

Martensite, however, is only a transitory state. At a sufficiently high temperature, or after a sufficiently long time, it assumes the shape of the true crystals (network *C*), with a tendency towards their natural forms, which is a development of the previous forms of martensite. Martensite is the crystallitic condition of *Gamma*-iron: it has a tendency to disappear in pure iron, while the presence of a little carbon favors its retention. It is an individual case of a general law: carbon plays the same part in the crystallization of iron as gum in the crystallization of saline solution: it interferes with and retards the formation of true crystals; it hastens and facilitates the formation of crystallites, not so much in opposing organisation under a crystalline condition, as in causing this organisation to take place too suddenly.

Finally during the cooling following the heating, iron assumes again the *Beta* state when the point Ar_3 is reached. When this temperature is reached the networks *B* and *C*, although they may leave some traces, no longer correspond to the actual condition, and the *Beta*-iron takes the shape of pseudo-crystalline polyhedric grains (network *D*). The transformation of *beta* into *Alpha*-iron does not modify the crystallization, since both varieties are isomorphous.

We have now to ascertain the relations which may exist between the last two networks, *C* and *D*.

It is the network *C*, itself an evolution of the network *B*, which marks the limit of the growth of the grains upon reheating above the point A_3. It would seem, therefore, as if the brittleness of steel after reheating to a sufficiently high temperature and the corresponding size of the grains, must be attributed to that network *C*. The micrographic examination, however, shows us that the joints of this network are simply the limits of various crystalline orientations, without real thickness, and which are not enlarged by an acid etching. This network *C*, therefore, does not exhibit the characteristics of a network of least resistance, and, moreover, it is no longer the actual network of the cooled metal.

It is the actual network *D*, sharply outlined and easily dug out by sulphuric acid, which appears to be the network imparting weakness to the metal.

Here, however, we meet with a difficulty: Why is the size

of the network *D* the same as that of the network *C*? Since it is always formed at the same temperature, how can it be influenced by what took place at a higher temperature, seeing that the passage through the point A_3 causes an independent rearrangement of the previously existing states?

The first explanation which suggests itself is that *Beta* and *Gamma*-iron crystallize with the same axes. This assumption, however, presents some difficulties. If it were true, the networks *C* and *D* should coincide exactly, while, as a matter of fact, if they occasionally have some common parts, they also cross each other as shown so clearly in Fig. 69.

Some of the observed facts, therefore, appear to indicate the independence, and others, the inter-dependence of the networks *C* and *D*.

These opposite indications might be explained as follows:

The crystallization under the *Gamma* state causes the formation of crystals of certain dimensions, which dimensions are a function of the temperature and the time. In the same crystal the molecular orientation is constant. During subsequent cooling, when *Gamma*-iron changes to *Beta*-iron, the accompanying change of axes, whatever it may be, will have a tendency to be the same wherever the molecular orientation was the same, i.e., there will exist a tendency to replace each crystal of *Gamma*-iron by a crystal of *Beta*-iron. Each of these molecular varieties, however, has its individual forms. *Gamma*-iron includes pieces of true crystals with numerous twinnings, while *Beta*-iron takes the shape of pseudo-crystals without twinnings, approaching, through mutual interference, the pentagonal dodecahedron. *Beta*-iron cannot, therefore, follow exactly the preëxisting forms of *Gamma*-iron; it accepts them as much as possible, but rectifies them according to its own crystalline requirements.

To sum up, this leads us to the notion of a limited dependence between the networks *C* and *D*, and if this notion be correct, the reheating operations will, hereafter, have to be guided accordingly.

General Conclusions

In the left column will be found the temporary conclusions reached by Mr. Osmond through the interpretation and the discussion of previously known facts, as stated in the *Annales des*

Mines for January, 1900.* In the other column, we have summed up the conclusions based upon the researches just described. They are perhaps of a less temporary character, but differ little from the first ones.

Iron, in its three states, crystallizes in the cubic system.	Iron, in its three states, crystallizes in the cubic system.
Gamma-iron, generally in octahedra, more or less perfect.	*Gamma*-iron assumes all the combinations of the cube and the octahedron; never any modifications on the edges; frequent twinnings with a_1 as junction face. In commercial iron, more or less carburetted, the crystallization is preceded by a more or less extended crystallitic phase.
Beta-iron and *Alpha*-iron in cubes. *Beta*-iron and *Alpha*-iron form isomorphous mixtures.	*Beta*-iron and *Alpha*-iron crystallize identically, which is a necessary condition, if not a sufficient one, of their isomorphism, which was, moreover, suggested by the progressive character of their mutual transformation. The cube is the crystalline form, with frequent modifications, probably b_1 or b_2, on the edges. No modifications on the angles.
Gamma-iron does not form isomorphous mixtures with *Beta*-iron.	*Gamma*-iron does not form isomorphous mixtures with *Beta*-iron.
The transformation of *Gamma*-iron into *Beta*-iron appears to include a change in the planes of symmetry, at least in the carburetted irons.	The transformation of *Gamma*-iron into *Beta*-iron appears to include a change in the planes of symmetry, at least in the carburetted irons. The size of the grain, however, acquired under a certain state, is not without influence upon the size of the grain acquired under the following state.

The crystalline forms which have been detected make possible the existence of cleavages p and a_1 in *Gamma*-iron, and of

* *The Metallographist*, Vol. iii, page 287, October, 1900.

cleavages p and b_1 in *Beta* and *Alpha*-iron. Our experiments, however, do not show this conclusively.

Summing up, it is seen that the crystallographic differences which divide *Gamma*-iron from *Beta* and *Alpha*-iron are such as are generally met with in the history of substances for which there exists no reason to suspect allotropic or isomeric modifications. If the allotropy of iron were not conclusively proved by other evidences, it would not be revealed by its crystallography. This suggests the thought that the signification of crystallographic differences, considered to-day as having no bearing upon the molecular state of substances, may, in certain cases, be misunderstood. On the other hand, it is not impossible that the differences of this nature which we have detected in the different varieties of iron, will some day be erased by the results of other experiments.

Appendix I.

Martensite and Austenite

The name of martensite was first applied to hardened steel considered as a mineralogical entity. Later, through a natural extension, it was also used to designate the condition of steel above the point A_3, admitting implicitly that the crystalline forms acquired at a high temperature were retained by quenching. Our experiments dealing with the reheating of polished surfaces have shown such extension to be justified, since it has been possible to cause the appearance on these sections of the same forms which characterize martensite in hardened steel.

These forms represent the crystallitic state of *Gamma*-iron, and since *Gamma*-iron holds the carbon in solution, there is no reason to suppose that martensite, above the point A_3, is not homogeneous. It does not follow; however, that it remains homogeneous, after cooling, even if we give to the word homogeneous the restricted meaning of *not resolved by the microscope.* Strictly speaking, it is evident that martensite is not homogeneous, since it is magnetic, which indicates the return of the iron to the *Alpha* condition, at least in part. The question has been asked, notably by Mr. Sauveur, whether it could not be resolved by the microscope into two constituents. It is, indeed, possible that the transformations do not take place with the same speed in all parts of the mass, starting, on the contrary, from certain points or certain lines, and being arrested when the corresponding change of volume has become sufficient. This question remains undecided.

As a matter of fact, martensite does not represent the *final* structure of *Gamma*-iron; it is confined to certain definite conditions of temperature and carbon content. The final structure, i.e., the structure towards which *Gamma*-iron tends, without being always able to assume it, is the crystalline structure, and this crystalline structure, when it is possible to retain it intact by quenching, as in the case of certain manganese or nickel steel and of hypereutectic carbon steel quenched in ice-water at about 1100° C., is represented by austenite. In the case of carbon steels, however, it has not yet been possible to obtain austenite in a pure condition; the largest proportion that it was possible to retain in the metal does not exceed 70 per cent, the balance being martensite, or ***hardenite,*** if we retain this term, proposed by Mr. Howe, to designate *saturated* martensite.

These mixtures of hardenite and austenite present a difficult problem. If it be assumed that the mixture existed previously to the quenching, it would follow that a hypereutectic steel containing, say, 1.60 per cent of carbon, would not be at 1100° a homogeneous solution, which is opposed by the appearance of the cooling curves and by the very definition of solution. If, on the contrary, the breaking up into constituents takes place during the quenching, this phenomenon occurring in a solution previously homogeneous cannot be readily explained.

Our experiments lead us to propose the following explanation:

Let Fig. 71 be a graphical representation of a mixture of austenite and hardenite. The dotted lines represent the axes of the sheets of hardenite, and these axes, as is often the case, are parallel to two rectangular directions or to their bisectrix.*

From what we know about the crystallography of *Gamma*-iron, such a disposition represents probably a face of a cube cut by the cleavages of an octahedron.

Previous to quenching, say at about 1100° C., the solid solu-

---

* These axes of hardenite are not imaginary; every leaf is actually divided into two parts by a median line.

tion is homogeneous, but the most favorable carbon content, about 1.80 per cent,* is not in itself sufficient to retain, through the osmotic pressure which it creates, the totality of the iron in the *gamma* state, however severe the quenching. The transformations begin, therefore, but they do not begin at random. They start from the cleavage planes p or a_1, because these cleavage planes correspond to the maximum molecular separation, and they proceed in a parallel direction until the increase of volume resulting from the partial transformation of *Gamma*-iron into *Beta*-iron has added to the osmotic pressure of the dissolved carbon a sufficient mechanical pressure to prevent the transformation of *Gamma*-iron in the balance of the mass. Hardenite represents iron partially transformed, and austenite, iron not transformed.

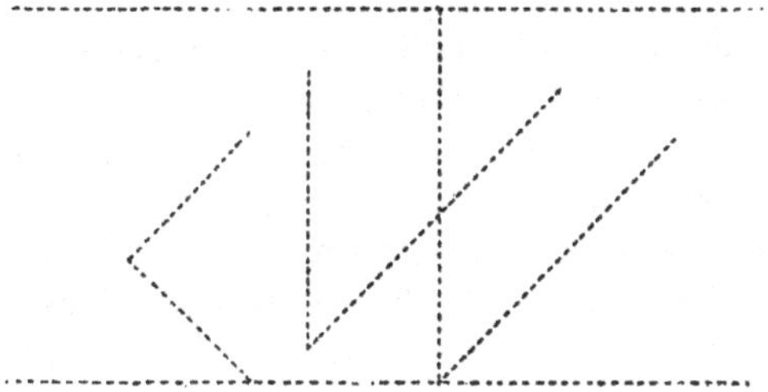

Fig. 71.

If now the carbon content is lowered below 1.80 per cent, the osmotic pressure decreases correspondingly; a greater mechanical pressure will be required, therefore, to retain a portion of *Gamma*-iron in a pure condition, that is the proportion of hardenite in the mixture will increase, and the complementary proportion of austenite will decrease correspondingly.

Finally below, say, 1.00 per cent of carbon, no austenite can be retained, which signifies that the osmotic pressure of the dissolved carbon and the mechanical pressure resulting from the transformation of austenite in hardenite are no longer sufficient to retain *Gamma*-iron, in a pure condition, in some parts of the mass.

* There is no interest in carrying higher the percentage of total carbon, because above 1.80 per cent of carbon some cementite segregates, no matter how rapid the cooling, and the proportion of dissolved carbon, therefore, does not increase.

APPENDIX II.

In one of our experiments, dealing with the crystallization of *Alpha*-iron by the Péligot method, we had introduced beforehand in the tube and in front of the boat containing the ferrous chloride, an electro-magnet made of a core of nearly pure iron, polished on one side, and of a solenoid of iron wire previously heated in hydrogen. The solenoid was isolated from the core by a porcelain support, and a current of two amperes was made to pass through it.

We wanted to ascertain where the deposit of reduced iron would take place, and whether the structure would be modified. No modification of structure was detected.

Some iron was found deposited, as a coating, with the characteristics of *Alpha*-iron, around the two poles of the core, and in the shapes of crystalline festoons between the two or three first spirals, and on the tube, bearing, however, no relation to the lines of force. No iron was found on the neutral part of the electro-magnet.

The formation of a deposit on the tube, a few millimeters from one of the poles of the electro-magnet, seems to prove that iron at the instant it leaves the molecule $FeCl_2$, is not magnetic, although iron as a mass is magnetic at that temperature. The union of two or more non-magnetic atoms is apparently required to form a magnetic molecule.

THE MICROSTRUCTURE OF HIGH CARBON STEEL QUENCHED NEAR ITS MELTING POINT

By ALBERT SAUVEUR

IN the July 1900 issue of *The Metallographist* some photomicrographs, taken by Mr. F. C. Lau, were reproduced showing the structure of a bar of steel containing 2.40 per cent of carbon, and quenched when the end of the bar was beginning to melt.

One of these photographs, magnified 1000 diameters, is again shown in Fig. 1. In describing this structure it was stated that the needles were martensite, and the large white areas, cementite. Prof. J. O. Arnold, in a letter published in the October 1900 issue of *The Metallographist,* took exception to these conclusions, arguing that the needles as well as the bright irregular areas were cementite. Mr. Osmond, in a private communication to the writer, expressed likewise his belief that the needles were cementite, and stated besides that the white areas were not pure cementite, but a eutectic mixture of cementite and a solid solution (martensite).

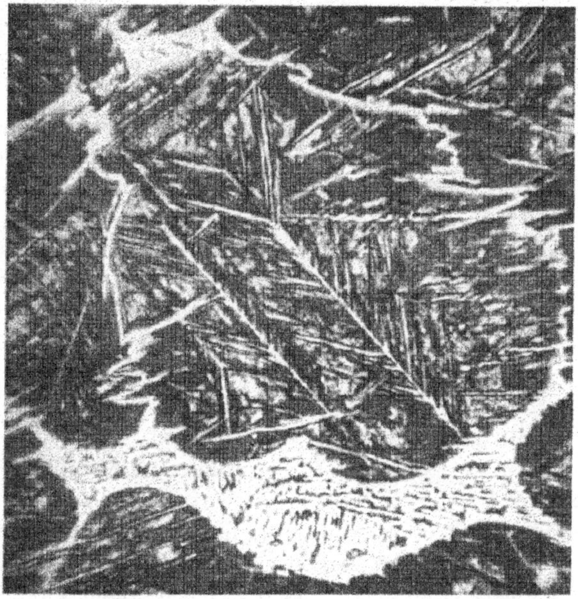

Fig. 1.

In a letter published elsewhere in the present number of this journal Mr. Lau maintains his contention that the needles are martensite, and the white areas cementite.

Let us see if we cannot throw some additional light upor this controversy by recalling what occurs when a steel of that carbon content (2.40 per cent) reaches its solidification-point.

Let us consider Prof. Bakhuis Roozeboom's cooling curves* as they are probably the most accurate of all the curves so far constructed.

It will be seen that iron, containing 2.40 per cent of carbon, begins to solidify a little below 1300° C., and that the solidification consists in the formation of a solid solution containing about 1.25 per cent of carbon. The amount of carbon in the portion remaining liquid is thereby increased, and the freezing-point of the mass correspondingly lowered. The solidification of the steel does not, therefore, take place at a constant temperature, but proceeds as the temperature is lowered from 1300° to about 1130°. The solid solution resulting from the solidification contains an increasing amount of carbon. The solid solution formed at the beginning of the solidification contains 1.25 per cent of carbon, while the solid solution formed at 1130° contains about 2 per cent of carbon. At 1130° the carbon content of the portion still remaining liquid has reached 4.3 per cent, that is the molten mass has now the composition of the eutectic alloy, whose two components are carbon and a solid solution containing 2 per cent of carbon. Like all eutectic mixtures, it will solidify at a constant temperature (1130°), giving rise to the formation of a mechanical mixture made up alternately of particles or plates of graphite, and particles of the solid solution.

The relative proportions of the solid solution and of the eutectic mixture which should be present, after solidification, in a steel containing 2.40 per cent of carbon, may be readily calculated.

We have seen that the carbon content of the solid solution varies continuously from about 1.25 per cent to about 2 per cent. The average carbon content of the solid solution is therefore 1.62 per cent, while the eutectic alloy contains 4.3 per cent of carbon.

Let x represent the percentage of solid solution in the steel after solidification, and y the percentage of eutectic alloy.

We have

$$x + y = 100 \quad (1)$$

And since the 2.40 per cent of carbon present in the steel are divided between the solid solution and the eutectic alloy in

* *The Metallographist,* Vol. iii, page 295, October, 1900.

the respective proportions of 1.62 per cent and 4.3 per cent, we have the second equation

$$\frac{1.62}{100}x + \frac{4.3}{100}y = 2.40 \quad (2)$$

The resolution of these two equations indicate that the steel under consideration should make up, after solidification, of 71 per cent of solid solution and 29 per cent of the eutectic alloy.

If the metal were now allowed to cool slowly to the atmospheric temperature, it would be finally made up of 73 per cent of pearlite, and of 27 per cent of cementite. Unless the heating had been prolonged, or the cooling extremely slow, it would not contain any graphite. If graphite was formed, therefore, during solidification, as indicated by Prof. Roozeboom's diagram, it must have been converted into cementite on subsequent slow cooling. It is quite probable, however, that the graphite eutectic just described will only be formed (in steel of that carbon content) on extremely slow cooling seldom realized in practice. The eutectic compound, formed under normal conditions at 1130°, is probably made up of *cementite* and solid solution.

The pearlite present in the slowly cooled steel is the final form assumed by the solid solution, and since pearlite contains only about 0.80 per cent of carbon, it is evident that the solid solution which contained initially 1.62 per cent of carbon, and the solid solution forming a component of the eutectic alloy and containing 2 per cent of carbon, must have rejected, during slow cooling, their excess of carbon in the form of cementite. This is supposed to occur in the zone SEE^1F_1K of Prof. Roozeboom's diagram.

The solid solution of carbon and iron is generally called martensite, but it would seem more rational, and it would avoid confusion, if the solid solution *resulting from the solidification* of the metal were called austenite, and the name martensite reserved for the solid solution after it had rejected its excess of carbon, that is, for the constituent found in steel after the quenching operation as carried on in practice.

It would seem, moreover, as if the transformation of martensite (solid solution containing 0.80 per cent of carbon or less) into pearlite, during the recalescence, were not an abrupt, but a gradual transformation, the martensite passing through some transitive forms, called by Osmond troostite and sorbite, before assuming the final pearlite condition.

To return to the composition of a steel containing 2.40 per cent of carbon after solidification; we have seen that it is theoretically made up of about 71 per cent of a solid solution containing on an average 1.62 per cent of carbon, which we shall call austenite, and of 29 per cent of a eutectic mixture whose theoretical constituents should be graphite and a solid solution with 2 per cent of carbon, but in which the first constituent is probably cementite instead of graphite. It can be readily calculated that this eutectic mixture should be made of about equal parts of cementite and solid solution, so that the steel after solidification would contain about 14 per cent of cementite.

If it were possible to retain by sudden cooling these constituents unaltered, such would also be the structural composition of the quenched metal. It would contain some 14 per cent of cementite associated "eutectically" with an equal amount of a solid solution holding 2 per cent of carbon, and the balance of the metal would consist of a solid solution (austenite) with an *average* carbon content of 1.62 per cent.

It is not, however, possible to prevent at least a partial transformation of the solid solution (austenite), that is, to prevent it from rejecting some of its carbon as cementite, that is, from approaching, if not from assuming the composition of martensite. We should, therefore, expect the quenched metal to contain more than 14 per cent of cementite, but less than 27 per cent (the amount which would be present after slow cooling). With our present knowledge of the phenomena it is not possible to state the amount of carbon which the solid solution should be able to retain under the stated quenching conditions. It must be somewhere between 1.62 (average content) and 0.80 per cent.

The 2 per cent solid solution forming a component of the eutectic mixture should also reject some of its carbon as cementite, modifying correspondingly the structural composition of that mixture.

Returning now to Mr. Lau's photograph, the appearances of the white areas certainly support Mr. Osmond's contention that they are not pure cementite, but a eutectic alloy of cementite and a solid solution, for they clearly exhibit the characteristics of all eutectic alloys. Moreover, such inference agrees with the indications of the cooling curves to which I have just referred.

With regard to the needles, if they are cementite, the proportion of cementite will seem excessive, greater than would have been expected from the considerations just outlined. If only the dark constituent corresponds to the solid solution (austenite-martensite), then that constituent is here quite structureless. The composition of the solid solution retained by quenching so near the melting-point, moreover, should approach the composition of austenite more than that of martensite.

METALLOGRAPHIC NOTES

The Value of Research. — *Extract from the Presidential Address of William Whitwell, Iron and Steel Institute, May* 1901 *Meeting.* — One of the leading features in the work of the Institute in the past lies in the fact that it has been the medium whereby much valuable research work has been placed at the disposal of its members and the whole metallurgical world. Up to the present we have not been able to offer any direct encouragement to workers in this line, but owing to the munificence of Mr. Andrew Carnegie we have now at our disposal a sum of £6500 for the founding of scholarships and medals for research work in the development of iron and steel manufacture. It is to be hoped that such prizes will encourage those who are already working, and induce others to join the same army, with considerable benefit ultimately to our industry. As to the value of research in the solution of metallurgical problems, and the improvement of our methods of manufacture, there can be no question. If we glance through the records of our Institute, we can find many examples that bear this out. Take for instance the basic process. The way to success was undoubtedly pointed out by the teachings of Percy and Grüner, followed up by the research work on dephosphorisation of Snelus, Edward Riley, Jordan of Paris, and Stead. Thomas himself admitted that these investigations had been of the greatest assistance to him. The Saniter desulphurising process, by which thousands of tons of good steel have been made from pig iron quite unfit if used without the desulphuriser, was founded on pure laboratory experiment conducted by Saniter himself. The research of Stead into the effect of arsenic on steel, proving this element not so pernicious as was at one time thought, has opened up the use of ores which were previously condemned on account of their arsenic contents. The investigations on the heat treatment of steel by Brinell, Tschernoff, Osmond, Howe, Sauveur, Roberts-Austen, Stead, Arnold, and Campion have given us the correct temperatures and treatment which steel must undergo to produce the best structure with the best mechanical tests. In the *Iron Age* for December 20, 1900, there was given an account of some radical changes in the arrangements of the Edgar-Thomson Steelworks of the Carnegie Company, carried out by Kennedy and Morrison to introduce a new method of rolling rails founded on the facts brought to light by the researches on heat treatment. The work of Hadfield on the effect of manganese produced that wonderfully useful material known as manganese steel. Many more illustrations of the value of research work in the preparation of new alloys of iron might be given. These examples, however, fully illustrate the value of research, and in the future developments of the manufacture of iron and steel, research will of necessity play an important part.

The Andrew Carnegie Research Scholarship. — It was announced at the May meeting of the Iron and Steel Institute that Mr. Carnegie had decided to double the amount of his gift, the interest of which is to be applied to original metallurgical research.

The council of the Iron and Steel Institute announced that the following candidates had each been awarded an Andrew Carnegie Research Scholarship of $500:

Alfred Stansfield, D.Sc., instructor in assaying at the Royal College of Science; John Alexander Mathews, Ph.D., of Brooklyn, U.S., and Julius Goldberg of Leoben, Austria.

The National Standardizing Bureau. — Director Stratton said to the correspondent of *The Iron Age* before leaving for Europe:

"I shall go first to England, and will there inspect the British National Physical Laboratory, the Standardizing Laboratory of the Board of Trade, the physical laboratories of the most important educational institutions and Kew Observatory. In France I will, of course, visit the International Bureau of Weights and Measures which has charge of the distribution of the standards of the metric system throughout the world, and will also inspect the principal physical and technical laboratories. In Germany, I shall devote a good deal of time to the famous Reichsanstalt and to the Normal-Aichungs-Kommission, which is the equivalent of our Bureau of Weights and Measures. I will also inspect the important Astro-Physical Laboratory at Potsdam, and the physical laboratories at several of the leading universities. In all these countries I will give special attention to the equipment and methods of testing laboratories.

"An important feature of my work abroad will be to ascertain the methods employed in the several countries visited to make the work of standardizing valuable to the great iron and steel industry. In Germany, especially, the Reichsanstalt has performed work of the greatest value to the iron and steel manufacturers, and we wish to learn just how this has been done with the hope of improving upon it if possible. As an illustration of the value of this great institution in Germany the fact might be cited that recently at the famous Krupp works a problem was encountered so perplexing that it was referred to the Reichsanstalt, which, instead of undertaking to develop a theory at long range, promptly sent its most expert specialist to the Krupp works with instructions to remain there until the difficulty was solved. It hardly need be said that such service is invaluable, but I trust that in a few years we may be in position to do as much for American manufacturers.

"The subject of high temperatures is a most important one in steel making and is one to which I shall give a great deal of attention while abroad. It is my expectation that when our scientific force is organized one of our most competent men will be a specialist in pyrometry. I hope to be able to bring back authentic information with regard to a number of subjects upon which we have more or less incomplete information, as, for example, the matter of the uniformity of screw pitch, which has been the subject of a good deal of agitation in Germany. It will not be the purpose of our bureau to suggest standards for screw threads, but simply to ascertain what the world at large is doing, and in case our manufacturers should adopt a uniform standard to assist them in promulgating and preserving standards, in distributing gauges, templates, etc., and in comparing them from time to time with the official standards in the custody of the bureau.

" Modern laboratories at big steel works are so complete in all details that these institutions will be interested in almost every branch of our work in addition to that which relates especially to manufacturing processes. We will be in position to standardize and calibrate all the philosophical apparatus of a large chemical laboratory, to standardize the weights of testing machines, to test weights for all kinds of scales, to standardize rules, tape lines, etc., and in a general way to assist in spreading scientific uniformity through every department of the great iron and steel industry. This, of course, is an ambitious programme, and cannot be realized at once, but it will be the mark toward which the bureau will constantly strive."

The Secretary of the Treasury within a few weeks will appoint a visiting committee to consult with the Director of the new bureau on important matters of equipment, practice, etc. The iron and steel industry will be represented by a prominent chemist attached to one of the leading steel works of the country, and an equally eminent electrical expert will represent that industry. In this way the institution will enjoy the advantages to be derived from the advice of men who not only possess high attainments in the scientific world, but who in addition are thoroughly practical. — *Iron Age,* April 25, 1901.

Distinctions for Metallographists. — The honorary degree of Doctor of Science has been conferred by the Victoria University upon Sir William Roberts-Austen, K.C.B. Mr. F. Osmond has been awarded the Vaillant Prize by the Paris Academy of Sciences for his researches on metals and alloys. The Swedish Technological Society has awarded the Polhem Prize to Mr. J. A. Binell. Prof. H. M. Howe has been created a Knight of the Legion of Honor.

Metallography at the Technical Schools. — *Harvard University.* — Mr. Albert Sauveur has been appointed Assistant

Professor in Metallurgy and Metallography. A new course in general metallography will be added to the curriculum of the Mining Department. It is described as follows: Metallurgy 14; Metallography; Lectures, laboratory work and reading.

The lectures of this course include a description of the operations required to make the structure of metals apparent under the microscope, and of the examination and photographing of the revealed structure. The microstructure of industrial metals and alloy will be fully described and illustrated, and the influence of chemical composition and treatment (both thermal and mechanical) upon the structure carefully considered, as well as the close relation existing between the structure and the physical properties.

The modern theory of metallic alloys, based in part upon their microstructure, will receive proper attention.

A practical knowledge of the technology of the subject will be imparted to the student by a series of laboratory experiments. Samples of the most important industrial metals and alloys will be polished, etched and otherwise prepared for microscopical examination and the resulting structures photographed. The influence of chemical composition and treatment upon the structure and the relation between the structure and the physical properties will be ascertained by properly selected experiments, which will include the use of the Le Chatelier thermo-electric pyrometer, the determination of the thermal critical points of iron and steel, etc.

The student should acquire in this course a good working knowledge of metallographic methods and manipulation.

CORRESPONDENCE

To the Editor of *The Metallographist*.

Sir: — Referring to Professor Arnold's letter, which appeared in the October 1900 number, and his discussion relative to the structure of steel containing 2.4 per cent of carbon quenched from its melting-point, a photograph of which appeared in the July number of *The Metallographist*, the steel experimented upon contained the following impurities: phosphorus 0.010 per cent, sulphur 0.014 per cent, manganese 0.15 per cent, silicon 0.081 per cent.

When I first examined the structure I concluded that the needles were striations of cementite, but upon a closer inspection it was found that the two structures did not stand on the same level, the cementite standing slightly in relief.

After a photograph had been taken, an oxidizing solution was made up of chloride of iron, chloride of antimony, and gallic acid, the polished surface was painted with the solution, the sample was then laid on its side to dry, the excess of solution draining off. The section was again examined, and it was found that the cementite had not been affected, or oxidized, remaining bright, in fact the method revealed minute membranes of cementite which

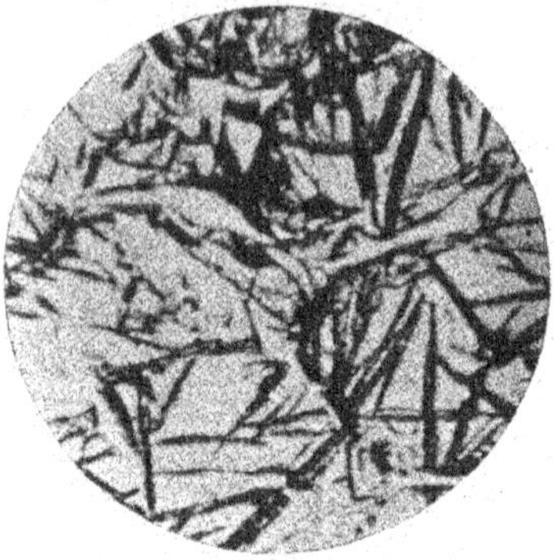

Fig. 1.

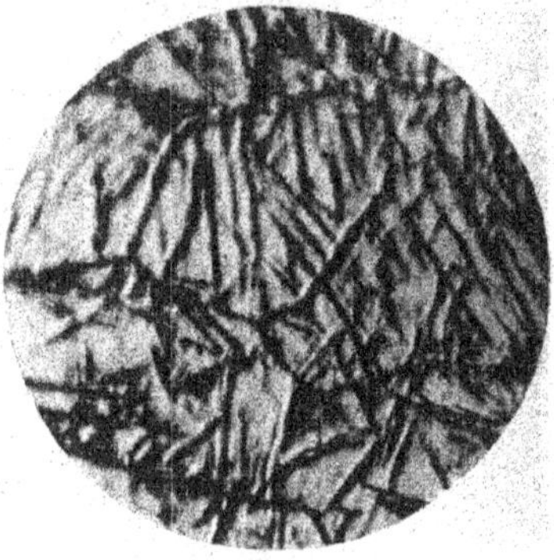

Fig. 2.

had not been detected in the polished surface; the needle structure had been obliterated; even after a brisk rubbing the white striations could not be detected.

The sample was again re-polished and reëtched, and again oxidized with a solution of glacial acetic acid, powdered nut-galls, water, and a few grains of nitrate of silver. This solution gave the same results as those just described.

The tinting method also failed to support the cementite theory.

I claim that Professor Arnold is in error in his interpretation of the structure, and in support of the statement I offer the following photographs:

Photomicrograph No. 1 exhibits the structure of self-hardening steel quenched from its melting-point. The chemical composition is the same as that given in the July number of *The Metallographist.* See also photograph in the same number; it will be noticed that a vast difference exists between the two structures. This is due to the fact that the former was quenched from its melting-point in a large section, while the later section was less than ½ inch square.

In No. 1 we also notice the white dots and striations in the needles of martensite; the dots begin to appear when 1200° C. have been exceeded.

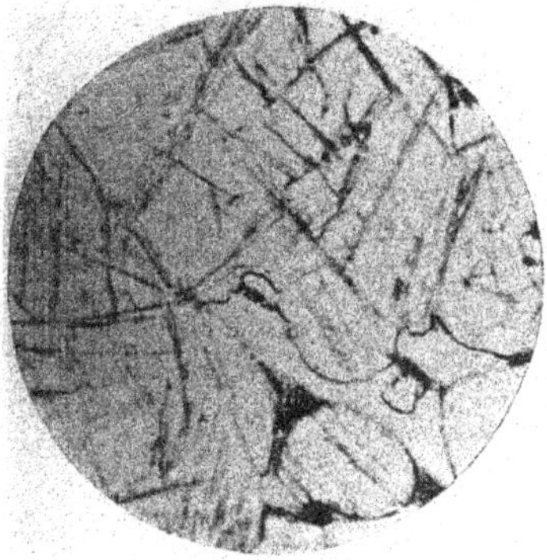

Fig. 3.

Fig. 4.

No. 2 shows the structure of a self-hardening steel containing about 2 per cent of carbon and considerable molybdenum quenched from near the melting-point.

No. 3 illustrates the structure of the same steel cooled quickly from about 1020° to about 600° C. and then allowed to cool slowly. Here the needles of martensite are rather indistinct owing to the slow cooling. The structure appears flat and of uniform hardness, making an excellent tool for heavy work. The structure of No. 2 appears wrinkled and is very brittle, and will not hold a cutting edge.

No. 4 exhibits the structure 1 mm. from the edge of a steel

containing 1.39 per cent of carbon, 0.071 per cent of silicon, 0.012 per cent of phosphorus, 0.011 per cent of sulphur, 0.017 per cent of manganese, quenched from 1303°. While No. 5 shows the structure of the same section 3 mm. from the edge.

No. 6 illustrates the structure of a steel quenched from about 850° C. containing 1.18 per cent of carbon, 0.175 per cent of silicon, 0.013 per cent of phosphorus, 0.009 per cent of sulphur, 0.29 per cent of manganese; here the martensite runs in veins

Fig. 5.

Fig. 6.

well interlaced. The sample was polished in relief, using a 2 per cent solution of sodium nitrate in place of water.

Professor Arnold also says: "For many years I have been in the habit of examining blades quenched by professional hardeners belonging to the Sheffield Blade Makers' Union, and had never noticed in such blades the needles form of martensite."

I agree with Mr. Arnold, I should not expect to find the needle form of martensite in any properly hardened blade or tool, as they are quenched from Ac_1 or slightly above. The structure remains granular up to about 825° C. and even above that if a forced heat is used or a large section.

F. C. Lau.

Chicago, Illinois, April 1901.

HENRI LE CHATELIER

The Metallographist

A QUARTERLY PUBLICATION DEVOTED TO THE STUDY OF METALS, WITH SPECIAL REFERENCE TO THEIR PHYSICS AND MICROSTRUCTURE, THEIR INDUSTRIAL TREATMENT AND APPLICATIONS.

Vol. IV OCTOBER 1901 No. 4

HENRI LE CHATELIER

HENRI LE CHATELIER, so well known to the readers of this journal and, indeed, to all students of applied science, entered the Ecole Polytechnique at the head of his class in 1869. After graduating from this school in 1871, he spent three years at the Paris School of Mines, and upon leaving the School of Mines was sent to Algeria as a member of a commission whose mission was to ascertain the existence of a sea which was thought to have existed, in the past, in the deserts of Southern Algeria. Upon his return he practiced for two years the profession of engineer at the " Corps des Mines " of Besançon.

In 1877, Mr. Le Chatelier was appointed Professor of general chemistry at the Paris School of Mines, and in 1884 " interrogateur " at the Ecole Polytechnique, a position which he resigned in 1897. In 1884 he was the first choice of the Board of Improvement of the Ecole Polytechnique for the chair of Chemistry. In 1887 he gave up the chair of general chemistry at the Ecole des Mines to occupy that of industrial chemistry at the same school, a position which he still holds.

In 1892, he was awarded the Jerome Ponti prize, and in 1895 the La Caze prize.

In 1898, Mr. Le Chatelier was appointed Professor of mineral chemistry at the Collège de France, where he is the colleague of Professor Berthelot, who occupies the chair of organic chemistry. Professor Le Chatelier is vice-president of the Chemical Society and of the Mineralogical Society; a member of the Committee of Applied Chemistry of the Société d'Encouragement, foreign member of the Société des Sciences of Holland, and member of many other scientific and technical societies. He is secretary of the Alloy Committee of the Société d'Encouragement, and the success of this important enterprise is due in a great measure to his untiring energy. He is also the French member of council of the International Association for the testing of materials.

As Professor Le Chatelier himself says, the starting-point of all his scientific investigations is to be found in his desire to apply scientific methods to industrial problems. Even his researches in pure science were suggested by industrial operations. His intellectual activity was at an early age directed along scientific lines. His father, who was an eminent inspector general of mines, devoted most of his time to the solution of problems connected with railroads, but his leisure moments were given to the study of metallurgy. He was a friend of Sir Williams Siemens and greatly contributed to the introduction and extended use, in France, of the regenerative heating system. He was the first metallurgist to apply this system to the production of steel. The trial conducted under his direction at the Fourchambault Works proved a failure owing to the melting of the furnace which had not been constructed of sufficiently refractory material. He was also a friend of St. Claire Deville, and he worked with him on the extraction of aluminium and on the various applications of that metal.

Professor Le Chatelier's maternal grandfather, Mr. Durand, was a friend of Vicat and devoted the last years of his life to repeating and developing the work on lime and cement of this eminent engineer.

Professor Le Chatelier's scientific investigations and writings may be classified in two groups which at first sight appear quite unrelated: Chemical Mechanics and Industrial Chemistry. These two departments of chemistry, however, as the author himself says, possess numerous points of contact. The laws of chemical mechanics control the phenomena of industrial chemistry as rigorously as they do the reactions studied in scientific laboratories.

His first researches dealt with hydraulic materials, and his numerous and valuable contributions to this important subject have given him a world-wide reputation among constructing engineers. The study of the dissociation of carbonate of lime was the starting-point of his researches in chemical mechanics as well as of those dealing with high temperatures. His admirable investigations and dissertations in chemical mechanics resulted in the enunciation of eight laws which threw much light upon a subject in which there existed much obscurity. It is interesting to note that Professor W. Gibbs had, before the publication of Professor Le Chatelier's work, arrived at some algebraic formulas in which these laws were implicitly contained. The latter, however, was not at the time acquainted with Professor Gibbs's results. In no other line of research and study has Professor Le Chatelier shown greater erudition and keenness of mind than in this masterly work in a field into which only those gifted with exceptional intellectual power dare to venture.

Professor Le Chatelier's eminently successful researches dealing with the measurement of high temperatures are too well known to the readers of this journal to demand more than a passing notice. His thermo-electric couple is now in use in numerous industrial establishments where its usefulness could hardly be overestimated. This admirable pyrometer has made possible investigations of the greatest commercial and scientific importance in the physics of metals and alloys. The discovery by Osmond of the upper thermal critical points of iron and the scientific and systematic study of all the critical points of iron and steel, which have thrown so much light upon the constitution of steel, have been made possible by the use of the Le Chatelier thermo couple. Without it the construction of the fusibility curves of many metallic alloys, so invaluable in ascertaining the constitution of these alloys, would be a matter of great difficulty and much uncertainty. The discovery of the simple law connecting the temperature of the junction of the two metals and the electric current generated, which was the result of systematic and strictly scientific work is a monument which would satisfy the ambition of most physicists. Had Professor Le Chatelier done nothing more than this in the domain of pure and applied science, it would have sufficed to secure the gratitude and admiration of the scientific and industrial world. This discovery, however, which he has so generously presented to the world, constitutes only a relatively small part of his contributions to the advancement of science.

Professor Le Chatelier's investigations dealing with the setting of hydraulic material were the starting-point of his theoretical studies of the laws of solubility. These, in turn, suggested to him his experimental researches on the fusibility of mixtures of salts, and finally led him to investigate the constitution of metallic alloys, thus joining the ranks of those engaged in metallographic investigations. And we all know what valuable services he has already rendered both to the technology of the subject and to the interpretation of the results. Professor Le Chatelier showed that the fusibility curves of metallic alloys were in every way similar to the freezing curves of aqueous saline solutions or of mixtures of melted salts from which he inferred a similarity of constitution between these two classes of substances. These conclusions were in every particular confirmed by microscopical examination. According to the appearance of the curve of fusibility, he was led to classify all alloys into three groups, similar to those he had previously proposed for mixtures of salts, namely: (1) Alloys forming neither definite compounds nor isomorphous mixtures, (2) alloys forming definite compounds and (3) alloys forming isomorphous mixtures. To these three groups should be added a fourth one, including all alloys with abnormal curves of fusibility, and which future investigation will probably further subdivide. This modern theory of the constitution of alloys has been fruitful in deductions of industrial importance and will undoubtedly result in the abandonment of the empirical methods still in such extended use in the manufacture of metallic alloys.

After having thus studied the constitution of metallic alloys in general, it is quite natural that Professor Le Chatelier should have been attracted by the fascinating problem connected with the constitution of steel. To investigate this subject, he naturally adopted microscopical methods as one of the most promising means of advancing our knowledge of the true nature of steel. He entered this field exceptionally well equipped for effective experimental work. Fully aware of the importance of time-saving appliances in laboratory work as well as in industrial operations, he endeavored to shorten the required manipulations and especially to obtain some polishing powders which would greatly reduce the length of the polishing operation. The results of his investigations were published in full in the January, 1901, issue of *The Metallographist.* He also devised a special microscope described in the same journal and recommended the use of the monochromatic light of a mercury arc vacuum lamp for photomicrography. Professor Le Chatelier's discussion of the data upon which the modern theory of the constitution of steel has been erected is most valuable, notably his paper on "The Present Condition of the Theories of Hardening" and on "Iron and Steel from the Point of View of the Phase Doctrine," both of which were published in full in *The Metallographist.*

His experimental work dealing with the electrical resistance and with the dilatation of steel, undertaken with a view of throwing additional light upon the allotropic transformations of iron discovered by Osmond, have yielded very important results. They have shown that the upper thermal retardation of iron (Ar_3) occurring at about 850° Cent., while often hardly noticeable by the calorimetric method, is on the contrary very marked when determined by the electrical method. This upper critical point always corresponds to an abrupt and very great alteration in the electrical resistance of iron. The second critical point on the contrary, occurring at about 750° Cent., and the point of recalescence have hardly any effect upon the electrical resistance. His investigations on electrical conductibility were extended to nickel steel and to manganese steel, as well as to various non-ferrous alloys. They revealed the existence of some allotropic transformations in several alloys and notably in brass (at 730°), in aluminium bronze (at 500°), in zinc (at 350°) and in an alloy of copper, iron and nickel (at 690°).

I can only mention here two other directions in which Professor Le Chatelier has shown much activity, as they fall so far outside the scope of this journal: (1) His researches on marsh gas conducted in collaboration with Mallard, with a view of preventing mine explosion, and which was the starting-point of their scientific investigation of the temperatures of combustion of gases, their specific heat and the theory of explosives; and (2) his many and important experimental studies of the properties of ceramic products, most of them conducted at the Ecole des Mines with the assistance of some of his students in industrial chemistry.

It will be seen that besides being a distinguished metallographist, Professor Le Chatelier possesses many other titles; he has won an eminent position in several departments of pure and of applied science, through his experimental skill, scientific methods, erudition, and intellectual power.

Professor Le Chatelier has published over one hundred and twenty-five papers. He also wrote a book, in collaboration with Mr. Boudouard, on the "Measurement of High Temperatures," and translated into French the important work of Professor Gibbs on the equilibrium of chemical systems. The greatest number of his writings, by far, were published in the Comptes Rendus of the French Academy of Sciences, the others mostly in the Annales des Mines, the Revue générale des Sciences and the Bulletin of the Société d'Encouragement.

A complete list of Prof. Le Chatelier's publications dealing with metallographic and related subjects is given below in their chronological order.

On the Measurement of High Temperatures with Thermo-Electric Couples. *Journal de Physique,* VI, January, 1887.

On the Determination of Coefficients of Dilatation at High Temperatures. *Comptes Rendus,* CVII, p. 862.

On the Dilatation of Metals at High Temperatures. *Comptes Rendus,* CVIII, p. 1096.

On the Electrical Resistance of Iron and its Alloys. *Comptes Rendus,* CX, p. 283.

On the Electrical Resistance of Metals. *Comptes Rendus,* CXI, p. 445.

Influence of Quenching on the Electrical Resistance of Steel. *Comptes Rendus,* CXII, p. 40.

Various Molecular Conditions of Metals according to their Electrical Conductibility. *Journal de Physique,* 1891.

On the Temperatures developed in Industrial Furnaces. *Comptes Rendus,* CXIV, p. 470.

On Manganese Steel. *Comptes Rendus,* CXIX, p. 272.

On the Definite Compounds of Metallic Alloys. *Comptes Rendus,* CXX, p. 835.

On the Definite Compound of Copper-Aluminium Alloys. *Comptes Rendus,* CXX, p. 1050.

Metallic Alloys. *Revue Générale des Sciences,* June, 1895.

On Microscopic Metallography. *Bulletin de la Société d'Encouragement,* April, 1896.

Actual Condition of the Theories of the Hardening of Steel. *Revue Générale des Sciences,* January, 1897. *The Metallographist,* Vol. I, p. 52.

On the Electrical Resistance of Steel. *Comptes Rendus,* CXXVI, p. 1709.

Influence of Quenching upon the Electrical Resistance of Steel. *Comptes Rendus,* CXXVI, p. 1782.

A New Microscope for the Study of Opaque Objects. *Revue Générale des Sciences,* 1897, p. 43. *The Metallographist,* Vol. I, p. 83.
Influence of the Casting Operation upon the Structure of Metals. *Bulletin de la Société d'Encouragement,* 1897, p. 441.
On the Dilatation of Metallic Alloys. *Comptes Rendus,* CXXVIII, p. 1444. *The Metallographist,* Vol. II, p. 334.
On the Transformations of Iron and Steel. *Comptes Rendus,* CXXIX, p. 279. *The Metallographist,* Vol. III, p. 38.
On the Dilatation of Iron and Steel at High Temperatures. *Comptes Rendus,* CXXIX, p. 331. *The Metallographist,* Vol. III, p. 152.
On the Points of Transformation. *Comptes Rendus,* CXXIX, p. 497.
The Law of Phases. *Revue Générale des Sciences,* 1899, p. 759.
Application of the Law of Phases to Alloys and to Rocks. *Comptes Rendus,* CXXX, p. 20. *The Metallographist,* Vol. III, p. 290.
On the Angular Points of Solubility Curves. *Comptes Rendus,* CXXX, 1900.
Notes on the Technology of Microscopic Metallography. *The Metallographist,* Vol. IV, p. 1.
Iron and Steel from the Point of View of the Phase Doctrine. *Bulletin de la Société d'Encouragement,* Nov. 30, 1901. *The Metallographist,* p. 161.

THE PROGRESS OF METALLOGRAPHY IN 1900

By ALBERT SAUVEUR

I. Preparation of Samples of Metals for Microscopical Examination

IN the following pages will be found a short, critical review of the progress of Metallography in 1900.

Polishing.—In order to examine the structure of metals through the microscope, it is necessary to polish a surface of the fragment until it is almost absolutely free from the minutest scratches. To accomplish this, it is quite evident that the sample must be rubbed successively over various abrasive substances of increasing fineness. Supposing the surface to be examined has been filed with a smooth file, the tool marks would not be removed and a perfectly specular condition could not be imparted to the surface in a single operation. The transformation must be gradual. The file marks must be effaced by rubbing the sample over a properly selected polishing substance, and replaced by finer markings. These, in turn, must be removed by a second rubbing with a finer abrasive agent, being replaced by still finer marks, and so on, until finally, the last operation removes the very minute marks from the previous treatment and leaves the surface absolutely free from scratches or nearly so. Emery powder of various degrees of fineness is the abrasive substance which naturally suggests itself, at least for the first treatments. The powders may be used in the shape of emery wheels, emery cloths or papers, or even spread loose over a suitable support, in which case it is kept wet during the rubbing. Emery may, of course, be replaced by carborundum for these operations. The polishing powder known as jeweler's or gold rouge suggests itself for the final polishing. It is generally spread over a piece of wash leather or other soft and smooth texture, which in turn is fastened to a block of wood. The powder is generally kept wet during the rubbing. The details of the manipulation vary greatly with different operators, but it may be said that they all use emery (or carborundum) in some form for the rough polishing and jeweler's rouge for the final treatment.

The markings left even by the finest emery powder, however (French emery paper, No. 000 and No. 0000), cannot always be removed readily by the rouge, and the operation is frequently a long and tedious one. The need of a suitable polishing powder to be used after the last emery treatment and before the final rouge polishing was soon made evident and many substances, such as water of Ayr stone, crocus, tripoli, diamontine, "brillant Belge," etc., have been tried. It is in this particular that the polishing methods of the various experimenters engaged in metallographic work vary the most. Somewhat roughly stated, the method at present in use for the preparation of metallic samples, is in most cases as follows:

Three treatments with emery (or carborundum) of increasing fineness.

One treatment with one of the powders mentioned above.

One treatment with jeweler's rouge.

It was also found that the emery powders and emery papers on the market are seldom of sufficient uniformity. They frequently contain some coarse particles which will scratch the specimens deeply and greatly lengthen the polishing operation.

This fact led some experimenters to prepare their own emery powders and papers. Commercial emery powder is allowed to settle slowly through a column of water and is divided into several grades, according to the time required for settling. The resulting powders are more uniform and by their use the polishing operation is shortened and better surfaces are obtained. Emery papers can readily be prepared with these various grades of powder. The same treatment is sometimes applied to commercial jeweler's rouge.

The polishing may be done entirely by hand or it may be hastened by the use of some simple power-driven machine consisting of revolving emery wheels and discs upon which the various powders are spread. Until quite recently, however, and even with all the refinements outlined above, the number of treatments was seldom less than five, and the average time required for the preparation of a surface measuring 0.25 sq. in., was not far from 30 minutes when polished by hand and 20 minutes when a polishing machine was used. Occasionally a sample was prepared in a much shorter time, but others would require considerably more time, and the average figures just given, represented very good practice. The polishing of a larger surface will generally demand a correspondingly longer time.

It was, of course, very desirable to reduce both the time and the labor required for the preparation of the samples, and it is with that end in view that Professor H. Le Chatelier conducted some very valuable and fruitful experiments described at length in his paper published in our January, 1900, issue.

The levigation method of preparing polishing powders, as usually carried on, does not yield perfectly homogeneous products. The pulverulent matters in suspension in water contain some salts, especially of calcium, which have a tendency to cause the formation of lumps containing both coarse and fine grains. A very small amount of salt is sufficient to produce this result, the carbonate of lime present in ordinary water being more than sufficient. These considerations led to the application of Schloesing's method for the analysis of kaolins to the preparations of the polishing powders. The method consists in adding one part of nitric acid to every one thousand parts of water used in the operation, in order to dissolve the calcium carbonate and sulphate as well as other salts which might be present. After a few hours, and occasional stirring, the mixture is allowed to settle. The powder falls rapidly to the bottom, and the clear liquid above it can be easily decanted. This liquid is now replaced by distilled water, and the mixture stirred; it is again allowed to settle, and is then decanted. After a few similar operations, when all the acid has been removed, the settling takes place more slowly, and the liquid remains milky. The conditions are now favorable to proceed with the levigation, and the operation is further facilitated by adding 2 c.c. of ammonia to each liter of water, as it helps the suspension of the finest particles. Decantation is now resorted to, at stated intervals, by means of a siphon forming a hook at its extremity, so as to avoid disturbing the portions which have settled. The decantations are made at the following intervals: 15 minutes, one hour, four hours, 20 hours, eight days. The deposit left after the first decantation contains all the coarse grains unsuitable for polishing. The second deposit, resulting from the first decantation after one hour's settling, yields a substance which is not yet very homogeneous, but which may serve to start the polishing. The third deposit constitutes a good polishing powder for hard metals such as iron. It is, however, the deposit collected between the first and the eighth day which constitutes true polishing powder. Instead of waiting eight days for the formation of the last deposit, it may be precipitated immediately after the removal of the 24 hours' powder, by adding some acetic acid, in order to saturate the ammonia. The totality of the powder, still remaining suspended, is then deposited after a few hours.

To preserve the powders thus prepared and prevent all possibility of their being contaminated by dust or other foreign matter, they are mixed with some shavings of castile soap, and sufficient water to make a thick paste when heated, which is preserved in tin tubes similar to those used for oil paints, paste, etc.

This levigation method has been applied to a number of substances, such as alumina produced by the calcination of ammoniacal alum; commercial emery powder; chromium oxide obtained by the combustion of ammonium bichromate; and iron oxide resulting from the calcination in air of iron oxalate. The alumina yielded by far the best results.

Professor Le Chatelier uses now the following method for polishing samples of iron and steel, after they have been previously rubbed upon commercial emery papers, including the finest grades:

1. An emery paper prepared with albumine, according to Osmond, with the deposit obtained in between a quarter of an hour and one hour in the ammoniacal washing of flour-emery.

2. A felt disc, covered with some soap paste prepared with the deposit of alumina or of emery, obtained in between one and three hours.

3. A flat disc made of wood, metal or ebonite, covered with cloth, velvet or leather which is strongly glued upon it. Upon this covering the soap preparation, obtained with the deposit of alumina after 24 hours, is spread.

The last two discs are rotated by some mechanical device producing a great speed.

Under these conditions the polishing proper may be finished in about five minutes. To this must be added the time required for preliminary treatment with commercial emery.

Mr. J. E. Stead has described and illustrated the polishing machine which he has finally adopted.

After much experimenting, I have devised an arrangement (Fig. 1) by which the whole polishing operation is greatly shortened and simplified. My outfit consists of a grinding machine of the usual style, carrying four discs revolving in a vertical plane and covered with suitable cloths — a simple arrangement permits the quick removal of torn cloths whenever necessary. In this way four polishing surfaces are obtained always ready for use with the additional advantage that they may be used simultaneously. This is decidedly better than Mr. Stead's arrangement in which one block only can be made to rotate, necessitating a frequent handling of the blocks and a greater consumption of time and labor. The four polishing surfaces of my outfit are fed automatically and simply. Water is fed near the center but the rapid revolution of the discs causes it to be immediately distributed over the entire surface. Shields are provided for catching the water thrown off the discs when at work. By properly regulating the feeding, the proportion of water, etc., the discs are at all times in perfect condition, and require but little attention. A paste made essentially of tripoli and castile soap has been found very effective. With most samples the rouge polishing may be omitted, the entire polishing operation consisting of

Fig. 1.

three treatments. Indeed, with some samples, two treatments suffice in which cases the operation lasts but three or four minutes. With the above outfit 5 or 6 minutes is a high average for polishing a surface measuring 0.25 sq. in. Two specimens may readily be prepared at the same time by one operator holding one sample in each hand and pressing them upon the two sides of the same disc. If the samples be delivered to him after filing them with a smooth file, one operator might readily prepare 15 samples per hour and probably more. The industrial importance of being able to prepare daily a large number of samples for microscopical examination will not escape the reader. At the rate mentioned above one operator could polish in one day's work a section from each blow of a three-vessel Bessemer plant, making an average of 150 blows per 24 hours.

II. Development of the Structure

Iron and Steel.—Many treatments have been tried to develop the structure of polished samples of iron and steel, i. e., to make their structure apparent when examined under the microscope. To do this, it is necessary to impart unlike appearances to the various constituents. Of all the methods tried, it may be said that five only have been retained:

1. Etching with nitric acid (both concentrated and diluted).
2. Etching with iodine tincture.
3. Combined polishing and etching with a solution of ammonium nitrate.
4. Polishing in relief.
5. Heat tinting.

The proper way of conducting these various operations has been described on many occasions; I shall only indicate here the improvements which have been introduced during the past year in their *modus operandi*.

Professor H. Le Chatelier recommends applying the iodine tincture to the polished section with the tip of the finger, with constant rubbing; the etching is then more uniform, the constituents which are not acted upon remain brilliant, and the filling up of the minute parts, such as pearlite, by some deposit from the etching of other parts, is prevented.

With regard to heat tinting, Mr. Stead* has applied this method extensively and with much success to samples of pig iron. He has found it especially effective in distinguishing between the pearlite areas and those of the phosphide eutectic. His beautiful photographs testify to the excellency of the method for this special purpose.

The method which should be used for the development of the structure depends upon the nature of the metal, the treatment it has received and to a great extent, upon the characteristics of the structure which it is desired especially to reveal. It is not possible to state positively what developing method should be used in every case, so as to obtain the best resolution of the structure. No sharp lines of demarcation can be drawn in the application of the various methods.

Soft Alloys.—It is well known that the polishing of soft alloys presents much difficulty. For instance, in polishing an alloy of tin and lead, the lead spreads itself upon the tin in such a way that it is impossible to distinguish clearly the structure under the microscope. In order to obtain smooth surfaces, Messrs. Irving and Rosenhain proposed to cast such alloys upon plates of glass or of polished steel. The frequent breaking of the glass plates was the most serious objection to the method, and this induced Mr. H. J. Hanover† to replace the glass plates by plates of mica, which gave very satisfactory results.

Microscopical Outfit. — Professor H. Le Chatelier‡ has devised a special microscope for the examination of metals and opaque objects in general, which he has fully illustrated in his article.

Seeing that so-called achromatic and apochromatic objectives are never absolutely free from chromatic aberration, Professor Le Chatelier recommends the use of a monochromatic source of light, especially for photography. He finds that an excellent monochromatic light is obtained by means of a mercury arc lamp in vacuum. In the lamp he uses the vacuum space is very large, so as to increase the radiating surface, and thereby decrease the heating of the lamp. The central tube is covered with a refractory coating which greatly lengthens the life of the lamp. By

---

* *The Metallographist,* October, 1900, p. 261.

† *The Metallographist,* 1900, p. 29. ‡ *The Metallographist,* 1900, p. 1.

passing the light emitted from this lamp through suitable solutions, either the green radiation, desirable for ocular examination, or the indigo rays, so effective in photography, may be readily isolated.

Preparation of Alloys. — In studying the constitution of metallic alloys it is frequently necessary to examine a large number of alloys of various proportions of the same constituents, the object in view being to ascertain what proportions of the constituent metals correspond to critical structural changes. Hence the necessary preparation and examination of numerous alloys which have only a negative value. To obviate this Professor Le Chatelier suggests superposing the two metals so that they will mix only in the vicinity of their contact. By preparing then a vertical section of the resulting mass, the metals may be observed alloyed in all proportions from one pure metal to the other.

Troostite and Sorbite. — In 1895 Mr. Osmond* described two constituents of steel, whose existence had not been previously discovered, and which he called respectively " troostite (in honor of the eminent French chemist, Troost), and " sorbite " (in honor of the distinguished English microscopist, Sorby, the pioneer of microscopic metallography). Mr. Osmond described troostite as being a transition form between martensite and pearlite, while he considered sorbite as a form of transition between ferrite and cemetite. As Mr. Osmond himself now admits,† these two constituents had not been sufficiently clearly defined; many metallographists failed to identify them satisfactorily, and some, no doubt, even questioned their existence; all steel microscopists are, therefore, greatly indebted to Mr. Osmond for the clear description he now gives of these constituents. Their existence as structural entities, at least in the case of troostite, can no longer be contested. In order to make the matter clear, it will be necessary to consider briefly the transformations which take place spontaneously when some carbonized iron is allowed to cool slowly from a high temperature to the atmospheric temperature. Steel, or rather carbonized iron, is composed after solidification of mixed crystals of *Gamma*-iron holding the carbon in solution

---

* *Bulletin de la Société d'Encouragement*, May, 1895, p. 480.
† *The Metallographist*, Vol. IV, p. 150.

to which, considered as a structural entity, Mr. Osmond has given the name of austenite. It is well known that as the iron cools slowly it undergoes at certain temperatures, at least two allotropic changes which are accompanied by structural changes as well as by changes in the physical and chemical properties of the metal, and that the cold metal will be made up of ferrite (Fe) and of cementite (Fe_3C) both in the free state and forming a mechanical mixture (pearlite). During slow cooling then, austenite has changed to ferrite and cementite (pearlite). This transformation of austenite, however, does not take place abruptly at a certain temperature, but gradually, assuming some forms of transition more or less stable at certain temperatures. These forms of transition are in their order, according to Osmond: martensite, troostite and sorbite. To retain any of these forms in the cold metal it will be necessary to cool the steel at a rate which will be sufficiently slow to allow the desired constituent to be formed and sufficiently rapid to prevent its further transformation. The tendency of austenite to pass to martensite when the proper temperature is reached (i. e., the temperature at which austenite ceases to be stable), is so great that it is only by quenching high carbon steel very suddenly that it can be retained in the cold metal, the carbon present assisting in its retention. Fig. 2, taken by Osmond, shows the structure of a steel with 1.50 per cent C quenched in ice water. The light portions correspond to austenite, the dark to martensite. Part only of the austenite could be retained as such, the balance passing to martensite. By the water-quenching operations as generally carried on in practice it is not austensite, but martensite which is retained. The cooling is not rapid enough to prevent the austenite-martensite transformation. In low-carbon steel, moreover, austenite cannot be retained even by extremely quick cooling, for lack of carbon, which, as already stated, assists in retaining it. Fig. 3 illustrates the structure of steel containing 0.45 per cent C, and quenched in water; it is made up entirely of martensite.

To retain the constituent troostite in the cold, the metal should be quenched in water while it is passing through its critical range, or quenched in oil from a temperature superior to that of the transformation, or it may be produced by proper reheating (tempering) of quenched samples, the martensite being thereby partially converted into troostite. Fig. 4 shows the structure of steel

Fig. 2. Steel containing 1.5 per cent C, quenched in ice water. Magnified 1000 diameters.

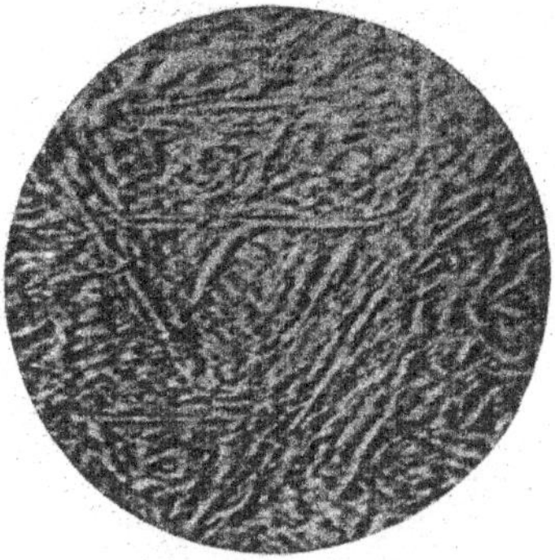

Fig. 3. Forged Steel. Reheated to 825° C. and quenched at 720° C. Combined polishing and etching. Magnified 1000 diameters.

Fig. 4. Steel containing 0.45 per cent C. Heated to 730° C. and quenched in water during critical range. Magnified 1000 diameters.

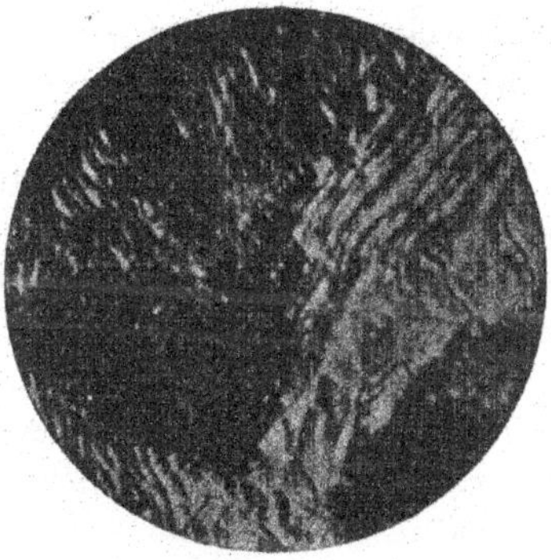

Fig. 5. Steel containing 1.24 per cent C. Reheated to 860° C. and slowly cooled in the furnace. Magnified 2000 diameters.

containing 0.45 per cent C heated to 730° C. and quenched in water during the critical range. The dark constituent is troostite, the light background martensite with here and there a ferrite grain.

To produce sorbite, the rate of cooling should be still further reduced (quenching in molten lead, for instance), or the quenching take place at the end of the critical range, or it may be formed by reheating (tempering) quenched samples to a purple color. According to Osmond the cooling in air of pieces of small sections is sufficiently rapid to retain some sorbite in the metal, i. e., to prevent the complete transformation of austenite into ferrite and cementite.

Fig. 5 illustrates the structure of a steel containing 1.24 per cent C, reheated to 860° C. and slowly cooled in the furnace. The dark areas are sorbite, the light lamellar region, pearlite.

The Crystallography of Iron. — Mr. Osmond has published two scholarly and exhaustive memoirs (the last one in collaboration with Mr. Cartaud),* on the crystallography of iron. It is now well accepted that iron may exist in at least three allotropic states, *Gamma, Beta,* and *Alpha,* each stable at a certain temperature, but the crystalline forms belonging to these different allotropic conditions were only imperfectly known. If it could be shown that iron assumes distinct crystalline forms, when allowed to crystallize at temperatures at which the various allotropic conditions are stable, it would constitute an irrefutable proof of the existence of these allotropic conditions. Such proof, however, is not needed. On the other hand should it be found that the three allotropic varieties assume exactly the same crystalline forms, it would in no way disprove their existence. Allotropy does not by any means imply a change of crystalline forms of the allotropic substance. In his first memoir Mr. Osmond reviews critically all the available data bearing upon this interesting subject. The conclusions which he reached are in the main identical with those based upon some experiments of his own described in his second memoir. These experiments consisted essentially in the reduction of some ferrous chloride either through the action of hydrogen or of zinc vapor at temperatures at which the various allotropic conditions of iron are stable, i. e.:

---

* *The Metallographist,* 1900, p. 181.

Below 700° C. to cause the iron to crystallize in the *Alpha* state. Between 750 and 860° C. to cause the iron to crystallize in the *Beta* state. Above 860° C. to cause the iron to crystallize in the *Gamma* state.

The transformations which take place during the further cooling of the metallic masses produced by the last two experiments will not affect the external forms of the crystals, and it will then be possible to study the crystalline forms assumed by iron under its three allotropic conditions. These experiments conducted with great skill led the authors to the following conclusions:

"Iron in its three states, crystallizes in the cubic system.

"*Gamma*-iron may assume all the combinations of the cube and the octahedron; never any modification along the edges; numerous *macles* with a′ as junction. In industrial iron more or less carburetted, the crystallization is preceded by a crystallitic phase more or less wide.

"*Beta*-iron and *Alpha*-iron crystallize identically, which is a necessary condition, if not a sufficient condition, of their isomorphism, which, moreover, was further indicated by the progressive character of their mutual transformation. The crystalline form is the cube with frequent modifications, probably b^1 or b^2 along the edges. No modification on the angles.

"*Gamma*-iron does not form an isomorphous mixture with *Beta*-iron.

"The transformation of *Gamma* into *Beta*-iron appears to be accompanied by an alteration of the positions of the planes of symmetry, at least in the case of carburized iron, but the size of the grains assumed by one of the allotropic varieties exerts a certain influence upon the dimensions of the grains resulting from the passage of the iron to the next allotropic state."

Iron and Phosphorus. — Mr. Stead has embodied the results of his experiments in a most important paper,* the substance of which I shall endeavor to present here as concisely as possible.

Hundreds of samples containing from a trace up to a very large amount of phosphorus, and practically no carbon, were first prepared and examined through the microscope, their structure being in most cases made apparent by heating the polished samples as described in a previous section of this review. From the

* *The Metallographist,* Vol. IV, p. 84.

microscopical and other evidences described at length in the paper, the author is led to the following interesting and valuable conclusions:

1. When phosphorus is present in amount varying from traces to 1.70 per cent, it forms a phosphide corresponding to the formula Fe_3P, which is held in solution by the iron (in solid solution, of course). All the metals used commercially, such as wrought iron and steels containing practically no carbon, may be included in this class.

2. When the metal contains from 1.70 to 10.2 per cent P, it consists of a saturated solution of Fe_3P in iron (1.70 per cent P) and of a eutectic alloy containing about 10.2 per cent P and made up of about 61 per cent Fe_3P and 39 per cent of the saturated solution of Fe_3P in iron. To account readily for this structure and that of the following groups, it is only necessary to consider these metals as alloys of two constituents; one the phosphide Fe_3P and the other a saturated solution of Fe_3P in iron. It is well known that a certain class of binary alloys when solidifying give rise to the formation of a eutectic alloy, that is, to a mechanical mixture made up in definite proportions, of extremely small plates alternately of one and the other constituents, the balance of the mass consisting of that constituent which is present in excess over the amount required to form the eutectic alloy. It is precisely what happens in the case of iron containing over 1.70 per cent P. It is also what occurs in carburized iron considered as an alloy of iron and the carbide Fe_3C, giving rise to the eutectic alloy known as pearlite ($Fe+Fe_3C$), whose structure like that of all eutectic mixtures is so strikingly similar to the phosphide eutectic discovered by Mr. Stead. It would seem highly appropriate to designate this constituent by the name of *steadite*.

3. When the iron contains from 10.2 per cent to 15.58 per cent P it is composed of crystals of Fe_3P surrounded by the eutectic mixture just described.

4. With a phosphorus content varying between 15.58 and 21.68 per cent, Mr. Stead finds a phosphide of the formula Fe_2P associated with Fe_3P, the former increasing in amount and the latter decreasing as the phosphorus increases. Fe_3P, the author says, is easily attracted by the magnet, while Fe_2P is only slightly attracted.

The second part of Mr. Stead's paper is devoted to a study

of the influence of carbon upon the structure of iron containing phosphorus and his conclusions are in part as follows:

"On melting saturated solid solutions of phosphide of iron in iron with carbon, the latter causes a separation of the phosphide near to the point of solidification, which appears in the solid metal as a eutectic in irregular-shaped areas, if the carbon present is small, and in envelopes, increasing in thickness with the amount of carbon present, but is incapable of throwing the whole of the phosphide out of solution even when 3.5 per cent C is present. A residuum always remains in solid solution. This residuum is smallest, however, when the carbon is at a maximum."

He also finds that "a portion of the phosphide of iron in steels containing under 0.10 per cent P is thrown out of solution by carbon when it exceeds 0.9 per cent, and the phosphide so separated is liable to form a brittle cell structure enveloping the grains, yielding a more or less fragile mass. In such cases, however, the larger part of the phosphide remains in solid solution."

In the third part of his paper he deals with the structure of phosphoretic pig iron, his conclusions being as follows:

"1. There are fusible eutectics in all pig irons containing phosphorus, and which have solidifying points of about 900° C.

"2. The free phosphide of iron can be left insoluble from such irons by dissolving them in cupric chloride (Schneider's method), or by nitric acid of specific gravity 1.20 in the cold. By attracting the magnetic portions of phosphide from the residue with a magnet, and further purifying what is attracted, almost pure Fe_3P is obtained.

"3. With the aid of the microscope it is possible to detect the phosphide eutectic in pig iron, even when the amount is as low as 0.03 per cent.

"4. In white irons it is necessary to use the heat-tinting process to enable the cementite to be distinguished from the phosphide.

"5. The eutectics from such irons have a very complex constitution, containing phosphide of iron, carbide of iron in plates, pearlite, and silicide of iron. They deserve more careful study.

"6. The reason why phosphoretic pig irons are not more brittle than they are is because the eutectic separates into isolated segregations, and does not form continuous cells round the crystalline grains. When the phosphorus does not exceed 1.7 per cent the metal is comparatively strong, but an addition of 0.3 per cent reduces the strength materially, the reason of which is that the eutectic brittle areas in metal with 2 per cent P approach each other closely, leaving less of the strong ground mass intervening."

Professor Arnold and Mr. McWilliams had in a previous paper asserted that solid iron phosphide diffuses, under certain conditions, into solid iron, and to this subject the fourth part of Mr. Stead's paper is devoted. His results confirm the above conclusions and show that the quantity of phosphide which will diffuse is very considerable, amounting to about 1 per cent of phosphorus. The temperature at which diffusion begins, however, has not been ascertained.

NOTES ON THE CHEMICAL CONSTITUTION OF CAST IRON AND STEEL *

By A. CARNOT and E. GOUTAL

OUR knowledge of the constitution of cast iron and steel has made much progress in recent years, thanks to micrographic studies, in which Sorby and Osmond have been the pioneers, and to the chemical and physical researches carried on simultaneously in France, England, Germany, Holland, Austria, Russia and in the United States.

Through purely chemical methods, four conditions of carbon had been detected:

1. *Graphite* or free crystallized carbon, not acted upon by acids even when concentrated.

2. *Graphitic carbon or temper carbon,* which is also in a free state and unaffected by acids, but presenting an amorphous appearance.

3. *The carbon of the carbide of iron* (cement carbon), forming a definite compound, acted upon by concentrated acids, but insoluble in diluted, cold acids. This carbide answers to the formula Fe_3C.

4. *Hardening carbon,* which appears to form a solid solution with the iron (or which is perhaps present as a compound, itself

---

* *Annales des Mines*, October, 1900.

dissolved in the iron), and which is acted upon by cold, dilute acids.

Micrographic methods have made it possible to carry these investigations further, and to detect, in carefully polished and etched sections, the various constituents of iron and steel. Four constituents are to-day known in carbon steel:

1. *Ferrite* or nearly pure malleable iron.

2. *Sorbite,* or iron slightly carburetted, not yet well defined, but which is acted upon by dilute, cold acids.

3. *Pearlite,* having a pearly or watered appearance and composed of alternate plates of ferrite or sorbite, and of the carbide Fe_3C.

4. *Cementite* or the carbide Fe_3C generally present in plates, standing in relief after etching with nitric acid.

The indications furnished by the microscope have been confirmed by chemical tests. Abel, Weyl and Müller succeeded in completely isolating the insoluble cementite plates from the easily soluble ferrite.

5. Finally three other constituents, *martensite, troostite* and *austensite* forming the bulk of hardened steel, are not yet clearly defined from a chemical standpoint. The carbon which they contain is changed to gaseous hydrocarbons by the action of diluted hydrochloric or sulphuric acid; it remains as a black residue when acted upon by nitric acid of 1.20 specific gravity, which is readily dissolved, imparting a brown coloration to the solution.

Steel, cast iron and alloys of iron in general contain other elements besides carbon, but their chemical condition was in most cases a matter of mere speculation.

It appeared to us that chemical analysis could give important information concerning the condition of the elements and the composition of the constituents that may be present in the metallic mass.

Our purpose was, therefore, to undertake a sort of *proximate analysis* of metallurgical products, by means of various reagents, capable of dissolving the ferrous matrix without affecting the foreign elements or constituents, or which could dissolve some of the constituents to the exclusion of those which it was desired to study. After having thus isolated them, they were subjected to an ultimate analysis, so as to ascertain as exactly as possible the combinations formed by the various elements.

In order to be able to isolate successively each of the constituents in sufficient quantity and in the best possible conditions, so as to avoid serious difficulties of interpretation, it was necessary to use metallurgical products of special quality, which contained large proportions of the elements to be studied, and small proportions of all other elements. It was, for instance, found desirable to employ for our researches such products as ferrosilicons, ferrochromes, silicospiegels, etc.

The samples which we used, however, were in every case industrial products or at least manufactured in metallurgical furnaces, for it is quite possible, owing to the very high temperature prevalent in electrical furnaces, that some compounds are produced in these furnaces different from those obtained at lower temperatures. We shall have to call attention on several occasions to the different combinations obtained through the influence of the electric arc. In our personal researches strictly metallurgical products only have been used, in order that our results might be applied industrially.

We must here express our thanks to the managers of steel works who have had the kindness to furnish us, and sometimes to prepare especially for us, the products which it was desired to examine. Special mention should be made of the works of Saint-Jacques (Montluçon), of Saint Louis (Marseille) and of Unieux (Loire).

We shall divide our investigation in two parts:

The first one will deal with the ordinary elements, exclusive of carbon, found in steel and cast iron.

The second part will deal with the rarer elements, found especially in special steels.

In the first part we shall endeavor to define the condition of the ordinary elements, silicon, sulphur, phosphorus, arsenic and manganese in several common metallurgical products. We shall illustrate how the results make it possible to figure the complete chemical constitution of cast iron.

In the second part we shall deal with the rarer elements which have in recent years been introduced in steel, to improve it, and which in some instances have yielded alloys possessing exceptional industrial qualities. The elements examined are chromium, tungsten, molybdenum, titanium, copper and nickel.

Part I

ORDINARY ELEMENTS, EXCLUSIVE OF CARBON, CONTAINED IN STEEL AND CAST IRON (SILICON, SULPHUR, PHOSPHORUS, ARSENIC AND MANGANESE).

SILICON. — 1. Some ferrosilicons containing as little manganese as possible (about 1%) and from 10 to 14% of silicon, were first experimented upon.

The metal was dissolved in dilute hydrochloric acid in an atmosphere free from air; 250 cubic centimeters of dilute hydrochloric acid (7 parts acid in 100 parts of water) were poured in a glass flask through which a current of carbonic acid gas was made to pass. The acid was boiled for a few minutes and a few grammes of very finely ground ferrosilicon was then introduced. The current of carbonic acid was again started and the temperature of the solution maintained at about 40° Cent., shaking the flask from time to time. After all effervescence had ceased, the clear solution holding in suspension a few flakes of silicic acid ($H_2Si_2O_3$), was decanted and the residue poured on a filter, washed with dilute hydrochloric acid and then with alcohol in order to prevent subsequent oxidation. After drying the residue, a black powder easily attracted could readily be separated from it by means of a magnet.

This powder contains, together with a little graphite, a silicide of iron, which when moist or dry (provided it is not calcined) is rapidly decomposed in contact with alkalies, and slowly decomposed even by pure water, with evolution of hydrogen, a property which recalls those of silicic acid and of hydrogen silicide.

The composition of the silicide of iron is readily determined by treating it with hydrochloric acid and bromine, evaporating to dryness, separating the resulting silica and precipitating the oxide of iron by ammonia.

These analyses lead us to assign to this compound the formula FeSi, the results obtained being as follows:

In 100 parts of metal:

	I	II	III
Silicon	0.0607	0.1474	0.1936
Iron	0.1260	0.2905	0.3790

or in 100 parts of silicide:

	I	II	III
Silicon	32.50	33.65	33.81
Iron	67.49	66.35	66.19

The formula FeSi calls for 33.33 per cent of silicon and 66.67 per cent of iron.

2. The same samples were exposed for fifteen days to the action of cold and very dilute sulphuric acid. (5 parts of acid in 100 parts water. A complex residue was obtained made up principally of carbon, various silicides and of some silicic acid (itself due, undoubtedly, to the decomposition of certain silicides).

By treating this residue with a hot, dilute solution of caustic potash both the silicic acid and the ferrous silicide were removed, while the hydrates of iron and manganese were dissolved by dilute sulphuric acid, either alone or mixed with a few drops of oxygenated water. The residue is then washed and gently ignited. Finally a magnetic powder is separated from the residue.

The analysis of this magnetic part shows that it consists of a compound having the formula Fe_2Si (20 per cent of silicon and 80 per cent of iron), four experiments having yielded the following results:

In 100 parts of metal:

	I	II	III	IV
Silicon . .	0.0473	0.0902	0.1197	0.1410
Iron . .	0.1918	0.3625	0.4886	0.5665

or in 100 parts of silicide:

	I	II	III	IV
Silicon	19.75	19.86	19.68	19.94
Iron	79.92	79.85	80.36	80.07

3. Some samples of silico-spiegels containing 20 per cent of manganese and 12 per cent of silicon were treated in the same way. A residue was obtained consisting of free carbon, and of a double silicide of iron and manganese, which in five experiments was found to have the following composition:

	I	II	III	IV	V
Silicon .	15.40	15.80	14.10	14.04	14.00
Iron .	68.25	68.60	68.80	69.15	71.22
Manganese	15.10	14.70	16.30	16.80	14.58

These various compositions answer fairly well to the general formula M_3Si, in which M represents the sum of both metals: iron and manganese.

The presence of a large proportion of manganese, appears to cause, therefore, an important modification in the composition of the metallic silicide.

4. Upon pulverizing several samples of ferromanganese, whose mass was not sensibly magnetic, it was found that it contained, nevertheless, some particles attracted by the magnet. These particles were relatively poor in manganese, but much richer in silicon and iron than the metal as a whole. Three samples of ferromanganese, for instance, gave the following composition:

	I	II	III
Iron	40.30	42.60	68.83
Manganese	51.60	51.15	27.10
Silicon	0.55	1.04	0.95

It was possible by means of a magnet to separate from the powder about 3 per cent of magnetic particles which were found to have the following composition:

	I	II	III
Iron	88.20	84.02	89.80
Manganese	7.20	9.45	2.90
Silicon	3.95	6.44	6.50

These particles were treated successively with very dilute acid, caustic potash and again dilute acid, and left a residue which proved to be the silicide Fe_2Si, previously found in ferrosilicons containing little manganese.*

It may be concluded from the above experiments that ferrosilicons contain two combinations of iron and silicon, corresponding respectively to the formulas FeSi and Fe_2Si. They may also contain, when they are sufficiently rich in manganese, like silico-spiegels, a silicide answering to the formula Mn_3Si, calling for quite a large proportion of manganese.

The compound Fe_2Si, prepared for the first time by Hahn,† had been obtained by Mr. Moissan in an electrical furnace.‡ It has also been very recently isolated by Mr. Lebeau from industrial ferrosilicons prepared in an electric furnace and containing from 10 to 20 per cent of silicon.§

* *Comptes Rendus,* CXXVIII, page 933.
† *Annalen der Chemie und Pharmacie,* CXXXI, page 57.
‡ *Comptes Rendus,* CXXI, page 621.
§ *Comptes Rendus,* CXXXI, page 933.

The silicide FeSi, which had been obtained by Fremy and by Hahn has been prepared, since our last experiments by Mr. Lebeau* in an electrical furnace, but in presence of silicide of copper, whose possible effect may have been to lower the temperature of the reaction, bringing it nearer that of metallurgical furnaces.

It should, however, be noted that the properties of these silicides are not absolutely identical, those produced in the electrical furnace being unaffected by acids, while those obtained in the blast furnace, are, on the contrary, quite readily acted upon by warm acids, even when diluted.†

5. We tried to isolate the above silicides from ordinary cast irons, but without success. We are led to believe that cooling causes the decomposition of the silicide FeSi, or else that the silicide forms with the excess of iron a solid solution or homogeneous mixture.

The absence of free silicide of iron in steel appears also to be shown by the experiments of Mr. H. Le Chatelier concerning the electrical resistance of steel,‡ for they showed that the resistance increases proportionally with the amount of silicon, from which it must be inferred that silicon exists in steel in the condition of a solid solution or homogeneous mixture.

The analysis of the residues resulting from the treatment of cast iron by dilute sulphuric acid without admission of air, led to the detection of a small quantity of a silicide of manganese corresponding apparently to the formula MnSi. The proportions of manganese found in the residues after separation of the light parts were as follows:

	a	b	c
Manganese	0.108	0.102	0.134
Silicon	0.053	0.047	0.071

or in percentage:

	a	b	c
Manganese	67.08	68.45	65.87
Silicon	32.92	31.54	34.62

* *Comptes Rendus*, CXXVIII, page 933.

† Two other iron silicides, Fe_3Si_2 and $FeSi_2$, were detected in some ferrosilicons containing from 25 to 50 per cent silicon and produced in an electrical furnace by the Wilson Aluminium Co. The samples were treated by dilute hydrofluoric acid (American Chemical Society, XXI, page 59).

‡ *Comptes Rendus*, June 13, 1898.

The formula MnSi calls for 66.26 per cent of manganese and 33.74 per cent of silicon.

This silicide of manganese, however, which corresponds only to a small part of the silicon, is always mixed with an excess of carbide of iron. We have been unable to isolate it and in order to ascertain its formula, it was necessary to make numerous analyses of a great number of samples.

SULPHUR. — In order to isolate the sulphuretted compounds of cast iron and steel, the method used in the case of silicides cannot be applied, for the ferrous products treated by dilute acids, give up nearly their totality of sulphur as sulphuretted hydrogen.

In dissolving them in neutral chloride of copper and potassium, on the contrary, the totality of the sulphur remains in the residue. This residue contains, sometimes, some sulphide of iron, FeS, but it was generally found to be made up mostly of sulphide of copper, CuS.

It was found through some direct experiments, that sulphide of iron is not decomposed by the double chloride. In melting, for instance, a mixture of pure iron and of sulphide of iron, and treating the pulverized mass by some double chloride, a residue was left having the following composition:

Iron	0.1542 or 63.80 per cent
Sulphur	0.0875 or 36.20 " "

The formula FeS corresponds to 63.63 per cent of iron and 36.37 per cent of sulphur.

We inferred from this that sulphur must be present in most metallurgical products, combined with some element other than iron forming a compound easily decomposed by copper chloride. This element is manganese.

We confirmed this conclusion by subjecting to the same treatment a mixture of pure iron, sulphide of iron and ferromanganese. The result was a sort of cast iron containing 1.55 per cent of sulphur and 0.82 per cent of manganese. The residue left by the solution of 5 grammes of the metal in the double chloride contained:

Copper	0.0480
Iron	0.0932
Sulphur	0.0773

The residue was made up, therefore, of a mixture containing 0.1465 grammes of the sulphide of iron FeS and 0.0720 grammes of the sulphide of copper CuS. The 48 milligrammes of copper are precisely equivalent to the 41 milligrammes of manganese which were present in five grammes of the cast iron tested. Our inference, therefore, that cast iron rich in sulphur and also in manganese, contains the compound MnS rather than FeS, is well founded. Under the influence of copper chloride, the sulphide of manganese is transformed into chloride of manganese and copper sulphide. This conclusion, moreover, agrees perfectly with the fact well known to metallurgists, that the sulphur content of cast iron may be lowered through the addition of spiegeleisen or ferro-manganese. It accounts for the segregation of manganese and sulphur, so frequently detected, in the upper part of ingots and castings slowly cooled.

It seemed to us interesting to ascertain whether quenching modified the condition of sulphur in steel. To that effect, quenched steel containing 0.47 per cent of carbon, 0.65 per cent of manganese and 0.18 per cent of sulphur was treated with neutral chloride of copper.

The insoluble residue, after being carefully washed, contained nearly the totality of the sulphur, combined with copper in atomic proportions. Ten grammes of steel have given a residue containing 16.8 grammes of sulphur and 34.1 grammes of copper.

It may, therefore, be admitted that quenching does not modify the condition of the sulphur in steel containing manganese.

Phosphorus.—In treating some cast iron rich in phosphorus with some double chloride of copper and potassium absolutely neutral, the phosphorus remains in its entirety in the insoluble residue, as phosphide of iron, occasionally mixed with a very small quantity of phosphide of manganese, and also with some silica, silicic acid, carbon and sulphide of copper.

As phosphide of iron is only attracted very slightly by a magnet, it was not possible to separate it by this means, and in order to ascertain its formula, it was necessary to compare and discuss a great many analyses of the total residues left by cast iron and steel rich in phosphorus.

To analyze these residues, they were dissolved with nitric acid and bromine; the filtered solution was divided in two parts; in one of them, the iron was precipitated by ammonia, redissolved in dilute sulphuric acid, reduced with metallic zinc and titrated with permanganate; in the other portion the organic matters were destroyed by chromic acid and the phosphoric acid precipitated with ammonium molybdate.

We give here the results obtained in three such analyses:

	I	II	III
Phosphorus	0.0068	0.0336	0.0030
Iron	0.0360	0.0180	0.0152

or in percentage:

	I	II	III
Phosphorus	15.9	15.7	16.5
Iron	84.1	84.3	83.5

The phosphide of iron present in cast iron and steel, corresponds, therefore, to the formula Fe_3P, whose theoretical composition calls for 15.60 per cent of phosphorus and 84.40 per cent of iron.

Leopold Schneider had obtained, under slightly different conditions, a nearly identical result.* It is also the formula found long ago by Schrötter, Hvoslef and Percy for the phosphides prepared in the laboratory.

In order to ascertain whether this compound was still present in quenched steel, a sample of quenched steel containing 0.35 per cent of carbon, 1.48 per cent of manganese and 0.37 per cent of phosphorus, was treated in the same way. Ten grammes of the metal yielded an insoluble residue containing 0.0332 grammes of phosphorus and 0.165 grammes of iron and free from manganese. The condition of the phosphorus is not therefore modified by quenching in the case of steel containing manganese.†

Arsenic. — In dissolving cast iron or steel containing much arsenic in double chloride of copper and potassium, the whole of

---

* *Oesterreichische Zeitschrift für Berg- und Hüttenwesen*, page 735, 1886.

† The present paper was already printed and ready for circulation, when we noticed in *Engineering* for October 19, 1900, the first part of a very interesting investigation on the phosphides of iron, recently presented to the Iron and Steel Institute by Mr. J. E. Stead.

The authors review briefly here the results obtained by Mr. Stead. The reader is referred to the last three numbers of *The Metallographist* in which Mr. Stead's paper was published in full. — Ed.

that impurity passes into the solution, the residue being absolutely free from it.

If the same metal be treated with weak hydrochloric acid, with the precautions described in the case of ferrosilicons, the whole of the arsenic remains in the insoluble residue;

Faint traces of the impurity only can be detected in the gaseous product of the reaction and in the solution.

A sample of steel containing 4.25 per cent. of arsenic was experimented upon. Two grammes of this steel, containing, therefore, 0.085 grammes of arsenic, left an insoluble residue weighing 0.0852 grammes, entirely free from iron and made up exclusively of arsenic. It is, therefore, inferred that arsenic does not form any compound in slowly cooled iron and steel, but is simply dissolved in the metal.

It is different in the case of hardened steel. Ten grammes of hardened steel containing 1.12 per cent of arsenic, and 0.09 per cent of carbon was dissolved in a 7 per cent solution of nitric acid, and left a residue weighing, after being washed with alcohol and dried, 0.265 grammes and containing nearly the whole of the arsenic. Upon being heated in a stream of pure and dry nitrogen, to about 250° C., the residue lost 0.126 grammes made up of arsenic and water. The composition of the remaining residue was ascertained by treating it with nitric acid and evaporating in presence of an excess of sulphuric acid, until white fumes were evolved. The solution was diluted and the silica separated by filtration. The arsenic acid was then reduced by sulphurous acid and the sulphide of arsenic precipitated by sulphuretted hydrogen. The arsenic sulphide was freed from a small amount of copper sulphide also formed by means of ammonium carbonate, finally the iron after being oxidized was precipitated with ammonia.

The results obtained were as follows:

Arsenic	45.2 per cent
Iron	40.0 " "
Copper	11.6 " "
Silica	2.4 " "

from which the following proportions between the atomic weights may be calculated

Arsenic	0.7
Iron	0.8
Copper	0.2

Leaving aside the silica, which may be considered as an impurity, the formula of this complex arsenide would approach M_3As_2. The presence of some copper in the sample used, however, left some doubt concerning the true composition of the arsenide of iron. It was desirable to repeat the experiment with a sample free from copper. For that purpose the sample of steel containing 4.25 per cent of arsenic which we used in our first experiment was quenched at different temperatures. Twenty grammes were treated with sulphuric acid as stated and gave the following results:

Quenching temperature .	800°	1100° Cent
Weight of arsenide . .	0.325	0.839 grammes
Weight of arsenic . .	0.1308	0.338 "
Weight of iron . . .	0.1920	0.499 "

or in percentages:

Arsenic	40.2	40.3 per cent
Iron . , ...	59.0	59.5 " "

If we neglect a small amount of carbide of iron difficult to expel, the proportion of the atomic weights leads almost exactly to the formula Fe_2As, which corresponds to 40.1 per cent of arsenic and 59.7 per cent of iron.

Contrary to the results obtained with the steel containing copper, an appreciable quantity of arsenic entered the solution upon dissolving the sample in dilute acid.

To sum up, it seems certain that slowly cooled steel contains only uncombined arsenic, while quenched steel contains also an arsenide of iron. We shall adopt for this compound the formula Fe_2As as being probably the correct one, admitting, however, that in steel rich in copper, a single or double arsenide richer in arsenic may be formed.

Arsenic acts, therefore, much like carbon; it forms, under the influence of sudden cooling, some definite compounds, while under the influence of slow cooling it remains free.

It is interesting to note how different are the parts played by arsenic and phosphorus. The latter is always wholly combined, while arsenic, which is partially combined in hardened steel, is absolutely free in slowly cooled steel. This difference in the chemical condition of these two elements, otherwise so similar in many respects, appears to afford a ready explanation of the very great difference noted recently in the influence exerted by these elements upon the properties of steel.

MANGANESE. — The experiments described in the foregoing pages have already shown that manganese possesses a special affinity for silicon and sulphur. When it is present in small quantity in cast iron, it may exist wholly as a silicide or as sulphide of manganese.

When it is present in larger proportions, it forms some double carbides, whose study presents much difficulty, owing to the fact that several such carbides may exist together, possessing nearly the same properties.

In order to distinguish between them, it was necessary to subject to the action of various reagents, over forty samples, whose manganese content varied progressively from 5 to 85 per cent.

The results obtained are summed up below:

The richest ferromanganeses are acted upon by boiling water. They yield gaseous and liquid hydrocarbons mixed with some free hydrogen. This fact had been detected by Cloëz,* in 1878, and had suggested an hypothesis concerning the mineral origin of petroleum, at the time when Mendeelejeff proposed a similar explanation.

The action of cold water is generally very slight, even when the metal contains 85 per cent of manganese. The resulting gases are a complex mixture in which hydrogen predominates.

It appears logical to infer from these results that ferromanganese does not contain, at least in appreciable quantity, the single carbide of manganese Mn_3C, discovered by Messrs. Troost and Hautefeuille, in 1875,† for Mr. Moissan has shown that the decomposition of this carbide by cold water yields a mixture of equal volumes of methane and hydrogen, free from liquid hydrocarbon.‡

Ferro-manganese containing less than 74 per cent of manganese is not appreciably acted upon by boiling water, but it is affected even in the cold, by a five per cent solution of acetic acid. The reaction is the more pronounced, the more manganese

* *Comptes Rendus,* LXXXVI, page 1248.
† *Comptes Rendus,* LXXX, page 909.
‡ *Comptes Rendus,* CXXII, page 421.

present; the insoluble residue amounts to one-half of the total weight when the metal contains 35 per cent of manganese, and only to one-fourth when it contains 55 per cent, while rich ferromanganeses are entirely dissolved. They are, likewise, strongly acted upon by a solution of ammonium chloride, even when an excess of ammonia is present.

They dissolve slowly, on the contrary, in a solution of acetate of ammonium slightly ammoniacal and boiling. It is with this last reagent that we succeeded in isolating, through a slow action lasting several days, the carbides present in the richest ferromanganeses.

1. The samples used in our experiments contained (a) 84 per cent and (b) 79 per cent of manganese.

The treatment with boiling ammoniacal solution of ammonium acetate, with exclusion of air, left some crystalline, non-magnetic residues not acted upon by boiling water, but soluble in cold and very dilute acetic acid. These residues consist of a double carbide of iron and manganese corresponding to the formula $Fe_3C.4Mn_3C$, for their analysis showed the following composition:

	a	b		a	b
Manganese . . .	0.3908	0.5580	or in per cent	71.05	74.40
Iron	0.0937	0.1380	" "	17.041	8.64
Carbon and oxygen	0.0523	0.0515	" "	9.51	6.87
Silica	0.0132	0.0670	" "	2.40	0.09

The theoretical composition of the formula $Fe_3C.4Mn_3C$ calls for:

Manganese	74.32	per cent
Iron	18.92	" "
Carbon	6.76	" "

The presence of some silica in (a) is evidently accidental; it could not be removed either by repeated washings, or by methyl iodide. Such is also the case with a certain quantity of oxide of manganese, set free by the ammoniacal reagent, and this difficulty to purify the residue, accounts very well for the slight difference between the composition of (a) and the theoretical composition. The analysis of the residue (b) gave results agreeing perfectly with the formula.

The manganese content of 74 per cent, which corresponds to the formula $Fe_3C.4Mn_3C$, is precisely the percentage below which ferromanganese ceases to be acted upon by boiling water. Like the residues left by richer alloys, they are, moreover, deprived of magnetic properties, decomposable by copper chloride, by very diluted mineral acids, even by a 5 per cent solution of acetic acid.

2. Ferromanganeses whose manganese content varied between 74 and 60 per cent, when treated in the cold with a 5 per cent solution of acetic acid, left a residue consisting of a double carbide, crystalline, non-magnetic and whose composition corresponds to the formula $Fe_3C.2Mn_3C$.

Here are for instance, the results obtained with two samples (a and b) containing respectively 70 and 65 per cent of manganese.

	a	b		a	b
Manganese	0.3035	0.3944	or in per cent	60.70	61.65
Iron	0.1606	0.2006	" "	32.12	31.55
Carbon	0.0337	0.0429	" "	6.74	6.70
Silica	0.0025	0.0024	" "	0.51	0.38

The theoretical composition of the formula $Fe_3C.2Mn_3C$ is:

Manganese	61.82
Iron	31.46
Carbon	6.73

3. The products containing from 60 to 30 per cent of manganese when they are treated in the same way leave a residue consisting of a mixture of double carbides, in the shape of a crystalline powder. One of these carbides is the one just described, while the other which is also non-magnetic corresponds to the formula $2Fe_3C.Mn_3C$.

This last double carbide is affected like the preceding one by the various reagents, boiling water, diluted, cold, mineral acids, acetic acid and copper chloride. It is, therefore, very difficult to obtain it pure, from the treatment of ferromanganese containing over 30 per cent of manganese. In dissolving a sample of cast iron contains 35.70 per cent of manganese, 57.40 per cent of iron and 6.90 per cent of carbon, however, a residue was obtained containing:

Manganese	0.2576	or in per cent	30.31
Iron	0.5297	" "	62.32
Carbon	0.0612	" "	6.96
Silica	0.0024	" "	0.28

The composition of the formula $2Fe_3C.Mn_3C$ is as follows:

Manganese	30.72
Iron	62.57
Carbon	6.70

This last double carbide shows a tendency to segregate during the slow cooling of large castings of ferro-manganese. A large sample from the old works of Terrenoire exhibited a crystallized portion made up of large plates (d), while the balance of the piece had a crystalline structure (c). One gramme of each portion was analyzed with the following results:

	d	c
Manganese	30.45	46.19
Iron	62.42	47.14
Combined C	6.80	5.98
Graphitic C	0.10	0.25
Silicon	0.02	0.19

The lamellar portion, therefore, had exactly the composition of the double carbide $2Fe_3C.Mn_3C$.

Such sharp separations are seldom found, but appreciable segregations frequently occur in ferromanganese, probably owing to the difference between the melting-points of the various carbides.

4. The metallurgical products containing between 30 and 18 per cent of manganese can no longer be dissolved by acetic acid, whose action is then very slight. They are, on the contrary, too easily affected by sulphuric acid, even very diluted and cold, which causes a nearly complete solution, only a small quantity of silicide and of carbide of iron remaining undissolved.

5. Finally, in the case of a manganese content lower than 18 per cent the free carbide Fe_3C begins to be found in appreciable quantity in the residue left by the treatment with very dilute and cold sulphuric acid. In treating 10 grammes of powdered cast iron containing respectively 10, 5 and 2.4 per cent of manganese with 200 cubic centimeters of diluted, cold sulphuric acid, with exclusion of air, the amounts of Fe_3C left as a residue weighed respectively, 0.808, 1.130 and 1.720 grammes.

The results of the analysis of these residues confirmed the accuracy of the formula Fe_3C previously found by several investigators for the normal carbide of iron or cementite.

It is inferred from the last two experiments that the double carbide with the smallest amount of manganese, contains about 18 per cent of that element; both carbides are united in the proportion of one atom of manganese and four atoms of iron, and the composition would be expressed by the formula $4Fe_3C.Mn_3C$. This double carbide, which it was not possible to separate from the excess of Fe_3C, present in the samples experimented upon, seems to be the same as that detected by Behrens, in a steel containing 1.73 per cent of manganese and 0.55 per cent of carbon, and to which he assigned the formula $Fe_7Mn_2C_3$, or in multiplying the coefficients by 5-3, a formula approaching closely the one adopted by us: $Fe_{12}Mn_3C_5$ or $4Fe_3C.Mn_3C$.

Conclusions.—The results of the proximate analysis of metallurgical products just described, permit us to arrive at the following general conclusions, regarding the chemical condition of the elements commonly found in cast iron and steel.

1. The sulphur is found combined with manganese nearly completely, as the sulphide MnS, which is soluble in acids and in cupric chloride; the excess only, if there is any present, forms the sulphide of iron FeS.

2. Phosphorus is combined with iron as a phosphide Fe_3P, soluble in acids and insoluble in absolutely neutral chloride of copper and potassium.

3. Arsenic is nearly always in a free state or forming a solid solution with the iron. In hardened steel, it is found partially combined as an arsenide, Fe_2As, soluble in cupric chloride and insoluble in very dilute acids.

4. Silicon exists usually free in cast iron. It may unite with manganese and iron to form the silicides MnSi and FeSi. The latter compound, however, is dissociated through the action of slow cooling, or at any rate, it cannot be isolated by solution in dilute acids. The silicide MnSi remains after slow cooling; it is insoluble in dilute acids and also in cupric chloride. Ferrosilicons contain both the compounds FeSi and Fe_3Si. The silicospiegels may contain a double silicide M_3Si.

5. Carbon is combined with iron and with manganese. In ferromanganese it forms double carbides whose composition and properties vary with the manganese content of the alloy. The following carbides were detected:

$Fe_3C.4Mn_3C$	for a manganese content varying	between	85 and 74	per cent
$Fe_3C.2Mn_3C$	" "	"	74 and 60	"
$2\ Fe_3C.Mn_3C$	" "	"	60 and 30	"
$4\ Fe_3C.Mn_3C$	" "	below	18	"

It is this last carbide, containing the smallest amount of manganese, which seems to be present in cast iron and spiegeleisen. The carbide is insoluble in acetic acid but soluble in dilute sulphuric acid and is decomposed by cupric chloride.

6. The excess of combined carbon is found combined with iron, principally as cementite Fe_3C, insoluble in dilute acids, and to a small extent, as an ill-defined sub-carbide, acted upon by acids. The carbon, which is dissolved at a high temperature, separates during cooling in the graphitic state, leaving part of the iron free or possibly in combination with hydrogen.

Chemical Constitution of Ordinary Cast Iron

As an example of the application of the principles just described let us propose to ascertain the constitution of a sample of cast iron whose ultimate analysis is known.

The composition was as follows:

Iron	94.00
Silicon	0.63
Phosphorus	0.15
Arsenic	0.05
Sulphur	0.12
Manganese	2.00
Combined carbon	2.45
Graphite	0.60
	100.00

The solution in dilute sulphuric acid left a residue consisting of carbide of iron, silicide of manganese and graphite, which represented, after washing with alcohol and drying at 180° Cent., 25.92 per cent of the weight of the cast iron.

The analysis of the residue indicated the following composition:

Silicon	0.21	or 0.62 per cent MnSi
Manganese	0.41	
Graphite		0.60
		1.22

Fe_3C (by difference) = 25.92 — 1.22 = 24.70 per cent.

0.12 per cent of sulphur entirely combined with manganese give 0.33 per cent of sulphide of manganese; 0.15 per cent phosphorus combined with iron give 0.96 per cent of phosphide. The excess of the manganese, which is not combined either with sulphur or with silicon, exists as a double carbide corresponding to the formula $4Fe_3C.Mn_3C$.

The chemical constitution of the cast iron is, therefore, as follows:

Sulphide of manganese, MnS	0.33	Manganese Sulphur	0.21 0.12
Phosphide of iron, Fe_3P	0.96	Iron Phosphorus	0.81 0.15
Silicide of manganese, MnSi	0.62	Manganese Silicon	0.41 0.21
Double carbide of iron and manganese, 4 Fe_3C. Mn_3C	7.48	Manganese Iron Carbon	1.38 5.60 0.50
Carbide of iron, Fe_3C	24.70	Iron Carbon	22.07 1.63
Carbon as ill defined subcarbide	0.32		
Graphitic carbon	0.60		
Free silicon	0.42		
Free arsenic	0.05		
Iron free or combined with hydrogen, or as undefined carbide	64.52		
	100.00		

Part II

Rarer Elements Found in Special Steels (Chromium, Tungsten, Molybdenum, Titanium, Copper, Nickel).

Chromium.—In our first investigations we used some chromium steel owing to the difficulty of dissolving ferro-chromes in acids. Later, however, we succeeded in dissolving them.

1. Two samples of ferro-chromes of the following composition were used:

Chromium	57.6	59.1
Iron	32.6	32.3
Carbon	9.9	9.1

In subjecting them to the action of pure hydrochloric acid for two days at a temperature of 60° Cent., some insoluble residues were obtained, which were purified by means of a heavy liquor of tungstate of boron and then treated with sulphate of potassium as previously described.*

The residues showed the following composition:

	I	II		I	II
Chromium	0.5216	0.6213	or in per cent	65.2	65.4
Iron	0.1884	0.2195	" "	23.6	23.1
Carbon	0.0904	0.1083	" "	11.3	11.4

This composition corresponds almost exactly to the formula $Fe_3Cr_9C_7$, which calls for 65.2 per cent of chromium, 23.2 per cent of iron and 11.6 per cent of carbon.

This formula may also be written $Fe_3C.3Cr_3C_2$, and we must here note that Fe_3C represents the composition of cementite and that Cr_3C_2 is the formula found by Mr. Moissan for the carbide of chromium, which he prepared in an electrical furnace. The compound found in ferro-chrome from metallurgical furnaces appears to be, therefore, a combination of these two carbides.

Messrs. Behrens and Van Linge had before us examined a sample of ferro-chrome containing 50 per cent of chromium.† In treating it with concentrated hydrochloric acid a residue was left having the following composition:

Chromium	65.2
Iron	24.6
Carbon	10.4

from which they inferred that the residue was a compound of the formula Cr_3FeC_2. We are inclined to think that this compound is identical to the one we found. The results of both analyses are indeed slightly different. Our results lead exactly to the formula $Fe_3C.3Cr_3C_2$, which we prefer as corresponding to the combination of two well-defined carbides.

A short time after our first publication, Mr. Williams obtained, in an electrical furnace, a double carbide belonging to the same formula, for its composition would be represented by the

* Ad. Carnot, "Méthodes d'Analyse des Fontes, des Fers et des Aciers," *Annales des Mines,* October, 1895.

† "Recueil des Travaux Chimiques des Pays-Bas," XIII, 1894.

formula $3Fe_3C.2Cr_3C_2$* made up of the two single carbides, Fe_3C and Cr_3C_2.

More recently, Mr. Williams prepared in an electrical furnace two new double carbides, also belonging to the same series: the carbide $Fe_3C.2Cr_3C_2$ and the carbide $2Fe_3C.3Cr_3C_2$.†

We see in these facts additional reasons to assign the formula $Fe_3C.3Cr_3C_2$ to the carbide found in commercial ferrochrome.

2. Chromium steels when they are highly carburetted are easily dissolved in acids, even when diluted and cold, but much less readily when they contain only a small proportion of carbon. On the other hand if the steel be treated by double chloride of copper and potassium, the whole of the chromium remains in the insoluble residue, when the steel contains much carbon, while if it contains only a small amount of carbon, a portion of the chromium is found in the solution. This constitutes a very great difficulty in comparing the residues left by chromium steels containing various amounts of carbon.

Fairly good results were obtained, however, under the following conditions:

Two samples of steel A and B, containing:

	Chromium	Carbon
A	2.35	2.00
B	2.01	0.57

were treated with cold and very dilute hydrochloric acid (7 per cent acid). The residues were found to have the following composition:

	A	B		A	B
Chromium	0.1253	0.0931	or in per cent	21.60	21.16
Iron	0.4066	0.3126	" "	70.10	71.05
Carbon	0.0470	0.0334	" "	8.10	7.60
Silicon	0.0012	0.0014	" "	0.14	0.32

The composition of residue A corresponds almost exactly to the formula $Fe_9Cr_3C_5$ whose composition is as follows:

Chromium	21.8
Iron	69.9
Carbon	8.3

* *Comptes Rendus,* October, 1898.

† These carbides, like the preceding one, were exhibited at the Exposition (Class of Electrochemistry).

This formula may also be written $3Fe_3C.Cr_3C_2$.

The composition of residue B differs little from that formula, and the difference may be explained by the presence of silicon, which indicates the presence of a silicide in the products analyzed. If the silicide has the formula Fe_2Si, it will be necessary to subtract 1.04 per cent of iron (required to combine with 0.31 per cent of Si), and the remaining composition will then correspond very closely to that of the formula.

3. A sample of ferromanganese containing 18 per cent of manganese and 0.97 per cent of chromium was dissolved in double chloride of copper and potassium. Seven grammes of the metal yielded a residue weighing 0.2982 grammes, in which was found nearly the totality of the chromium (64.4 out of 67.9 grammes).

The results of the analysis of the residue was as follows:

Chromium	0.0645	or in per cent	21.63
Iron	0.2093	" "	70.18
Carbon	0.0248	" "	8.33

These results also lead to the formula $Fe_9Cr_3C_5$ or $3Fe_3C$.-Cr_3C_2.

It is interesting to note that the whole of the chromium is in this sample in the condition of carbide united to the carbide of iron.

The composition of the insoluble residues after solution of the iron, in the case of steel containing a small amount of chromium, indicate, therefore, the existence of the same single carbides as in the case of ferro-chromes, but the carbide of iron and that of chromium appear to be united in different proportions, varying with the amount of chromium present in the metal.

In the case of ferro-chromes three molecules of carbide of chromium are united to one molecule of iron carbide. In the case of steel with a small percentage of chromium, one molecule of carbide of chromium is united to three molecules of iron carbide.

It is interesting to note that the double carbides of chromium and iron may constitute, like those of manganese of iron, a series of numerous compounds varying with the amount of chromium and manganese of the metallurgical product.

Tungsten.—In dissolving a sample of steel containing 6 per cent of tungsten and from 0.4 to 0.6 per cent of carbon, in very dilute hydrochloric acid, and heating gently without admission of air, a residue is left consisting of a compound of iron and tungsten, whose composition corresponds to the formula Fe_3W. Hardly a few milligrammes of tungsten are dissolved.

Two different samples have given the following results:

Tungsten .	0.1385	0.1526	or in per cent	51.58	51.73
Iron . .	0.1300	0.1424	" "	48.42	48.27

The formula Fe_3W calls for 52.27 per cent of tungsten and 47.73 per cent of iron.

In treating this alloy with double chloride of copper and potassium, it is decomposed leaving a residue made up of free tungsten and of oxide of tungsten. Mr. Behrens detected in ferro-tungsten, (containing 50 per cent tungsten), the presence of a crystallized alloy made up of octahedra, whose formula would approach Fe_2W.

2. In treating in the same way a steel more highly carburetted a double carbide was isolated, magnetic and corresponding to the formula Fe3C.CW. The samples of steel experimented upon contained (A) 6.1 per cent tungsten and 2.00 per cent carbon, and (B) 7.8 per cent tungsten and 2.2 per cent carbon.

After being crushed and treated with a 10 per cent hydrochloric acid solution, insoluble residues were left which after being purified with a heavy liquor, were found to have the following composition:

	A	B
Tungsten	48.28	48.17
Iron	45.13	45.56
Carbon	6.59	6.26

The formula $Fe_3C.WC$ calls for

Tungsten	48.88
Iron	44.73
Carbon	6.39

Mr. Moissan had previously obtained in an electrical furnace the carbide W_2C; while Mr. Williams had prepared, in presence of an excess of iron, the carbide WC, both in a forge fire and in an electrical furnace.* As in the case of silicide of iron, the for-

* *Comptes Rendus,* June 13, 1898.

mation of tungsten carbide, less rich in metal, is favored by a lowering of the temperature.

More recently, Mr. Williams produced in an electrical furnace a mixture of the carbide WC and of a double carbide to which he ascribed the formula $2Fe_3C.3W_2C$.**

MOLYBDENUM.—1. Molydenum steel containing little carbon, treated with dilute acids with exclusion of air leaves a residue made up of iron and molybdenum, whose proportions correspond exactly to the formula Fe_3Mo_2.

Two samples containing respectively 2.5 and 3.4 per cent of molybdenum gave the following results:

	A	B		A	B
Molybdenum .	0.1690	0.2441	or in per cent	53.48	53.34
Iron . . .	0.1470	0.2135	" "	46.52	46.66

The formula Fe_3Mo_2 corresponds to 53.33 per cent Mo and 46.67 per cent Fe.

In using the double cupric chloride for the solution the whole of the molybdenum is not found in the insoluble residue.

2. With samples of steel highly carburetted, the double carbide $Fe_3C.Mo_2C$ was isolated. The same compound has been prepared by Mr. Williams in an electrical furnace.†

The steel experimented upon contained (A) 3.9 per cent of molybdenum and 1.7 per cent carbon, and (B) 3.5 per cent molybdenum and 2.3 per cent carbon. They were dissolved in hydrochloric acid and the residues purified by means of a heavy liquor of methyl iodide. The analysis of the residues indicated the following compositions:

	A	B
Molybdenum	50.57	50.48
Iron	43.30	43.48
Carbon	6.13	6.09

The composition of $Fe_3C.Mo_2C$ is as follows:

Molybdenum	50.77
Iron	43.08
Carbon	6.15

Our conclusions concerning the chemical condition of chromium, tungsten and molybdenum agree exactly with those of

* *Comptes Rendus,* June 13, 1898.

† *Comptes Rendus,* September 12, 1898.

Mr. H. Le Chatelier based upon the electrical resistance. The slight influence exerted by these metals upon the tenacity of the metal, lead us to infer that they are isolated in the metallic mass in the condition of definite compounds,* and that they are not present as solid solution like silicon.

TITANIUM. — Ferro-titanium may be dissolved either by acids or by the double chloride of copper and potassium. In both cases the residue consists of titanium free from iron.

Two grammes of a ferro-titanum containing 48.6 per cent of titanium upon being dissolved in hydrochloric acid left a residue consisting of 0.968 grammes of pure titanium; five grammes of steel containing 4.62 per cent of titanium yielded as a residue 0.230 grammes of that element. Titanium does not appear, therefore, to be combined; it remains unacted upon while the whole of the iron is dissolved.

COPPER.—The copper present in steel cannot be isolated by the action of double chloride of copper and potassium. This separation may be accomplished, however, by means of a weak acid, such as a 5 per cent solution of hydrochloric acid, but it is necessary that the air should be excluded during the operation. Two grammes of steel containing 3.4 per cent of copper yielded in this way a residue consisting of 0.066 grammes of pure copper. The copper remained unaffected, therefore, without retaining any iron, and it might be inferred that it was simply dissolved in the metal. It might be asked, however, whether there had not been a precipitation by the iron of the copper in solution. This led us to use another reagent, which would not allow such uncertainty, and we selected, for that purpose, a solution of ammonium chloride containing oxygenated water, carefully neutralized.

The iron is converted to ferric oxide, which can be separated through decantation. During the first days no copper is dissolved. When traces of that metal begin to be found in the water from the decantation, the residue is carefully washed, dried with absolute alcohol, then with ether, and examined through the microscope. Some fine threads of pure copper may then be detected entangled in the mass not yet decomposed and sometimes extending quite far out of that mass. The appearance is alto-

* *Comptes Rendus*, June 13, 1898.

gether different from that of copper precipitated on·iron in the same conditions. It corresponds well to the appearance presented by a pure metal disseminated, through previous solidification, throughout the mass of another less fusible metal.

In treating this residue with ammonia mixed with a small amount of oxygenated water, the free copper is dissolved and the remaining mass may then be treated with ammonium chloride, as before. After several such treatments, a very small residue remains, which is non-magnetic and contains still some iron and some copper. Is it a carbide or an alloy? The small amount obtained does not permit us to answer this question positively.

It may on the other hand be noted that when some fine drillings of a copper steel are acted upon by diluted ammonia mixed with a small proportion of oxygenated water, the copper is partially dissolved, imparting a light blue coloration to the solution.

From the above observations it may be inferred that the bulk of the copper does not form any definite compound with the iron, in the case of commercial steel containing small proportions of copper, whether the steel be or not quenched.

NICKEL.—Nickel is entirely dissolved by neutral double chloride of copper and potassium. It was not possible to obtain it in the residue left by an acid even very diluted and with exclusion of air.

The use of ammonium chloride alone or mixed with oxygenated water did not prove satisfactory. Both metals are dissolved simultaneously, in the case of unhardened as well as in the case of hardened steel.

We did not find any carbide of nickel, moreover, in the slightly carburetted samples, hardened or not, which were examined, and it is our belief that the nickel is present in a free state, as a solid solution or a mixture and not as a definite compound. This conclusion agrees with the results obtained by Mr. Le Chatelier and based upon the electrical resistance of special steels.

Note on the Analysis of Ferromanganese

It has been stated in the portion of this paper dealing with silicon, that finely pulverized ferromanganese may be divided into two parts, one attracted by the magnet and the other not attracted. These two parts differ greatly in composition. The portion which is attracted is much richer in silicon than the other. The rough crushing of ferromanganese followed by sifting in order to separate the finer particles, results also in a division in two parts of different composition and which may be separated by physical means.

Three samples of ferromanganese containing on an average:

	I	II	II
Manganese, per cent	29.0	47.9	53.5

were crushed and passed through a fine sieve (No. 120). The portion that passed through the sieve and the one remaining on the sieve contained the following amounts of silicon and manganese:

		Fine particles	Coarse particles
Sample I .	Manganese	28.1 per cent	29.4 per cent
	Silicon	0.33 "	0.21 "
Sample II .	Manganese	47.0 "	48.4 "
	Silicon	0.20 "	0.14 "
Sample III .	Manganese	52.6 "	54.5 "
	Silicon	1.40 "	0.50 "

In treating in the same way a sample of ferromanganese containing a small amount of chromium, the following results were obtained:

	Fine particles	Coarse particles
Manganese	44.6 per cent	45.8 per cent
Chromium	1.58 "	1.16 "
Silicon	c.24 "	0.12 "

The fine particles, therefore, always contain less manganese and more silicon (and in the last case more chromium) than the coarser particles which failed to pass through the sieve.

If this phenomenon be compared with that of the segregation in castings of ferromanganese slowly cooled, previously mentioned, we may infer that in order to arrive at the average composition of ferromanganese, it is absolutely necessary to break a number of fragments from various parts of the sample, to crush them finely until the *whole* will pass through the sieve, and to mix the siftings carefully before weighing the sample for analysis.

It might also be noted that in the case of ferromanganese containing over 80 per cent of manganese, the double chloride of copper and potassium is not suitable for the determination of carbon, as it may result in a serious loss of carbon through the formation of hydrocarbons.

Direct combustion of the fine powder in a current of oxygen yields much more accurate results.

THE TEMPERATURE LIMITS FOR THE SEPARATION OF GRAPHITE FROM MARTENSITE IN PURE CAST IRON *

By HUGH P. TIEMANN

THE object of this thesis is to determine in the case of a pure cast iron the temperatures and conditions (if any) where the combined carbon is transformed into graphite, whether in its ordinary form or as temper carbon (*Temperkohle* of Ledebur). By *pure cast iron* is meant a highly carburized iron as free as possible from all other elements (impurities). The complete absence of silicon is necessary on account of its well known effect in producing gray cast iron.

The result aimed at is of importance (a) from a purely scientific standpoint as tending to confirm or disprove the theories on the subject advanced by various scientists, particularly Roberts-Austen and Roozeboom, which are discussed below; and (b) from a commercial point of view as having a direct bearing on the treatment of white cast iron in the manufacture of malleable castings.

Fig. 1 is the diagram presented by Roberts-Austen to illustrate the results of his investigations as described in the " Fifth Report of the Alloys Research Committee."†

Fig. 2 shows the results obtained by Roozeboom‡ from his

---

* The following experiments represent work done in the Metallurgical Laboratory of Columbia University for the degree of A. M. under Prof. H. M. Howe, 1900-1901.

† Read before the Institution of Mechanical Engineers, February 9, 1899; *The Metallographist*, 1899, pages 186-222.

‡ *Zeitschrift f. phys. Chem.*, Vol. XXXIV, 1900. Article read before the Iron and Steel Institute. *The Metallographist*, 1900, pages 293-300.

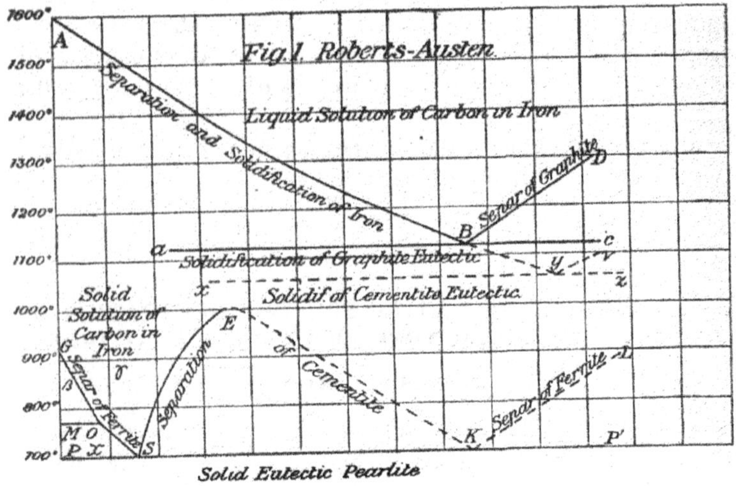
Fig. 1. Roberts-Austen.
1600°
1500°
1400°
1300°
1200°
1100°
1000°
900°
800°
700°
Liquid Solution of Carbon in Iron
Separation and Solidification of Iron
Separ. of Graphite
Solidification of Graphite Eutectic
Solidif. of Cementite Eutectic
Solid Solution of Carbon in Iron
Separ. of Ferrite
Separation of Cementite
Separ. of Ferrite
Solid Eutectic Pearlite

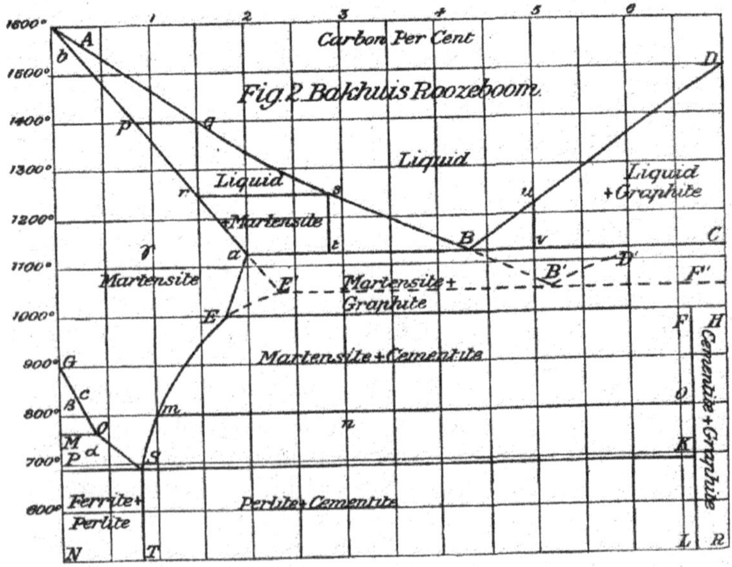
Carbon Per Cent
Fig. 2. Bakhuis Roozeboom.
1600°
1500°
1400°
1300°
1200°
1100°
1000°
900°
800°
700°
600°
Liquid
Liquid + Graphite
Liquid + Martensite
Martensite
Martensite + Graphite
Martensite + Cementite
Ferrite + Perlite
Perlite + Cementite
Cementite + Graphite

application of the phase-rule to Roberts-Austen's researches, the most marked difference from Fig. 1 being the line *SK* which marks the critical temperature between pearlite and cementite on the one hand and martensite and cementite on the other. The following extracts are taken from Roozeboom's paper to help the elucidation of the case in question: "As regards the iron-carbon alloys with more than 2 per cent of carbon, my results are more at variance with those of Sir W. Roberts-Austen. Beneath the line *aBC* at 1130° in the normal case there should be nothing but a solid conglomerate of two phases, viz., graphite and martensite with 2 per cent of carbon. Firstly I have traced the line *aE* to indicate that the amount of carbon in the martensite falls simultaneously with the temperature, and this would induce a further separation of graphite. Then at about 1000° the line *aE* meets the cementite line *SE* of Roberts-Austen. This can have no other meaning than that at 1000° there is in the normal state of stable equilibrium an abrupt transformation of the graphite into carbon (cementite), according to the equation—

"Martensite (1.8 per cent C) + Graphite = Cementite (Fe_3C)

"But then the temperature of 1000° is a transition point; that is, only at that temperature can the three phases coexist, and therefore the transformation of all the conglomerates of martensite and graphite must take place at this temperature; the border line *EFH* must be horizontal.

"It is impossible to state briefly here all the arguments which led me to accept this line in preference to the lines *EK* and *KL* of Roberts-Austen and of Stansfield, nor can I sum up the considerations which I have adduced to explain the many anomalous results obtained with highly carburized iron. A prominent part is played by the retardations which are possible in all the processes of transformation which occur during cooling. There is also to be taken into account the influence which retardation has upon the production of cementite during the period of solidification, and during the two following periods above and below 1000°."

The following table is a concrete example of an iron containing 4.5 per cent of carbon, and shows the structural composition during its various changes. (See Note A.)

Table Showing Structural Composition at Different Temperatures of an Iron Containing 4.5 per cent Carbon.

	Above 1150°	1130°+	1130°−	1000°+	1000°−	Below 690°
Fe	95.5	95.5	95.5	95.5	95.5	95.5
C (total)	4.5	4.5	4.5	4.5	4.5	4.5
Graphite	—	0.4	2.6	2.8	—	—
Martensite	—	—	97.4	97.2	40.8	—
Cementite	—	—	—	—	59.2	65.8
Pearlite	—	—	—	—	—	34.2

In examining the literature on the subject in hand a curious state of affairs is discovered. In spite of the fact that it is generally conceded that silicon (leaving sulphur and manganese out of the question) is a factor in determining whether white or gray cast iron is to be produced, this seems to have been either totally forgotten, or else deemed of such small importance that in researches on cast iron its presence is neglected. In a good many cases it is not even stated whether it is present, or, if mentioned, it is merely in an analysis along with the other impurities, and the subject dismissed without further notice. To give an illustration of the way in which results are presented the following researches of Stansfield* may be cited:

"The fact that slowly cooled iron, when rich in carbon, contained very little carbon in the combined state has long been known in the case of pig irons, in which, however, the simultaneous presence of silicon somewhat complicates the problem. Perhaps the earliest experiment in which pure materials were used was one described by Dr. Percy,† in which he melted pure iron (note how pure?) in crucibles brasqued with lampblack (see NOTE D for results with lampblack), and obtained a cast iron containing a little more than 4 per cent of graphite, and with no sensible amount of combined carbon."

Farther on in the same article he discusses the relative formations of graphite and cementite, and then mentions some experi-

* "The Present Position of the Solution Theory of Carburized Iron," Part II. By A. Stansfield. Iron and Steel Institute, Paris meeting, September, 1900. *The Metallographist,* 1900, pages 300-314.

† Percy's "Metallurgy, Iron and Steel," 1864, page 113.

ments of his own, where the iron "was carburized by melting in a graphite crucible, packed in charcoal powder, and was subsequently annealed for nearly two days at a temperature below that of its melting point, followed by very slow cooling." The question here is, How much silicon was taken up from the graphite crucible and the charcoal? In this example there is only a trace of combined carbon. Another example is given of a "more highly carburized iron, containing, when fluid, 7 per cent carbon and 3 per cent silicon. Combined carbon 0.22 per cent. The section [a reproduction of a photograph] shows only ferrite and graphite."

In a discussion of Roozeboom's paper on the phase-rule, Osmond* cites "an experiment of Mr. Royston,† who heated rapidly to 1030° C. and quenched some cast iron of the following composition:

Graphite carbon	trace
Combined carbon	3.85 per cent
Silicon	0.29 "
Sulphur	0.03 "
Phosphorus	0.02 "
Manganese	0.15 "

"After quenching, the metal contained 2.35 per cent of graphite and 1.50 per cent of combined carbon.

"Since the quenching treatment suppressed the transformation *EF* (see Roozeboom's diagram) of the system mixed crystals + graphite into mixed crystals + cementite, the composition of this quenched cast iron would also represent its composition at a temperature of 1030° C.: the agreement with our provision is qualitatively complete and quantitatively sufficient. The same experiment, however, in which the rapid cooling was replaced by slow cooling in the air, yielded the same proportion of graphite, but as it is accepted by all that the transformations graphite + mixed crystals into cementite must be slow, this fact is not abnormal and could have been anticipated."

In the above example Osmond takes no notice of the silicon (0.29 per cent). From my results, which I will describe later on,

* *Bull. de la Soc. d'Enc.*, November 30, 1900; *The Metallographist*, 1901, pages 150-61.

† *Journal Iron and Steel Inst.*, 1897, I, page 166.

I find two objections to Osmond's deductions: (a) the fact that he considers the temperature alone responsible for the formation of graphite, without considering the impurities present; and (b) that because a transformation proceeds in one direction during cooling it must necessarily proceed in the other upon heating. In some other experiments, published in the same article, he neglects to mention whether there is any silicon present in the samples tested, and, consequently the results are not considered in the present discussion. However, Osmond, corrects, to a certain extent, the impression given above, by saying at the end of his paper, "We must, of course, take into consideration, not only the time, but also the presence of certain impurities. It is known, for instance, that when silicon or aluminum are present in sufficient proportions, they cause the formation of graphite as soon as solidification begins. A very small quantity of the same elements might, if they are given time, act in the same way. A single atom of silicon represents a high proportion of that substance in a small mass of iron surrounding it; the proportion increases as we consider a smaller surrounding mass, and may be as high as we wish. If this atom be mobile, as might be expected in a solution, and if the redissolution of the precipitated graphite be slow, a state of equilibrium may be reached between the causes inducing the formation of graphite and those inducing the formation of cementite."

Now that a brief review of the situation has been made, an account of the present work will be given.

The iron used was supplied through the courtesy of Mr. H. G. Stoddard, Assistant Manager of the Worcester District of the American Steel and Wire Company, and had the following composition:

Swedish Iron

Phosphorus	0.031
Sulphur	0.002
Silicon	Nil
Manganese	0.040
Combined carbon	0.030

For work of this nature a condition of prime importance is the entire absence of silicon, and from the above analysis it will be seen how suitable was the iron employed.

The composition of the carburized iron (see NOTES B, C, D) with respect to the carbon and silicon, was

Combined carbon	4.271
Graphitic carbon	0.255
Total carbon	4.526
Silicon	Nil

For the sake of clearness, it may be mentioned again that the melted iron was poured into a large vessel of water at about 15°.

In the experiments the following method of procedure was followed: the heats were all made in a magnesia crucible which was put in a small, experimental, electric furnace (NOTE E) designed by Prof. H. M. Howe. The temperatures were ascertained by means of a Le Chatelier thermo-electric pyrometer (NOTE F) which was placed in direct contact with the pieces of iron being treated.

In all cases (except the last) the iron was quenched in iced brine (—5°) without any preliminary cooling. The quenching was done as soon as the iron had reached the desired temperature or after it had been held at that temperature for a definite length of time. The pieces of iron were packed around with magnesium oxide (c.p.), and a thin layer of sugar charcoal was laid over the top. Notwithstanding this precaution, the surface was somewhat oxidized, but all scale was carefully removed before analysis.

In the table below are recorded the data obtained under the given conditions:

Treatment of Cast Iron Free from Silicon

Marked	Temperature	To maximum temperature	At maximum	Temperature of quenching water	Graphite	Total carbon
A	Original sample			15°	0.255	4.526
B	906°	30 min.	0	−5°	0.260	not det.
C	1006°	10 min.	0	−5°	0.261	not det.
D	1115°	10 min.	0	−5°	0.263	4.254
E	996–1010°	10 min.	1 hour	−5°	0.252	not det.
F	1056–1061°	15 min.	1 hour	−5°	0.268	not det.
G	1113–1116°	50 min.	1 hour	−5°	0.262	not det.
H	1053°	10 min.	0	−5°	0.254	not det.
I	998–1012°	55 min.	3 hours	−5°	0.264	1.972

From an examination of the above it will be seen that the graphite remains practically constant throughout. As this is the reverse of the method employed in the experiments which supplied the data for Roozeboom's diagram, i.e., heating instead of cooling, another sample was cooled before quenching, as shown below :—

Treatment of Cast Iron Free from Silicon

Marked	Temperature		Thermal treatment			Temperature of quench. water	Graphite	
	Max.	When quenched	To maximum	At maximum	Fr. max. to temp. when quenched		Before	After
J	1300°	1040°	50 min.	0	8 min.	24°	0.255	0.677

This last result would tend to prove Roozeboom's deductions, while the preceding go to show that the reverse action does not take place. Perhaps silicon exerts some catalytic action, examples of which are fairly numerous in chemical reactions (e. g., the presence of a trace of moisture being necessary for the combination of H and Cl.). The fact that a large portion of the carbon has been burned off in " I " would not necessarily show that the graphite rather than the combined carbon had suffered a loss. For in the preparation of malleable castings the pieces are packed in iron oxide, and graphite is formed in abundance, although, of course, the total carbon is not removed to quite the same extent. That the explanation for the graphite formed in the experiments quoted earlier in this paper would seem to be due largely, and in some cases solely, to the silicon present, is borne out by the following experiments made with an iron containing

Graphitic carbon	0.938
Combined carbon	3.438
Total carbon	4.376
Silicon	0.75

the other impurities being the same as in that marked " A." As in the case of " A," it was poured, when molten, into a large vessel of water at about 15°.

Treatment of Cast Iron Containing 0.75 *per cent of Silicon*

Marked	Temperature	To maximum temperature	At maximum	Temperature of quench. water	Graphite	Total Carbon
AA	Original sample			15°	0.938	4.376
BB	975°	15 min.	0	−5°	1.690	not det.
DD	1125°	30 min.	0	−5°	2.795	not det.

The fact that graphite was formed below the critical temperature of 1000° shows that, granted Roozeboom's deductions and critical points are correct (in the case of an iron containing nothing but carbon), some impurity present must be the cause, and since silicon is the only feature in which "AA" differs from "A" (except that it contains slightly less total carbon) this would seem to be it.

That the activity of the silicon is a function of the temperature (as shown by the increased amount of graphite in "DD" over "BB") seems perfectly plausible owing to the greater plasticity of iron at the higher temperature.

The silicon present would then explain the results obtained by Royston (quoted above) and would also offer a reason why the graphite formed on heating was not converted into cementite on cooling, although Osmond's explanation is doubtless more nearly correct.

A cooling curve was tried, but without any very satisfactory results. The transition point at 690° was found without any difficulty, but there were no other distinct retardations. Several explanations can be offered for the difficulty of finding the higher points: (a) because the cooling could not be made sufficiently slow in any of the furnaces in the laboratory; (b) as shown in the experiments, the process is not reversible, and consequently the transformations which occur during the first cooling remain unchanged when the iron is heated up, and, therefore, do not take place again on subsequent cooling.; (c) at very high temperatures the surface of the iron is rapidly decarburized, and, of course, this is the part with which the couple is in contact. The best means for getting the critical points (on the first cooling) would be by the use of a differential couple such as is described by Roberts-Austen in "The Fifth Report of the Alloys Research Committee."

Some work has been done with the Roberts-Austen recording pyrometer, with a view of obtaining a very delicate curve, but up to the present the instrument has not been gotten sufficiently into shape.

A piece from each heat of the iron free from silicon (except " I ") was polished and etched with a 2 per cent solution of nitric acid for from 5 to 15 seconds. The results are shown in the accompanying photomicrographs.

They all have a magnification of 140 diameters. As the

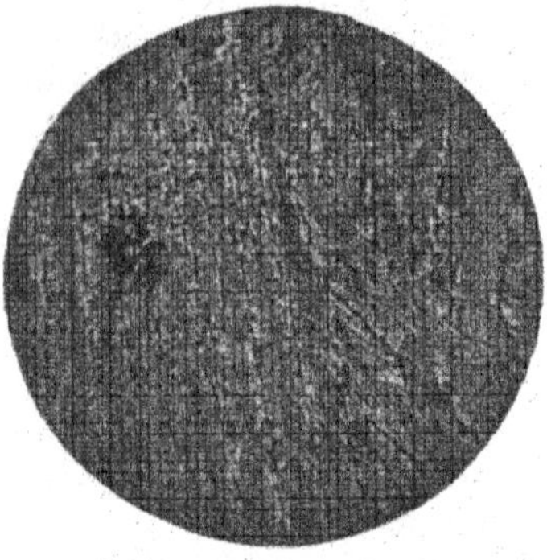

A1. Original Iron showing eutectic of cementite and martensite. Section perpendicular to cooling surface. Magnified 140 diameters.

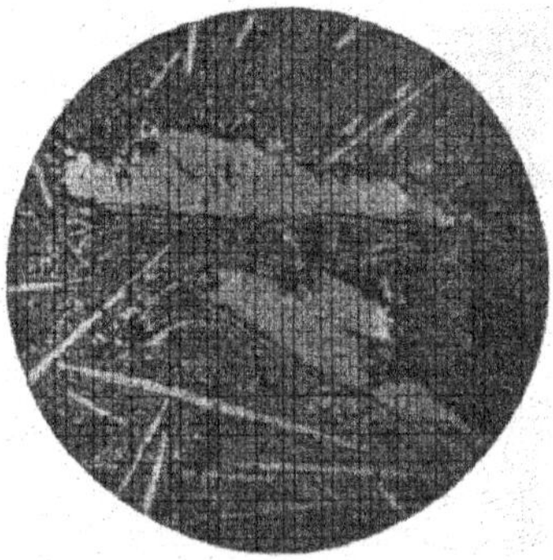

A2. Original Iron showing plates of cementite and eutectic. Section parallel to cooling surface. Magnified 140 diameters.

graphite present in all cases is only about 0.26 per cent, it is not distinguishable in any of the etched sections, although readily found before etching. The black spots are caused by minute blow-holes, and are not due to imperfections in the negative. All the photographs were made of sections taken parallel to the cooling surface with the exception of the first one shown, which was taken perpendicular.

All the photographs show a white constituent (cementite) and a dark one (martensite). Also the original melted iron was quenched in water at 15° and this would render the cooling slow enough to permit the development of the very large plates of cementite.

In all cases there is a eutectic, presumably of cementite and martensite. It has not been definitely proved that austenite is not present (at least in the specimens quenched above 1100°), as

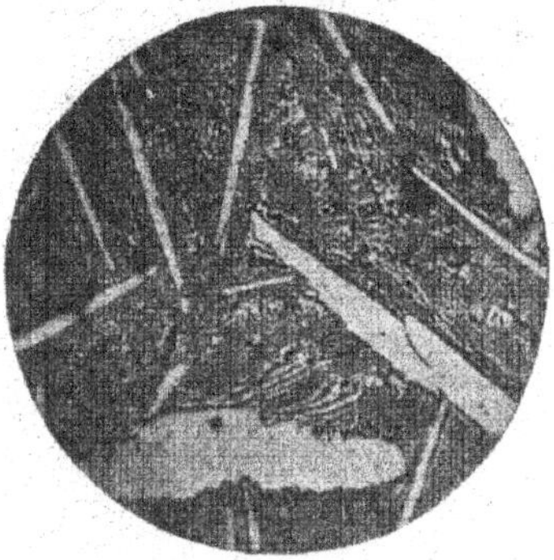

A3. Original Iron. Section parallel to cooling surface. Magnified 140 diameters.

B. Heated to 906° and quenched. Section parallel to cooling surface. Magnified 140 diameters.

C. Heated to 1006° and quenched. Section parallel to cooling surface. Magnified 140 diameters.

E. Heated to 996–1010° for 1 hour. Section parallel to cooling surface. Magnified 140 diameters.

the etching was done with nitric acid and not with iodine, as is necessary for the separation of cementite and austenite. This explanation will shortly be made, and the results published, together with some others bearing on the subject which were worked out this summer at the Boston Testing Laboratories, thanks, in large part, to the advice and suggestions of Mr. Sauveur. Judging by the lustre alone (cementite has a remarkable and peculiar lustre much more pronounced than in the case of austenite, as seen in other specimens) the uncolored constituent, both in the plates and the eutectic, would seem to consist entirely of cementite.

Note A

Calculations for the Table Showing the Structural Composition at Different Temperatures of an Iron Containing 4.5 *per cent of Carbon.*

The iron has the ultimate composition

Fe 95.5 per cent
C 4.5 " "

Above 1130° the carbon in excess of 4.3 per cent separates out as graphite,

$$\text{Graphite} = 4.5 - 4.3 \times 0.955 = 0.4 \text{ per cent.}$$

At 1130° the eutectic forms, consisting of graphite and martensite with 2 per cent of carbon, hence the additional graphite which separates out

$$= 4.1\ (4.5 - 0.4) - 2 \times 0.955 = 2.2 \text{ per cent.}$$

The total graphite$=2.2+0.4=2.6$ per cent and the martensite $=100-2.6=97.4$ per cent.

From 1130° to 1000° there is a decrease in the martensite from 2 per cent. to 1.8 per cent.

Hence there is a further separation of graphite which=

$$1.9\ (4.1 - 2.2) - 0.955 \times 1.8 = 0.2 \text{ per cent}$$

The total graphite is then

$$0.2 + 2.2 + 0.4 = 2.8 \text{ per cent.}$$

and the martensite

$$= 100 - 2.8 = 97.2 \text{ per cent.}$$

Below 1000° the martensite combines with the graphite to form cementite according to the equation

$$\text{martensite } 1.8\ C + \text{graphite} = \text{cementite } (Fe_3C).$$

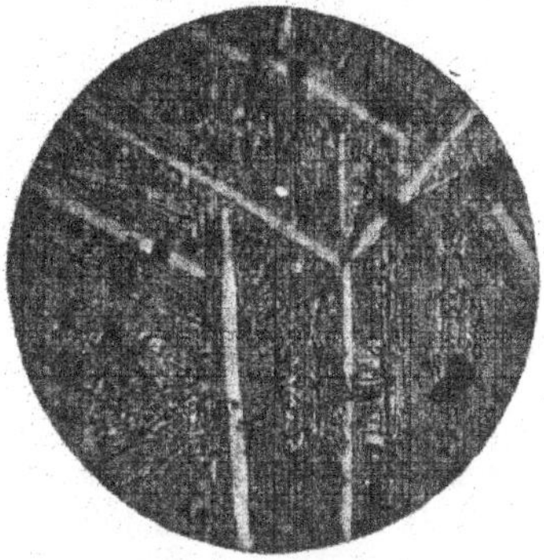

H. Heated to 1053° and quenched.
Section parallel to cooling surface.
Magnified 140 diameters.

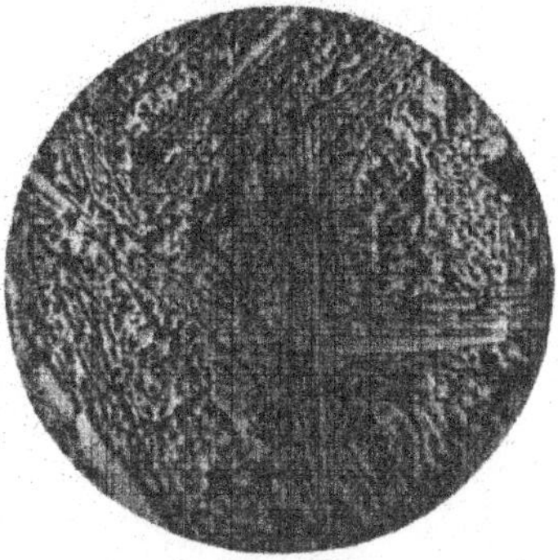

F. Heated to 1056–1061° for 1 hour and quenched.
Section parallel to cooling surface.
Magnified 140 diameters.

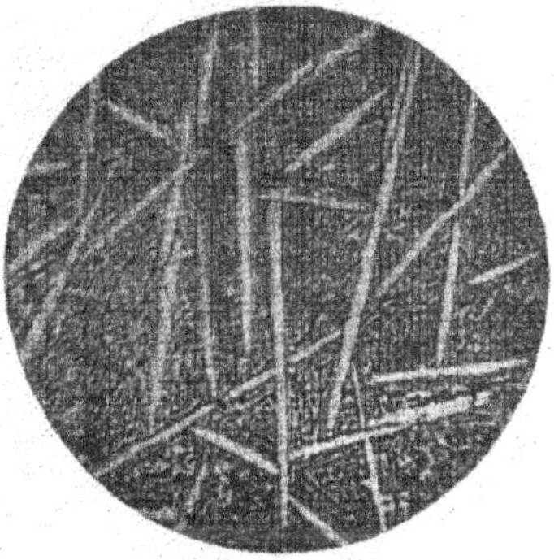

D. Heated to 1115° and quenched.
Section parallel to cooling surface.
Magnified 140 diameters.

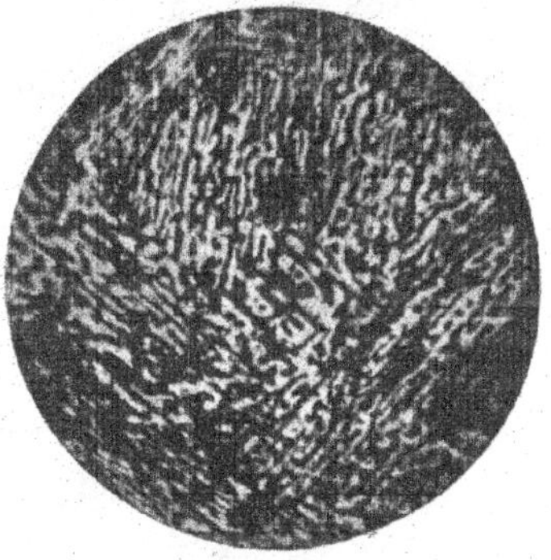

G. Heated to 1113–1116° for 1 hour and quenched.
Section parallel to cooling surface.
Magnified 140 diameters.

Cementite contains 6.7 per cent C, therefore, to convert all the iron into cementite, we should use $6.7 \times 0.955 = 6.4$ per cent C (making a total of over 100 per cent) ; the martensite, however, contains 1.8 per cent C. So the amount of graphite needed would be 6.4—1.8=4.6 per cent.

If x equals the amount of cementite formed from the graphite present we have

$$97.2 : x = 4.6 : 2.8,$$

whence, $x = 59.2$ per cent,

and the martensite remaining

$$= 100 - 59.2 = 40.8 \text{ per cent.}$$

Below 690° the martensite is converted into pearlite and cementite, hence there must be an increase in the cementite as the pearlite contains 0.85 per cent C. This is calculated as follows:

Pearlite $= x$ per cent
Cementite $= y$ " "

(1) $x + y = 40.8$

(2) $0.85\,x + 6.7\,y = 1.8 \times 40.8$

Multiplying (1) by 0.85 and subtracting from (2) we get

$$5.85\,y = 38.76$$
$$y = 6.6$$
$$x = 34.2$$

Hence the total cementite=

$59.2 + 6.6 = 65.8$ per cent,
pearlite $= 34.2$ " "

Note B

Preparation of Sugar Charcoal

A large iron dishpan, about 18 inches across and 5 inches deep, is supported on a suitable tripod, and two Bunsen burners are placed underneath. The burners are lighted, and 1 pound of granulated, white sugar is poured in. The mass soon melts and boils up with evolution of steam and small quantities of hydrocarbon vapors of characteristic odor and having the property of staining the skin yellow like nitric acid (xantho proteic acid?). When the liquid has become thin and is boiling vigorously, small portions of sugar are repeatedly stirred in (the mass being allowed to melt completely after each addition) until another pound has been added. This quantity (2 pounds) has been found to be the most convenient to handle in a dish of the given capacity, and it has been shown by testing that more charcoal can be produced in a given time from charges of this size than from larger or smaller ones.

When most of the water has been driven off (as shown by the increased viscosity), the hydrocarbon vapors are given off in increasing volume, and, being very inflammable, readily catch fire, being no longer diluted by the steam. The mass has to be constantly stirred at this point to prevent its caking on the bottom, and puffing up and over the sides. When the fumes persist in catching fire, a cover, made from a piece of asbestos board, is laid over the top to extinguish them. Occasionally the burners are taken away, the cover removed, and the mass broken up and stirred to bring a fresh portion in contact with the bottom. Complete charring is shown when no more fumes are given off, and when the charcoal merely smoulders on the admission of air.

The charcoal is then pulverized in a mortar to the required degree of fineness and sieved.

The treatment described took slightly less than one hour and the yield varied from 20 to 30 per cent (say, 6.5 to 10 ounces of charcoal in 2 pounds of sugar) depending on the care with which the operation was carried out. The theoretical yield as calculated from the formula $C_{12}H_{22}O_{11}$ is 42.1 per cent.

Note C

Brasquing the Crucible

As a binding material yellow, commercial dextrine was used, dissolved in a small quantity of water (say 500 grammes to the litre) — the yellow is as pure as the white for this purpose (neither containing any ash) and is more easily dissolved. This is best effected by allowing it to soak over night and warming up in the morning, whereupon everything goes into solution. If heat be applied at once the dextrine cakes on the bottom and floats on the top in balls, the centres of which are dry, and which are hard to break up and go into solution with difficulty.

The ground charcoal is put in a dish and enough of the dextrine solution stirred in to make the mass coherent without being soggy. If the charcoal be too moist it cannot be compressed by ramming, but remains in a condition resembling soft jelly, while if not soft enough it crumbles upon drying.

A layer is spread in the bottom of the crucible and tamped down well. Then a core consisting of a smaller-sized crucible is placed inside and the charcoal packed around as firmly as possible. The core is then gently removed and the lining dried. This drying takes several days and at first the temperature should not be above 150°. Too rapid heating at first causes the lining to crack and crumble. Before using, the crucible should be tested by heating to a bright, cherry red.

Note D

Carburizing the Iron

As is well known silica (SiO_2) is readily reduced at a fairly high temperature in the presence of iron which combines with it and thereby prevents its subsequent oxidation, except under special conditions. Considerable difficulty was experienced in preventing this reduction, as a very slight amount of silica in the crucible and other materials used made its appearance in the resulting iron. Ordinary charcoal was barred for this reason, and lampblack also, as a trial resulted in the iron showing a content of about 0.5 per cent silicon. Other commercial forms of carbon were not tested on account of their known impurities (principally SiO_2).

Sugar charcoal was finally adopted and with entire success. The ash produced on the combustion of purified cane sugar is inappreciable. This is the form of charcoal that Moissan* used in so much of his work. (For the method of carbonizing see NOTE B, and for brasquing the crucible NOTE C).

When this charcoal was first used, an unforeseen accident oc-

* "Le Four Electrique."

curred. The charcoal was tightly packed around the iron and up to the top in a crucible properly brasqued with the same charcoal. The cover was luted on with a mixture of fire-clay and borax water. This lute made a joint perfectly air-tight, but unfortunately some of it worked down into the melted iron which in consequence showed upon analysis 0.45 per cent of silicon.

To avoid the trouble of brasquing, a magnesia crucible was then used, sugar charcoal being packed around the iron as before. This time the cover was luted on with a mixture of magnesium oxide and powdered borax glass, so no trouble with silicon was experienced. The crucible was analyzed beforehand and showed:

SiO_2	2.0 per cent	(determined accurately)
(Fe_2O_3 and Al_2O_3) .	1.0 "	(determined approximately)
CaO	2.0 "	(determined approximately)
MgO	95.0 "	(by difference)

The iron was kept at a temperature of over 1500° for about 2 hours and was then poured slowly into a tall cylindrical vessel filled with cold water (15°). Owing to the size of the crucible a little less than a pound of the carburized iron was obtained.

Note E

Electric Furnace Used in Making the Heats

This furnace was designed by Prof. H. M. Howe for just such experimental purposes as the present. It is circular in cross-section and consists of two blocks of magnesium oxide which are held together by two iron hoops. The interior is cuneiform to make a rather loose fit for the crucible which is also of magnesium oxide. The heating is effected by 2 No. 18 platinum wires, twisted together and wound in a spiral groove cut on the inside of the furnace. A cap supported by two shoulders, all of magnesium oxide, fits over the crucible and makes it fairly air-tight. The pipe-stem holding the thermo-electric couple passes through a hole in the cap. The body of the furnace is 5 inches in diameter and 4 inches high. The shoulders make it ½ inch higher, and the top of the cap projects ½ inch above these. In order to retain the heat as far as possible the furnace was packed in lime. As resistance two rheostats in series with the furnace were used; (a) a fixed resistance consisting of iron wires submerged in water* and (b) a changeable resistance made by moving a sliding contact along a coil of iron wire wound on an insulated iron pipe.†

The two chief points aimed at and attained in the use of this furnace are high temperatures and close regulation. Iron contained in the crucible was several times heated up to a temperature of about 1400°. With a little practice the temperature could be kept indefinitely with variations of less than 5°; thus, a sample was kept for one hour at from 1021-1026°, a change of 5°, and another sample for one hour at 1113-1116°, a change of 3°. As the rheostats and furnace heated up, slight regulation was necessary to keep the current constant.

As very small changes in the current made very large changes in the resulting temperature, and as the heating of the circuit exerted a variable influence, no definite relation between ampères and temperature can be given; 31 to 33 ampères were found to be sufficient, in the heats made for one hour and three hours, to keep the temperature constant at points between 1000° and 1200°.

As was mentioned above, the platinum wire used was No. 18. Two lengths, each 36 inches long, were twisted together, thus making them the equivalent of one No. 15 wire. The object in using two wires was to present as much heating surface as possible.

A set of figures (from one of several preliminary tests made) may be of interest, as showing, principally, the resistance of the wire with increasing temperatures. In this case no attempt was made to keep the temperature constant. The wires can carry 53 ampères, but it was found best (and to answer all requirements) not to go above 45.

---

* "Immersed Rheostats vs. Water Rheostats," by Parks Rucker, *American Electrician*, February, 1901, pages 72-4.

† As designed and used by the Electrical Engineering Department of Columbia University.

Table Showing Variables in the Electric Furnace with Couple inside Crucible

Ampères	25	36	50	52	52.5
Volts	3.25	9.5	15.5	16.75	18.5
Ohms	0.13	0.26	0.31	0.33	0.35
Deflections (pyrometer) . .	0	6.7	11.7	13.1	16.6
Temperature (+ cold junction)	29°	621°	1029°	1126°	1401°
Time (minutes)	0	5	10	15	19

Note F

Standardizing the Thermo-Electric Couple

The pyrometer consisted of a couple formed from wires of platinum and of platinum with 10 per cent rhodium, with a Deprez-d'Arsonval galvanometer for measuring the current. It was calibrated in the usual way, the substances and fixed points with the corresponding deflections being as follows:

Substance	Point	Temperature ° C.	Cold function	Deflection ° scale
Water	Boiling	100	23	0.63
Naphthalene	Boiling	218	23	1.75
Zinc	Melting	420.6	23	4.0
Sulphur	Boiling	444.6	23	4.35
Aluminium	Melting	660	23	6.92
Zinc	Boiling	930	23	10.35

The method generally followed in the laboratory was to plot the curve directly, i.e., with deflections as ordinates and degrees of temperature between the hot and the cold junction as abscissæ. This method, after several trials, was found to produce invariably a slight curve.* In the reference given this is mentioned, and several other methods of plotting the results are shown. The logarithmic curve was the one chosen for the present work, and a perfectly straight line resulted. This method has the advantage over the direct of avoiding the error (possibly of several degrees)

* "Contributions from Phys. Lab. of Mass. Inst. of Tech.," XLIV. —*Thermo-Electric Interpolation Formulæ,* November 15, 1895, S. W. Holman.

due to the impossibility of accurately locating every point in the curve. The one drawback to the logarithmic curve is the fact that the size of the logarithms does not increase proportionately with that of the numbers, making the higher values somewhat close together. This trouble should not exist, provided the curve is plotted on a moderately large sheet of paper.

IRON AND PHOSPHORUS

By J. E. STEAD

(Concluded from page 236)

Part III

CONSTITUTION AND MICROSTRUCTURE OF PIG METALS CONTAINING IRON, CARBON, AND PHOSPHORUS

IT was (1) Sir Frederick Abel who first discovered that on breaking phosphoretted pig irons just after solidification sometimes a little fluid metal ran out from the central portion, which contained considerably more phosphorus than the average of the pig iron from which it liquated. Since that time many observers have confirmed this result.

In 1870 I subjected to hydraulic pressure in a properly prepared mould Cleveland pig iron after practically the whole of it had solidified. The git or runner was left open when the pressure was applied, with the result that a small quantity of metal was squeezed out. The original metal and fusible portion had the following analysis, viz. :—

	Original Metal No. 4 Brand	Portion Squeezed Out
	Per Cent	Per Cent
Carbon	3.00	1.75
Manganese	0.35	0.29
Silicon	1.63	0.79
Sulphur	0.12	0.06
Phosphorus	1.53	6.84

(15) Mr. Osmond has described certain experiments of his in which he proved the existence of a critical point in the solidification of phosphoretic pig iron at about 900° C. The analysis and critical points were as follows :—

Phosphoretic Metal

	Per Cent
Graphite carbon	1.50
Combined carbon	1.20
Manganese	0.74
Silicon	1.89
Sulphur	0.03
Phosphorus	1.98

Critical Points

	III	II	I	$Ar_{1,2,3}$
Cooling	1065	1025	895	698
Heating	1050	1040	910	800

Comparative analysis and critical points of hæmatite pig iron containing practically no phosphorus are given below. These are also by Mr. Osmond, viz.:—

Hæmatite Metal

	Per Cent
Graphite carbon	2.25
Combined carbon	1.04
Manganese	0.11
Silicon	2.45
Sulphur	0.06
Phosphorus	0.054

Critical Points

	Solidifying Point	II	I	$Ar_{1,2,3}$
Cooling	1240?	1140	1090	708
Heating	—	1155	1100	780

The characteristic difference between the two irons is, that in hæmatite pig iron there is an absence of a critical point at about 900° C. Mr. Osmond proved that this was peculiar to phosphoretic iron, and showed by experiments made by Mr. Lencauchez ([18]) that a compound high in phosphorus existed in pig iron which was more fusible than the greater mass of the iron. The results quoted are as follows:—

100 kilos. of Longwy iron were heated to 950° C. in a reducing gas for 100 hours. When cold, although the pieces had preserved their form, on the surface was observed a multitude of little spherical metallic particles like small lead shot. Some of these had joined together and run off the surface and had fallen on to the bottom of the retort, where they solidified. The following analyses give their composition, viz. :—

	Shot		Liquate	
	1	2	1	2
	Per Cent	Per Cent	Per Cent	Per Cent
Combined carbon	1.24	—	0.90	—
Graphite	0.75	trace	1.50	1.50
Silicon	0.82	0.63	1.41	2.17
Phosphorus	5.45	6.00	4.48	4.40

Previous to heating, the metal contained —

	Per Cent
Carbon	3.50
Silicon	2.60
Phosphorus	1.90

The analysis of the shot, it will be observed, closely resembles the analysis of the metal which in my own experiments I squeezed out of Cleveland pig iron by hydraulic pressure.

The phenomena observed by Mr. Lencauchez are peculiar to metals containing small quantities of a fusible eutectic, or a constituent which is more fusible than the greater mass, for I have found that it is only necessary to heat such alloys or compounds a little above the melting-point of the eutectic for small globular exudations to pass out of the metal, forming little shot-like bodies all over the surface. It is not difficult to understand how this occurs, for as soon as the temperature is raised slightly above that of the liquefied eutectic it expands, and a portion of it comes out of the metal in which it was imprisoned. There is also always a tendency in heterogeneous mixtures of metals for that part which is in greater mass to throw off the others in contact with it; hence small quantities of the eutectics, where it is physically possible, are thrown into large segregated areas or entirely outside the mass.

I have most carefully repeated Mr. Osmond's experiments, and find that in Cleveland iron, when it is slowly cooled from the molten state, there is a critical point at about 900° C., thus confirming his observation. The pure eutectic of phosphorus and iron solidifies at about 980° C., so it is clear the eutectic solidifying at 900° C. must have a different composition.

On melting together some of the phosphide of iron eutectic with carbide of iron, the eutectic formed solidified at 910° C. A similar result followed on cooling a mixture of silicious pig iron with the pure eutectic.

Leopold Schneider succeeded in isolating the free phosphide of iron from phosphoretic pig irons by dissolving them in cupric chloride, and found it to correspond to the composition expressed by the chemical formula Fe_3P.

In order to isolate the eutectic from grey Cleveland iron containing 1.56 per cent phosphorus, some lumps of it were dissolved in dilute acid. The residue, containing graphite and decomposition products of the phosphide in solid solution, and the eutectic and carbide of iron, had the two latter bodies removed from it by a magnet. The material so attracted was treated with cupric chloride to decompose the free carbide of iron, and the phosphide eutectic remaining was attracted from the residue magnetically. On testing a very small portion of what was obtained, it yielded about 0.70 per cent silicon. A second portion ground to powder in an agate mortar was treated with nitric acid of specific gravity 1.20 to remove the soluble iron and silicide. The residue contained 15.1 per cent phosphorus instead of 15.56 per cent. It may be considered to be nearly pure Fe_3P.

It would appear from this result that silicide of iron dissolved in iron is a constituent of the phosphide eutectic of grey silicious pig irons.

The last fluid portion squeezed out of Cleveland pig iron referred to on page 59, shows that from a forge iron containing 1.63 per cent of silicon, what was expressed contained 0.79 per cent silicon, and 1.75 per cent carbon, and nearly 7 per cent phosphorus.

The exact composition of eutectics in pig iron evidently requires more careful study. It is a somewhat difficult subject to master, but it is hoped that the foregoing remarks will lead others to endeavor to take up this interesting branch of research.

In view of these researches, it was naturally to be expected that the microscope would be able to detect the fusible phosphide eutectics in pig metals.

Practical observation of Mr. Osmond and myself, working independently, fully confirmed such a conclusion, for it has been proved beyond all doubt, that in grey pig iron containing only 0.03 per cent phosphorus the phosphide eutectic can be detected by the microscope, and with quantities greater than that it is in great evidence.

It is in grey irons free from massive cementite where the eutectic can be most easily detected, for then a simple acid etching process is sufficient to reveal its presence.

In following what occurs when grey phosphoretic pig iron solidifies, it would appear that the thick plates and semi-spherical shells of graphite first fall out of combination with the iron, leaving the remaining portion of the metal still in a plastic state; but this, as heat passes away, in time begins to solidify, commencing at the solid borders of the graphite.

This metal consists of all the iron, manganese, silicon, and phosphorus; it does not all solidify at the same time.

The crystalline combination of the greater part of the iron and manganese and all the silicon forms one part; the phosphorus and some of the iron and manganese the other part.

As the first part crystallises in advance, the remainder, consisting of the fusible eutectic, is driven before the metal, advancing in its growth on all sides of it; and finally, when they nearly approach each other, the eutectic (incapable of being forced further in any direction) is imprisoned in irregular-shaped cavities, where it remains fluid till the temperature falls to its freezing-point, when it solidifies, breaking up into its two constituents.

It follows, therefore, that the irregular-shaped solid masses of the eutectic must be found between and at a distance from the graphite plates, and this is actually where the greater number of them are located.

The photographs Nos. 14 and 15 illustrate the foregoing remarks. The first represents the structure of Cargo Fleet No. 3 iron, the second East Coast No. 1 hæmatite. They had the following composition:—

	Phosphoretic	Non-Phosphoretic
	Per Cent	Per Cent
Combined carbon	0.12	0.25
Graphite	3.59	3.86
Manganese	0.69	0.76
Silicon	2.76	2.83
Sulphur	0.05	0.04
Phosphorus *	1.49	0.04
Phosphorus in solid solution	0.29	0.03

* Equal to phosphide of iron, 9.50 per cent, 0.25 per cent.

The irregular white areas in the centre of the photographs are the phosphide eutectic.

When hæmatite metal is strongly etched, so as to darken the ground-mass, and is examined with vertical light, the specks of brilliant phosphide stand out in bold relief, and appear like the stars on a clear night, the number and size depending on the quantity of phosphorus present and the rate of cooling. When rapidly cooled, they are greater in number but are more widely distributed, whereas in the same slowly cooled material they are less numerous but of greater mass.

It is advisable to etch strongly when looking for phosphorus, and to examine with a high power, when the compound character of the phosphide eutectic will be easily identified.

The structure of chilled and white hæmatites has not yet been thoroughly studied; they are therefore reserved for future research.

In grey irons the amount of the eutectic is proportional approximately to the amount of the phosphorus, so that almost any one with a little practice, and with a standard series of sections, can now, in a few minutes, grind and polish a sample of metal, and approximately estimate the proportion of phosphorus present in it.

In grey pig iron cooled in not too large masses, even in the forge grey quality containing nearly 0.7 per cent combined carbon, the phosphide eutectic is invariably found in the same position.

When the metal is cooled very slowly, and when there is little silicon present and some combined carbon, the latter is found in segregated cementite or carbide of iron; but this, unlike the phosphide, is found surrounding and adhering to the graphite plates, but the phosphide eutectic is quite separate from the carbide and as far as possible from the graphite.

An interesting instance of this has been found in what was alleged to be a natural iron found on the top of a mountain in Norway. On breaking it, it was found to be a coarsely grained dark grey cast iron with a hackly fracture, the analysis of which, as made by one of my pupils, Mr. John Evans, was as follows, viz. :—

	Per Cent
Iron	95.192
Combined carbon	0.720
Graphite	3.030
Manganese	0.200
Silicon	0.520
Sulphur	0.016
Phosphorus	0.322
Nickel	Nil.
	100.000

The lump of metal was about 5 inches square and 12 inches long. Whether or not it was a natural product is doubtful, but that it had cooled exceedingly slowly in passing from the liquid state was evident, for the carbon which, under less slow cooling, would have existed as pearlite, was massive cementite enveloping the graphite.

White phosphoretic irons are not quite so easy to judge as to the proportion of phosphorus present.

The presence of large masses of carbide of iron makes it impossible to tell which part is phosphide and which carbide if the simple etching method is employed; it is necessary, therefore, to use the "heat-tinting" process.

By heating till the pearlite is blue, as a rule, the cementite will be found red and the phosphide yellow. It is then quite easy to see the proportion of each, and where the eutectic joins the carbide. It has been found better first to etch with iodine to darken the pearlite, and after wiping the specimen with a clean linen rag, to heat till the required tint is obtained. This method leaves the pearlite dark. If the structure is to be photographed, it is advisable to heat until the carbide has assumed a faint yellow tinge, when the phosphide will remain quite white

By using isochromatic plates, the prints show the phosphides white, the carbide in half-tone, and the pearlite dark.

The photographs Nos. 16 and 17 represent areas of white Clarence pig etched with iodine, and also heat-tinted in the manner described. It will be noticed how the heat-tinting in No. 17 shows up the phosphide, which is represented by the white areas.

Summary of Part III

1st. There are fusible eutectics in all pig irons containing phosphorus, and which have solidifying points of about 900° C.

2nd. The free phosphide of iron can be left insoluble from such irons by dissolving them in cupric chloride (Schneider's method), or by nitric acid of specific gravity 1.20 in the cold. By attracting the magnetic portions of phosphide from the residue with a magnet, and further purifying what is attracted, almost pure Fe_3P is obtained.

3rd. With the aid of the microscope it is possible to detect the phosphide eutectic in pig irons, even when the amount is as low as 0.03 per cent.

4th. In white irons it is necessary to use the heat-tinting process to enable the cementite to be distinguished from the phosphide.

5th. The eutectics from such irons have a very complex constitution, containing phosphide of iron, carbide of iron in plates, pearlite, and silicide of iron. They deserve more careful study.

6th. The reason why phosphoretic pig irons are not more brittle than they are is because the eutectic separates into isolated segregations, and does not form continuous cells round the crystalline grains. When the phosphorus does not exceed 1.7 per cent the metal is comparatively strong, but an addition of 0.3 per cent reduces the strength materially, the reason of which is that the eutectic brittle areas in metal with 2 per cent P approach each other closely, leaving less of the strong ground mass intervening.

Part IV

On the Diffusion of Phosphide of Iron in Iron

(8) Professor Arnold and Mr. MacWilliam have shown that phosphide of iron will diffuse from iron containing 1.36 per cent phosphorus into pure iron, when they are placed in close contact and are heated in vacuum to 1000° for ten hours.

The relative speed of diffusion of carbon and phosphorus was found to approximate to the ratio of 5 to 1.

That phosphide of iron does diffuse in iron has been fully confirmed by the observations made in my laboratory. One notable instance of diffusion was furnished in a remarkable sample of iron found in the hearth of one of the Ferryhill blast-furnaces after it had been blown out.

As a rule, the "bears" from furnaces are highly carburetted and are associated with much graphite, but this sample was quite an exception to the rule, for it was free from carbon, and, moreover, contained not a trace of silicon and manganese. The structure resembled that of the columnar iron already referred to, with this difference, that the columns were much larger and were greatly curved. The microscope showed that they were enveloped with thin layers of phosphide of iron, and in their centres there existed small globular masses of the same substance, together with globules of sulphide of iron.

The analyses of Ferryhill pig iron and of this curious formation are as follows:—

	Ferryhill Pig Iron		Sample from Furnace Hearth
	Grey	White	
	Per Cent	Per Cent	Per Cent
Iron	91.10	92.90	91.06
Graphite	3.36	—	Nil
Combined carbon	0.16	3.20	Nil
Manganese	0.66	0.25	Nil
Silicon	2.53	1.10	—
Sulphur	0.06	0.30	0.32
Phosphorus	2.01	2.10	1.98
Oxygen, etc.	0.12	0.15	0.64
	100.00	100.00	100.00

By crushing the sample to fine powder and removing the fine particles by sieving, the greater part of the brittle and easily crushed and free phosphides were removed. On treating the fine powder so removed with sulphuric acid, the sulphides were dissolved, together with that part of the iron which passed the sieve. The insoluble residue, consisting of a black powder and bright metallic particles, was separated into two parts by a magnet, the bright particles being attracted, the black powder remaining behind. The metallic powder had the following composition:—

	Per Cent
Iron	84.00
Phosphorus	15.41
Not determined	0.59
	100.00

It is practically pure phosphide of iron, Fe_3P, proving the correctness of the microscopic observation.

The iron portion which remained on the sieve had the following composition:—

	Per Cent
Iron	98.20
Sulphur	0.09
Phosphorus	1.47
Oxygen, etc.	0.24
	100.00
Phosphorus as Fe_3P	0.03
Phosphorus in the columnar masses	1.44

It is not certain that the iron which originally entered the hearth was grey or white, but it is immaterial in which state it was. It must have been fluid to get there at all, and if it was fluid at the temperature of the blast-furnace, after solidification it must have been carburetted, and contained not more than 0.5 per cent of the phosphorus in solid solution, and about 1.5 per cent as free or separate phosphide of iron. It is difficult to understand how the metal was oxidized, but it appears probable that it was subjected to the action of water vapour, which would oxidize the manganese and silicon, and probably a portion of the carbon, when the metal was still in the fluid state.

The last 1.5 per cent of the carbon, at least, must have been removed when the metal had become solid, for the temperature would not be sufficient to retain it in a fluid state. The greater part of the phosphide must still have existed in the separate state at the stage when 1.5 per cent carbon was present.

If these conclusions are correct, and I think there can be little doubt that they are, the carbon as it was removed from the solid solution must have been replaced by the phosphide which diffused into the iron from the segregated little masses of free phosphide, producing a final solid solution containing 1.44 per cent phosphorus, a diffusion equal to about 1 per cent, which had passed into the iron from what must originally have been free phosphide. The time occupied in effecting this result probably was very protracted. Possibly it took years.

It is evident, however, that although the change was very slowly effected, and that there was more than sufficient phosphide present to give a *saturated* solid solution of the phosphide, the amount which actually did diffuse into the iron was short of that quantity by about 0.30 per cent.

The peculiar columnar structure of the metal was no doubt produced when the *Beta*-iron dominated by the carbon, changed to *Gamma*-iron, when the carbon was replaced by phosphide of iron.

I am conscious, on comparing the results here recorded with those obtained by decarburizing phosphorus iron in the malleable castings furnace previously given, that I am at a loss to understand why there is such a wide discrepancy in the proportion of phosphorus left in solid solution in the two cases. I am, however, certain of the facts, and it remains to find a satisfactory explanation. The conditions, which of course must have been widely different, possibly may account for the anomalous results.

Many experiments have been made in my laboratory to ascertain how phosphide of iron penetrates into steel when the former is maintained in the liquid state. One result yielded most interesting results.

Some fluid eutectic in a crucible was maintained in the liquid state at 1000°, near its solidifying point, in contact with a half-inch soft steel bar containing 0.12 per cent carbon. The bar was placed vertically through the middle of the fluid eutectic. After keeping them in contact for about an hour, the crucible with contents was allowed to cool down. The compound mass was then sectioned for micro-examination. The bar was found partially dissolved in the eutectic, and was reduced in diameter in consequence to about ¼-inch. On polishing and heat-tinting, and then etching by iodine, the following interesting observations were made, viz.: —

The bar was surrounded by a bright, hard ring, apparently free from carbon, which was traversed by solidified " rivers " of Fe_3P, which appeared as if they had been flowing from the eutectic through the hard ring into or near to the iron core. This ring consisted of a solidified solution of phosphide of iron in iron. The temperature was never nearly high enough to melt this hard envelope. Between this hard bright ring and the bar there was a transition ring or envelope not so hard as the first, but slightly harder than the bar itself. This envelope was very thin; it did not apparently contain any carbon, and most probably was only partially saturated with the phosphide. At certain points the ferrite areas of the bar continued in unbroken tracks right up to and through the thinner ring into the hard ring.

It was observed that in the central portions of these there existed hard spines, consisting probably of the same constituent as one or both of the rings, and occasionally in the centre of a spine there appeared the fine tapering terminus of one of the streams of phosphide, which could be traced right through both of the rings to the eutectic. Although at intervals the ferrite of the bar joined the inner ring, the greater part of the circumference of the latter was in juxtaposition with pearlite, and where the barrier of pearlite faced them no phosphide penetration could be observed. (Photo No. 7.)

It is proved then that, under such conditions, liquid phosphide preferentially, actually flows into the iron along the junctions of the crystalline grains and then diffuses in the solid state into the iron, and apparently diffuses more readily into iron than into steel.

Summary of Part IV

1st. The observations of Professor Arnold and Mr. MacWilliam proving that solid phosphide of iron does diffuse when in the solid state into solid iron under certain conditions have been fully confirmed.

2nd. That if time is given, the quantity which will diffuse into iron is very considerable, amounting to what is equivalent to about 1 per cent of phosphorus.

3rd. It still remains to be proved at what temperature diffusion commences.

On a previous occasion I made the statement that at that time "they were simply in a fog at present, and could not explain the exact nature of the changes which took place when solids or liquids diffused into solids. Theory and discussion would probably direct their attention to certain grooves in which practical experiment should be made to enable them to explain it satisfactorily. They were not in a position as yet to dogmatize." More data and light were required on this obscure subject, before we could understand it thoroughly, and I think the little work I have accomplished more than justified the remark. A little more light has been thrown on a somewhat complicated question, but before correct generalizations can be arrived at more workers at the research are required and more of nature's revelations obtained.

In conclusion, I must acknowledge the assistance and co-operation of Sir William Roberts-Austen, Professor Bauerman, Professor Arnold, Mr. MacWilliam, Mr. F. W. Harbord, Mr. A. Richards, Mr. J. L. Potts and Mr. Duckinfeld of Messrs. J. H. Andrew & Co., Ltd. of Sheffield, Mr. A. Cooper, Mr. C. H. Ridsdale, Mr. T. W. Sorby, Mr. E. Saniter, Mr. David Evans, Mr. Horard Livens of Messrs. Ruston, Proctor, & Co., and my own assistants.

Without the assistance of these gentlemen I could have done little work.

Appendix I

As the words "eutectic" and "eutectic alloy" are now very generally used, it is important it should be clearly understood what is meant by the terms.

It was Guthrie who first applied it to the study of alloys.

Sir William Roberts-Austen has kindly given me his views on eutectics as follows:—

As regards the definition of a eutectic, reference to Guthrie's original paper, *Phil. Mag.*, June, 1884, shows that he intended the word "eutectic" to refer to a solidified "mother liquor," that is, to a mixture which has a lower freezing-point than any other of the series. The simplest case is a mixture of the two dissolved bodies A and B, the proportion of each that is present depending on their relative solubilities in each other.

It happens, however, that in certain series of alloys the two constituent metals form a true compound which acts like a third body, and this forms a separate eutectic mixture with each of the two constituent metals. Each of these mixtures has a definite composition and solidifies at a definite temperature, whatever may be the composition of the substances which "fall out" from it and the temperature at which they "fall out."

It is convenient to call all such mixtures "eutectic" alloys. A case is presented by the following system:—

The liquid which solidifies at the point C is a true eutectic, because the adjacent alloys, as indicated by the lines CA, CB, are less fusible than it is. The existence of other more fusible alloys

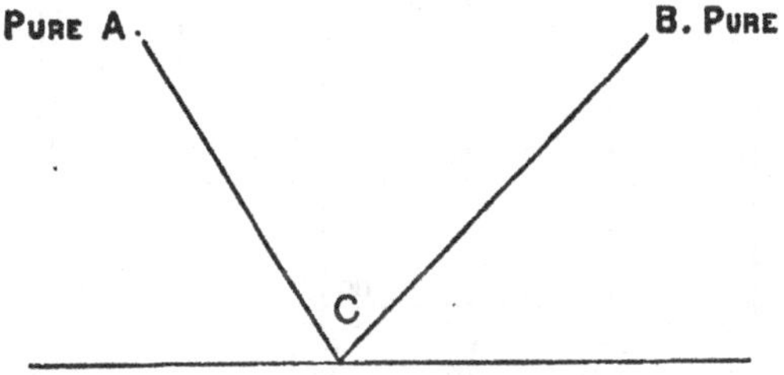

Mechanical mixture of A and B.

in the series would not affect its position as a eutectic alloy, provided the lines CA and CB both slope upwards. If, however, one of these lines does not slope up as in the lead-copper series, "Fourth Report to the Alloys Research Committee of the Institution of Mechanical Engineers," plate 7, at the point F, the alloy which solidifies is not a true eutectic, although it was first so called for want of a better term. In the copper-lead series the line DFC is horizontal, and represents the solidification of one of the two conjugate liquids into which the fluid alloy separates.

On reading the paper of Guthrie referred to by Sir W.

Roberts-Austen, it is clear that he recognized that the eutectic proportions of the metal present had not necessarily any relation to their atomic weights, and says, "Of course in many cases, metals can be fused together in simple multiples of their atomic weights, but in most such cases, one of the metals will separate out first and so destroy the ratio." . . . "When we are dealing with metals which, like antimony or arsenic, are themselves halogenous, or with such strongly chemico-positive metals as sodium, we may and do get alloys of atomic composition or metallo-metallic salts. But the temperature of fusion of these is never, so far as I am aware, lower than that of either (both) of the constituents; they are not eutectic."

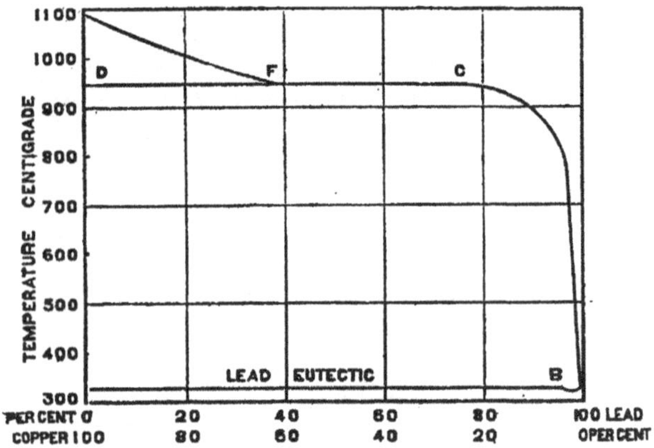

This last sentence makes it clear that one property of a eutectic, according to Guthrie, is that its melting-point must be lower than that of the mean of the separate constituents. All of the instances quoted in his paper illustrate this second principle.

Recognizing that certain metals may combine to form metallo-metallic salts, Guthrie, by inductive reasoning, concluded that in an alloy such salts might be formed, and that they would take the part of a free metal and form eutectic alloys with one or other of the constituents. He says, "The very bodies resulting from the chemical union of the two metals will possibly and probably furnish starting-points of new series of eutectic alloys, consisting of a single metal on the one hand and the chemical alloy on the other."

Professor Guthrie prepared any given eutectic by melting the metals and then allowing that metal which happened to be in excess to crystallize or fall out of solution, and when the greater part had become solid, the residual alloy which still remained fluid as a kind of "mother liquor" was poured off and allowed to set. This crude eutectic was remelted and allowed to partially solidify, and the mother liquor again poured off. The same treatment was repeatedly applied until the mother liquor invariably yielded the same constant composition. This alloy when solid, but for the contradiction in terms, might have been called a solidified mother liquor. Guthrie applied the term eutectic.

A eutectic, then, must have the following properties and components:—

1st. A solidifying and melting-point lower than that of the mean of the constituents.

2nd. A single melting and solidifying point.

3rd. It may consist of two or more metals which do not unite chemically, or of a metal and a definite chemical compound, and possibly of two or more definite chemical compounds.

Recent research has shown that it may be constituted as follows:—

4th. It may consist of a mixture of a solid solution of one metal in another and a free metal.

5th. It may contain a solid solution of a definite chemical metallo-metallic salt, and that same metallo-metallic salt in the free state.

6th. It may possibly consist of two solid solutions.

7th. Is an exception to the general rule, and relates to the eutectic pearlite which is formed in solid steel when it cools slowly from 700° C. to below 400° C. It is a solid solution at 700° C., but splits up into two constituents on cooling.

8th. A eutectic may contain two eutectics, one formed at its solidifying point, the other of pearlite formed on cooling below 700° C. The more fusible part of white pig iron, free from phosphorus and high in carbon, is an instance of this.

Preparation of Eutectics

The method of Guthrie is undoubtedly the best for making the eutectic mixtures, but it necessitates using large quantities of the metals. The second in value is that in which the metal is poured into a cylindrical cast iron mould lined with loam, and after solidification is nearly complete, it is compressed by hydraulic pressure, and the eutectic squeezed out of a suitable opening.

Sir W. Roberts-Austen has employed a similar plan to this. It is described in his Fourth Report on Alloys to the Institution of Mechanical Engineers, 1897.

"The alloy under examination is placed in a steel cylinder, which is fitted with two loose steel plungers, each about ½-inch diameter. The whole is placed between the jaws of a hydraulic press, and a pressure of about half a ton per square inch is maintained on the ends of the plungers. The cylinder is then slowly heated, the temperature at any moment being measured by a thermo-couple, which is placed in a hole drilled in the cylinder. At a certain definite temperature part of the alloy will liquefy, and can be squeezed out between the plungers and the wall of the cylinder. The temperature is noted at which the alloy is sufficiently liquid to be extruded. The heat is then raised, and a further portion of the alloy can sometimes be separated. Finally, a comparatively infusible residue is left. These several portions are then analyzed. In some cases it is found to be desirable to subject the extruded portion to a second treatment in the compressing cylinder."

It was the hydraulic method I used in expressing the liquid eutectic from Cleveland iron, and it is the best in cases where a small quantity of the eutectic is in presence of a large mass of metal.

A third method useful in such cases where the fusing-point of the alloy is under 500° C., and where only small quantities are available, has been employed with great advantage in my laboratory in preparing the eutectics of the fusible metals.

The melted alloy is poured upon the surface of a sheet of flexible asbestos paper. When the mass has solidified to the consistency of a thick paste, a second sheet of asbestos is placed on the top of the metal, and a wooden ruler is rolled over the plastic mass from the front in a forward direction. The eutectic is in this way squeezed out and runs as a fluid drop in front of the ruler leaving the solid metals behind. The crude eutectic so obtained is remelted and is again subjected to the same treatment. It is possible in this way to obtain the eutectic from 10 grammes of alloy.

The last method is that which I have used most extensively in the preparation of the eutectics containing two metals. It may be called the microscopic method. It is based upon the characteristic micro appearance of nearly all the eutectics, and on the statement of Guthrie, now generally recognized, that if one of the constituent metals is present in excess it will first solidify, leaving the residual eutectic to solidify at a lower temperature. The metal which falls out of the mother liquor is clearly seen in the micro section when it is properly polished and etched. It generally takes some clearly marked crystalline form.

Having prepared several mixtures synthetically by melting the metals in several proportions, and allowing them to solidify slowly, pieces of each are cut, polished, etched, and examined. As a rule, it does not take more than ten minutes to prepare a specimen. A glance through the microscope reveals at once which specimen contains the greatest proportion of the eutectic and which metal is in excess. Several further alloys are then made approximating to this, but with diminishing proportions of the metal known to be in excess. The resulting mixtures are examined, and probably one of them will be found to consist of pure eutectic, free from crystals or crystallites of either the constituent metals. A chemical analysis is then made to check the figures synthetically obtained.

It often happens that in a series of alloys of two metals more than one eutectic is formed. A very simple method of detecting these, and also of obtaining the complete series of alloys in one single specimen, has been used with most satisfactory results in my researches.

A small special mould of fire clay, having a cavity 2 c.m. deep and 1 c.m. square, is heated to a temperature a little below the melting-point of the metal which is most infusible. The two metals are melted in separate crucibles, and a portion of that which has the highest gravity is poured into the mould, and when the lower part has solidified the second metal is poured in drop by drop, the following drop being added when the metal in the mould has nearly solidified or become plastic. When the mould is filled, a cold metal plate is placed on the top to hasten solidification of the last portion of the metal added. When cold, if the experiment has been conducted properly, the one metal in the pure state will be found at the bottom and the other at the top, the metal between containing a complete gradation or series of alloys. One side of the little ingot is then polished, etched, and examined. The specimen so prepared constitutes a very perfect chart or map, and is the best guide to work from in making a systematic research of the alloys of any two metals, and if more than one eutectic is present it may be recognized by its characteristic appearance.

Appendix II

Solid Solution

The term solid solution has been applied to cases, when, in the solidifying of two or more metals from the molten state, one or more of the metals is retained in the others, but in such a form that the highest power of the microscope is unable to detect it or them. They crystallize in a form identical with or very closely approximating to that constituent which preponderates. In a true liquid solution of salt in water, no microscope can detect any of the constituents; it appears to be homogeneous. Solid solutions of one metal in another are similar in that respect.

Sir William Roberts-Austen has kindly given me his definition of solid solution, which is as follows:—

" A solid solution is a homogeneous mixture of two or more substances in the solid state. In metals, no one has as yet worked at non-crystalline mixtures, and solid solutions of metals when crystalline are solid 'isomorphous mixtures,' or 'mixed crystals.' "

This definition is a most concise one, and certainly appears to cover every kind of solid solution.

There are, however, several kinds of solid solution:—

1st. In which one metal, from an alloy in crystallizing, retains a portion of the other homogeneously diffused throughout its whole crystalline mass.

2nd. In which during crystallization the central portion of the crystals contains less of the dissolved metal than their external boundaries.

3rd. In which the metals form a definite chemical compound, which is retained in solid solution in the excess of the metal or metals.

4th. In which the non-metallic elements form definite chemical compounds with a portion of the dissolving metal and that these remain in solid solution.

The terms "isomorphous" and "mixed," as applied to crystals, have been and still are used synonymously and are synonymous with the term solid solution in metals. The term "mixed crystals" might lead to the erroneous idea that the crystals were mixed up together but remained independent. The term "isomorphous," according to ([20]) Professor Bauerman, can only be strictly applied to mixtures of substances crystallizing in the cubic system.

There seems to be less objection to the term "solid solution." It may be applied to crystallized or non-crystallized substances, and conveys the meaning that the bodies in solution are as intimately associated as if they were in liquid solution, and that there is no separation of constituents when the fluid solution passes into the solid solution.

The expression "mixed crystals" is particularly to be avoided in metallurgical nomenclature, for the eutectics are described as eutectic mixtures, the term mixture in this case meaning an alloy or compound separated into two or more parts, whereas a so-called mixed crystal consists of a homogenous whole.

As a rule, all metals and alloys are crystalline, therefore solid solutions of one or more metals in another may be accepted as crystalline solid solutions.

Appendix III

Method for Determining Free Phosphide of Iron

One gramme of drillings or powdered material is placed in a 10-oz. flask. This is placed in a large beaker containing water. About 70 c.c. of 1.20 nitric acid is suddenly poured upon it. The flask is then removed and shaken up until the action of the nitric acid upon the iron has raised the temperature until it is sensibly warm, i.e., not more than 20° C. The flask and contents during reaction must be violently agitated, so as to avoid any local heating, and as the reaction, unless checked, would result in temperatures considerably higher than 20°, the vessel must be agitated under a stream of running water so as to regulate it. A complete solution of the iron and dissolved phosphide is generally effected in about two minutes. The flask with contents, at a temperature not exceeding 20° C., is allowed to stand on a table until no more gas is seen to arise from the residue at the bottom. This does not take more than two minutes. The insoluble matter is then filtered off through filter paper, the residue rinsed on to the filter with cold distilled water, and the filter with contents is washed with cold water. The whole of the phosphorus existing in solid solution is by this treatment obtained in the filtrate, together with a small proportion of that existing as free phosphide of iron, whilst from 90 per cent to 97 per cent of the free phosphide of iron is left insoluble upon the filter. The filtrate is evaporated down to a small bulk or to dryness if silica is present, and the phosphorus determined in the usual way. The filter containing the insoluble phosphide is burned off in a platinum crucible, dissolved in nitro-hydrochloric acid, and the phosphorus determined in the solution so obtained.

On making practical trials to ascertain what effect 1.20 nitric acid has upon pure phosphide of iron, it was found that the proportion dissolved depends upon the fineness of the particles of phosphide, and the length of time the acid acted upon it, and the temperature of the acid. But on treating even the finest powder in the manner described above, not more than 10 per cent was dissolved. Whether the quantity of phosphide (of any given fineness) acted upon is great or small, provided there is always an excess of acid at the proper temperature, the proportion dissolved is constant. The following instances indicate this:—

0.1 and 0.4 grammes of phosphide were treated side by side under the conditions above named with 25 c.c. of 1.20 nitric acid at 20° C. In each case the result was the same, 0.095 was left insoluble in the one case and 0.385 in the other, or 95 per cent of the whole in each case. The influence of long-continued action is very pronounced, for 1 gramme of finely powdered phosphide left in contact with 1.20 nitric acid for twelve hours lost 20 per cent of its weight, but the residue was brightly metallic, and it was absolutely free from any decomposition products. An excessive quantity of acid, if employed, makes no difference provided the temperature and time elements remain constant.

In calculating the results of any analyses, allowance must be made for the solubility of the phosphide, and in the analyses given in this paper from 5 per cent to 10 per cent has been added to the phosphide actually found, according as the free phosphide was in a fine or relatively coarse state in the specimen examined.

Almost all other reagents, such as hydrochloric acid, sulphuric acid, chloride of copper or chloride of mercury, although they do leave more or less of the free phosphide insoluble, it is accompanied by a large proportion of the black decomposition product from the phosphide in solid solution.

Probably previous investigators have been misled in their researches by want of knowledge of this important fact.

In determining the free phosphide in presence of free carbide by this method, the latter will be more or less left in the insoluble condition, but that does not really matter, as the phosphorus in the insoluble matter will only exist as free phosphide, and is calculated into that substance.

Although the method leaves much to be desired and is far from perfect, it has rendered valuable aid in checking microscopic observations; for after considerable experience I have found, that whenever the microscope has shown that free phosphide of iron is present, even though there be very slight traces, the method has confirmed the observation.

It is important to ascertain, after the metals have been dissolved in nitric acid, that the residue is magnetic, — that is, capable of being attracted by a magnet. If it is not attracted it cannot be Fe_3P.

Appendix IV

"HEAT TINTING" METAL SECTIONS FOR MICROSCOPIC EXAMINATION

Etching polished metals with corrosive reagents is more or less liable to lead to the confusing of the constitutional and crystalline structures.

With very feeble etching, as a rule, the constituents are made visible. With long-continued or strong etching, the granular, and often the crystalline, structures are developed.

By heating polished sections of raw irons till they assume certain oxidation tints, their various constituents are differently colored. The coloring depends on the formation of oxidation films, which form at different rates on the different constituents.

Professors Martens and Behrens were the first to use this method, and Osmond and others have employed it in the development of the constitutional structure of many alloys. It does not usually give any indication of the granular and crystalline structure, and stands before all other methods as a means of detecting the separate and individual constituents. It may therefore be called the constitutional method. It has the great advantage of not removing any of the metal, so that when the oxidation coloring has been effected, the original flat surface still remains. The specimens, moreover, are much less liable to rust than when chemical reagents are employed. The relative rates at which cementite, sulphide of manganese, phosphide of iron, pure iron, and pearlite — containing 0.6 per cent phosphorus — assume various colors when heated, were carefully determined in the following manner:—

A slowly cooled sample containing all the above-mentioned constituents, excepting pure iron, had a fine hole drilled in it, and a piece of pure iron was inserted into this; the iron was then compressed so as to expand and perfectly fill the hole. After polishing it was bedded on sand in a porcelain capsule. The capsule was dropped into the circular opening on the stage of a Becks star microscope. A small spirit-lamp placed under the stage capsule was used to slowly heat the section. The changes of color were carefully noted, with the following results, viz.: —

No. 1 — Table of Heat Tints

With steadily rising temperatures 200° to 400° C.

Pearlite 0.6% P. 0.75% C.	Iron	Carbide of Iron	Phosphide of Iron	Sulphide of Manganese
White	White	White	White, tinted yellow	Pale lavender
Very pale yellow	Do.	Do.	Do.	Do.
Do.	Do.	Do.	Do.	Do.
Yellow	Very pale yellow	Do.	Do.	Do.
Do.	Do.	Do.	Do.	Do.
Yellow brown	Yellow	Very pale yellow	Do.	Do.
Do.	Do.	Do.	Do.	Do.
Brown	Yellow brown	Do.	Do.	Do.
Do.	Do.	Do.	Do.	Do.
Red purple	Brown	Yellow	Very pale yellow	Pale yellow
Purple	Red purple	Yellow brown	Do.	Do.
Blue	Purple	Do.	Pale yellow	Do.
Do.	Blue	Brown	Yellow	Do.
Pale blue	Do.	Do.	Salmon	Do.
Do.	Pale blue	Red brown	Heliotrope	Brownish white
Do.	Do.	Red purple	Greenish	Do.
Pale pea-green	Do.	Purple	Yellow	Do.
Do.	Pale pea-green	Blue	Do.	Do.
Pale yellow	Pale yellow	—	—	—
White	White	—	—	—

The above table gives a good idea of the changes in colour with steadily rising temperatures.

The results obtained by heating at fixed temperatures yielded still more valuable results. The following observations were made by heating the same piece of metal as was previously used after repolishing, together with pieces of carbide of iron, phospphide of iron, and pure iron. They were placed side by side on a fluid bath of metallic tin, the temperature of which was regulated by a pyrometer.

No. 2 — Table of Heat Tints

Temperature 232° C. throughout.

	Pearlite 0.6% P. 0.75% C.	Iron	Carbide of Iron	Phosphide of Iron
1 min. .	Very pale yellow	White	Very pale yellow	White
4 " .	Yellowish brown	Very pale yellow	Pale yellow	Do.
10 " .	Do.	Darker yellow	Do.	Do.

No. 3—Table of Heat Tints

At constant temperature of 280° C.

	Pearlite 0.6% P. 0.75% C.	Iron	Carbide of Iron	Phosphide of Iron
1 min. .	Brown	Yellow	Brownish yellow	White
2 " .	Purple	Red brown	Do.	Do.
3 " .	Deep purple	Brown	Do.	Do.
4 " .	Blue	Purple red	Do.	Very pale yellow
6 " .	Do.	Purple	Orange brown	Pale yellow
11 " .	Do.	Blue	Red brown	Pale salmon
15 " .	Pale blue	Do.	Burnt sienna red	Do.
30 " .	Paler blue	Paler blue	Purple brown	Heliotrope

No. 4—Table of Heat Tints

At constant temperature of 326° C.

	Pearlite	Iron	Carbide of Iron	Phosphide of Iron
1 min. .	Purple red	Red	Yellow brown	Very pale yellow
2 " .	Deep blue	Purple red	Red brown	Pale yellow
3 " .	Do.	Do.	Do.	Do.
4 " .	Do.	Do.	Do.	Do.
5 " .	Do.	Deep blue	Do.	Pale salmon
6 " .	Paler blue	Lighter blue	Purple red	Salmon
7 " .	Pale sea-green	Very pale blue	Purple	Do.
8 " .	Do.	Do.	Do.	Do.
9 " .	Pale yellow	Do.	Do.	Purple
10 " .	Do.	Do.	Do.	Heliotrope

In high carbon steels and white irons the free phosphide is always accompanied by cementite or carbide of iron, and it is easy to distinguish them by heating to 280° C. for ten to fifteen minutes. After such treatment the cementite will be red brown after each heating, whilst the phosphide will be pale yellow after ten minutes' heating, and salmon color after fifteen minutes. On continuing the heating, the blue at first assumed by the phosphide of iron is quite different from that of any of the other constituents. It may be described as heliotrope in tint.

When no cemetite is present, it is advisable to heat at 280° C. along with specimen chips of phosphide and white Swedish pig iron and pure iron, and watch the tints on these three indicators. As soon as the white iron has assumed a brown color, the phosphide a pale yellow or salmon tint, and the iron a fine blue, the specimen is at once removed and examined for the phosphide of iron.

It has not been exactly determined what the effect is of varying quantities of dissolved silicide of iron. That it does make a great difference has been proved, a fact which must be kept in view in examining high silicon pig irons.

The effect of chromium, aluminium, manganese, nickel, and tungsten in determining the tints has not yet been ascertained.

Some alloys contain a fusible constituent which melts and exudes from the surface before any oxidation can take place, therefore oxygen is not available in such cases. Different tints, however, can be obtained by heating these alloys at low temperatures in an atmosphere containing slight traces of iodine, bromine, or hydric sulphide. In using these reagents the specimen is heated in a box fitted with a glass lid, and air which has passed over iodine, or which contains traces of bromine or H_2S, is forced over the hot metal. The constituents then color differently, and may be distinguished from each other.

The general method of tinting the specimens when they are heated may conveniently be called the "Heat Tinting Process," and the chemical symbol of the particular reagent used may be added as prefix, thus:—

Heat tinting (O) oxidation films being formed.
" (I) iodide " "
" (Br) bromide " "
" sulphide " "

In marking the microscope slides the following abbreviations may be used:—

HT(O), HT(I), HT(Br), HT(H_2S).

Special precautions are necessary in preparing the polished surfaces. If, after leaving the polishing block, they are at once heated till the tints appear, they will not be true, as the moisture in the porosities, when it evaporates, will affect the surface. It is necessary to make a preliminary heating to about 120° C., and remove from the source of heat, and rub it whilst hot on a clean dry hair buff or piece of flannel. It is then at once re-heated to the proper temperature, and as soon as the tints have developed it is placed at once on a bath of mercury to cool it rapidly and avoid further change.

Special Electrical Apparatus for "Heat Tinting"

By arrangements such as have already been described, the tinting can be done under the microscope by the flame of a spirit-lamp, but the stage becomes highly heated, which is very objectionable. I have, therefore, devised a simple electric heater, which can be placed on the microscope stage, and the progress of the tinting observed. The following description, with the assistance of the accompanying sketch, will make the construction of the apparatus intelligible.

The bed is of ebonite, and on this two strips of brass are screwed, one on each side of the upper surface. To these are attached the binding screws AA.

The carrier consists of a small 1-inch porcelain capsule, on two opposite sides of which a closely fitting sheath of brass is secured. These, when it is placed in position, form the contacts with the two conducting brass strips on the bed.

The bottom of this capsule is filled with magnesia made plastic with a solution of chloride of magnesium. A platinum spiral is bedded upon this, and the terminals connected to the brass sheaths. Over the wire is painted a thin layer of silica paint, made by mixing silicate of soda and silica to the consistency of a thin paste, after which the apparatus is ready for use.

The porous magnesia under the wire is a good non-conductor, and practically no heat passes to the microscope stage.

The capsule is quite loose, and can be removed from its bed or swivelled to any angle, so that a specimen placed upon it can always be made to have its polished surface parallel with the stage.

The current can be switched off as soon as the proper tinting is obtained.

This simple apparatus will no doubt be found to be very useful where there is a supply of electricity.

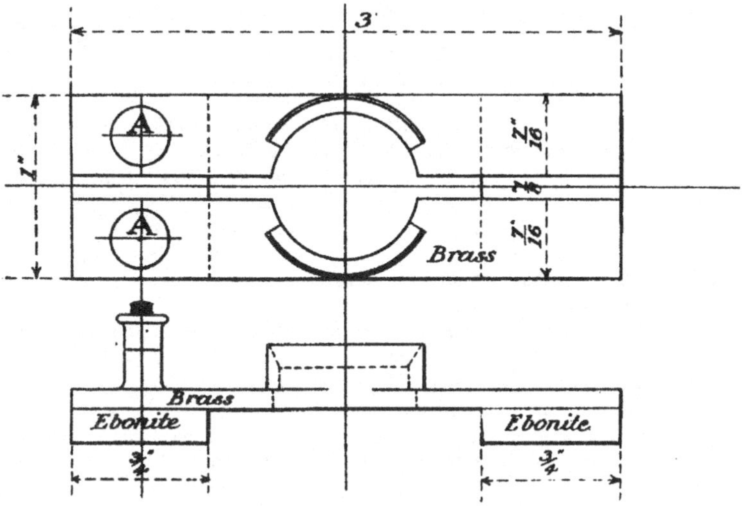

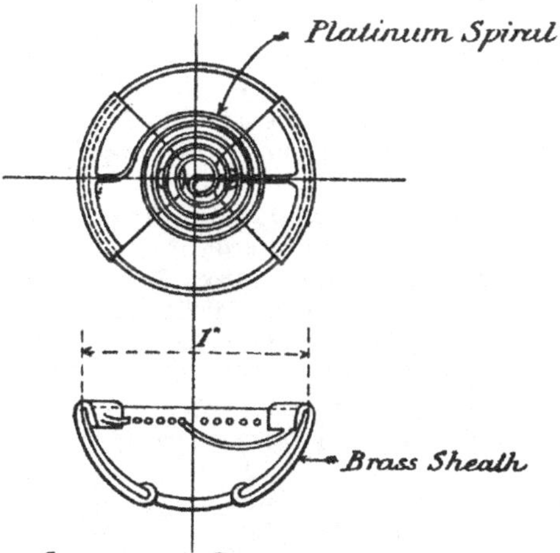

Sectional Elevation.

Appendix V

Bibliography

1. Sir F. Abel. — "On the Liquation of Phosphides from Solidifying Cast Iron." Percy's "Metallurgy of Iron and Steel," page 64, 1864.
2. Percy. — "Metallurgy of Iron and Steel," page 64, 1864.
3. Volhard. — *Oesterreichische Zeitschrift für Berg- und Hüttenwesen,* 1880.
4. Behrens and Van Linge. — *Zeitschrift für analytische Chemie,* Vol. XXXIII, page 513.
5. Leopold Schneider.—*Oesterreichische Zeitschrift für Berg- und Hüttenwesen,* page 736, 1886.
6. B. W. Cheever. — *Transactions of the American Institute of Mining Engineers,* Vol. XV, page 448.
7. Leopold Schneider.—*Oesterreichische Zeitschrift für Berg- und Hüttenwesen,* Vol. XXXI, page 361.
8. Arnold. — "The Physical Influence of Elements on Iron." *Journal of the Iron and Steel Institute,* 1894, No. I, page 107.
9. Stead. — "Phosphorus in Cleveland Ironstone and Iron." *Journal of the Cleveland Institute of Engineers,* 1875-76.
10. Baron H. Jüptner von Jonstorff. — *Oesterreichische Zeitschrift für Berg- und Hüttenwesen,* Vol. XVIII, page 209.
11. Baron H. Jüptner von Jonstorff. — *Journal of the Iron and Steel Institute,* 1897, No. I, page 224.
12. Arnold and MacWilliam. — *Journal of the Iron and Steel Institute,* 1899, No. I, page 85; *The Metallographist,* II, page 278.
13. Campbell and Babcock. — "On the Influence of Heat Treatment and Carbon upon the Solubility of Phosphorus in Steel." *Journal of the American Chemical Society,* Vol. XIX, No. 10, 1897.
14. Osmond and Werth. — "Théorie Cellulaire des Propriétés de l'Acier." *Annales des Mines,* 8th series, July to August, 1885.
15. Osmond. — "Etudes Métallurgiques." *Annales des Mines,* July to August, 1888.
16. Lencauchez. — *Mémoires de la Société des Ingénieurs Civils,* May, 1887.
17. Stead. — "Crystalline Structure of Iron and Steel." *Journal of the Iron and Steel Institute,* May, 1898; *The Metallographist,* I, page 289.
18. Stead. — "Note on the Structure of Phosphoretic Pig Metals." *Journal of the Cleveland Institute of Engineers,* February, 1900; *The Metallographist,* III, page 261.
19. Howe. — "Metallurgy of Steel." This work contains the most complete account of the effect of phosphorus on the physical properties of steel. 1891.
20. Bauerman. — "Systematic Mineralogy," 1881, page 336.
21. Hogg. — "On Segregation," page 238. *Journal of the Society of Chemical Industry,* March, 1893.
22. Sir W. Roberts-Austen. — Fourth Report on Alloys to the Institution of Mechanical Engineers, 1897.

INDEX

[The names of authors are printed in SMALL CAPITALS and the titles of articles in *italics*.]

Spotted something that isn't quite right?

Let us know and we'll fix it for future reprints. And as a thank you for your diligence, we'll send you a free Sothis book of your choice.

publisher@sothis.press

This page intentionally left blank

Made in the USA
Las Vegas, NV
13 April 2026

45318593R00218